The **BEST**
WRITING on
MATHEMATICS

2010

The BEST WRITING on MATHEMATICS

2010

Mircea Pitici, Editor

PRINCETON UNIVERSITY PRESS
PRINCETON AND OXFORD

Published by Princeton University Press, 41 William Street,
Princeton, New Jersey 08540

In the United Kingdom: Princeton University Press, 6 Oxford Street,
Woodstock, Oxfordshire OX20 1TW

press.princeton.edu

Library of Congress Cataloging-in-Publication Data

The best writing on mathematics 2010 / Mircea Pitici, editor.
p. cm.
Includes bibliographical references.
ISBN 978-0-691-14841-0 (pbk. : alk. paper)   1. Mathematics.   I. Pitici,
Mircea, 1965–
QA8.6.B476 2011
510—dc22   2010027717

British Library Cataloging-in-Publication Data is available

This book has been composed in Perpetua Std

Printed on acid-free paper. ∞

Printed in the United States of America

1 3 5 7 9 10 8 6 4 2

For my daughter, Ioana Emina

_____

Contents

Mathematics Alive

Mathematicians and the Practice of Mathematics

Mathematics and Its Applications

Mathematics Education

History and Philosophy of Mathematics

Mathematics in the Media

Foreword

WILLIAM P. THURSTON

Mathematics is commonly thought to be the pursuit of universal truths, of patterns that are not anchored to any single fixed context. But on a deeper level the goal of mathematics is to develop enhanced ways for *humans* to see and think about the world. Mathematics is a transforming journey, and progress in it can be better measured by changes in how we think than by the external truths we discover.

Mathematics is a journey that leads to view after incredible view. Each of its many paths can be obscured by invisible clouds of illusion. Unlucky and unwary travelers can easily stray into a vast swamp of muck. Illusions arise because we cannot directly observe each other think. It is even difficult for us to be self-aware of our thinking. What we cannot see we replace by what we imagine, hence the illusionary trap that can lead to nowhere.

People's inclination to disregard what they cannot easily see (or say) in order to focus on what can readily be seen (or said) results in predictable and pervasive problems in communicating mathematics. We have an inexorable instinct that prompts us to convey through speech content that is not easily spoken. Because of this tendency, mathematics takes a highly symbolic, algebraic, and technical form. Few people listening to a technical discourse are hearing a story. Most readers of mathematics (if they happen not to be totally baffled) register only technical details—which are essentially different from the original thoughts we put into mathematical discourse. The meaning, the poetry, the music, and the beauty of mathematics are generally lost. It's as if an audience were to attend a concert where the musicians, unable to perform in a way the audience could appreciate, just handed out copies of the score. In mathematics, it happens frequently that both the performers and the audience are oblivious to what went wrong, even though the failure of communication is obvious to all.

Another source of the cloud of illusions that often obscures meaning in mathematics arises from the contrast between our amazingly rich abilities to absorb geometric information and the weakness of our innate abilities to convey spatial ideas—except for things we can point to or act out. We effortlessly look at a two-dimensional picture and reconstruct a three-dimensional scene, but we can hardly draw them accurately. Similarly, we can easily imagine the choreography of objects moving in two or three dimensions but we can hardly convey it literally, except—once again—when we can point to it or act it out. It's not that these things are impossible to convey. Since our minds all have much in common, we can indeed describe mental images in words, then surmise and reconstruct them through suggestive powers. This is a process of developing mental reflexes and, like other similar tasks, it is time-consuming. We just need to be aware that this is the task and that it is important, so that we won't instinctively revert to a symbolic and denatured encoding.

When I grew up I was a voracious reader. But when I started studying serious mathematical textbooks I was surprised how slowly I had to read, at a much lower speed than reading non-mathematical books. I couldn't just gloss over the text. The phrasing and the symbols had been carefully chosen and every sign was important. Now I read differently. I rarely spend the time and effort to follow carefully every word and every equation in a mathematics article. I have come to appreciate the importance of the explanation that lies beneath the surface. I love to chase signals that direct me toward a theoretical background I don't know. I have little expectation for the words to be a good representation of the real ideas. I try to tunnel beneath the surface and to find shortcuts, checking in often enough to have a reasonable hope not to miss a major point. I have decided that daydreaming is not a bug but a feature. If I can drift away far enough to gain the perspective that allows me to see the big picture, noticing the details becomes both easier and less important.

I wish I had developed the skill of reading beneath the surface much earlier. As I read, I stop and ask, What's the author trying to say? What is the author *really* thinking (if I suppose it is different from what he put in the mathematical text)? What do I think of this? I talk to myself back and forth while reading somebody else's writing. But the main thing is to give myself time, to close my eyes, to give myself space, to reflect and allow my thoughts to form on their own in order to shape my ideas.

Studying mathematics transforms our minds. Our mental skills develop through slow, step-by-step processes that become ingrained into our neural circuitry and facilitate rapid reflexive responses drawing little conscious effort or attention. Touch typing is a good example of a visible skill that becomes reflexive. It takes a lot of work to learn to touch type, but during that process people can at least see what they need to learn. For me the process of typing is transparent. I think about what I want to say and I put it on the computer screen; I do not ordinarily attend to the typing at all. Although my moment-by-moment mental processing is complex, it seems seamless.

Now think of a skill similar to typing that involves no physical motion and no direct use of words or symbols. How do we observe and learn such a skill—and how do we communicate about it? This kind of skill is sometimes glossed as intuition—that is, a powerful insight, hard to explain. One of the main problems with doing mathematics is that we tend not to be aware that hardly definable skills *are* important, that they need to be nurtured and developed. Yet practicing mathematics requires them. Mathematical ideas *can* be transcribed into symbols organized into precise descriptions, equations, and logical deductions—but such a transcription is typically far removed from the mind process that generates the ideas.

These types of problems are pervasive but not insurmountable. Mathematics *can* be communicated without being completely denatured, if people are attuned to its mental dimensions. One important part of the remedy is reflective reading of thoughtful writing, such as the interesting and varied collection in this book.

For the last two years, Mircea Pitici has invited me to meet with the students in his Freshman Writing Seminar in Mathematics at Cornell University. I was struck by the imaginativeness and the intensity of the participants. It was immediately obvious that Mircea's writing seminars lead students to think about mathematics on paths different from those available in typical mathematics courses. It reminded me how compelling and interesting the journey of mathematics can be, at all its stages. When mathematics is rendered into technicalities, the story is removed. But discussions of mathematics without technicalities can evidence its poetry and bring it back to life.

We humans have a wide range of abilities that help us perceive and analyze mathematical content. We perceive abstract notions not just through seeing but also by hearing, by feeling, by our sense of body motion and position. Our geometric and spatial skills are highly trainable, just as in other high-performance activities. In mathematics we can use the modules of our minds in flexible ways—even metaphorically. A whole-mind approach to mathematical thinking is vastly more effective than the common approach that only manipulates symbols. And a collection like this, of writings on mathematics, opens up the whole mind toward a more comprehensive understanding of mathematics.

Happy reading!

Introduction

MIRCEA PITICI

This book offers a selection of texts on mathematics published during 2009 in a variety of professional and general-interest publications from several countries. The main goal of bringing together such diverse contributions is to make widely available representative texts on the role, importance, and dynamism of mathematics, applied mathematics, and mathematics instruction in contemporary society, as perceived by some of the notable writers on these topics. Our goal is to make accessible to a wide audience texts originally printed in publications that are often not available outside the scientific community or have limited distribution even inside it. We also intend to contribute to the dispersion of thinking on mathematics in the context of global competition in technical education, to illustrate the growing presence of mathematical subjects in the mass media, and to encourage even more and better writing of a similar sort.

Since the texts included here are not mathematics articles per se but writings on mathematics, the volume is not technical, as far as mathematical symbolism is concerned. It should appeal to a general audience but particularly to undergraduate and graduate students preparing for mathematical, scientific, and technical careers. The book is of special interest to mathematics and science teachers, as well as to instructors and researchers. But any person with a mathematical background equivalent to or better than high school mathematics will be able to read and understand most of the content. We hope the readers of this collection of texts will become familiar with some of the topics involving mathematics currently discussed in the specialized literature and with mathematical ramifications in the mass media.

Good writing on mathematics is important for at least three reasons. First, it is a means of intradisciplinary growth and can serve as a stimulant for starting on a research path in mathematics. Many professional mathematicians trace their passion for mathematics to some well-written slim book they read in early adolescence—whether it was G. H. Hardy's *A Mathematician's Apology*, Poincaré's *Science and Hypothesis*, Pólya's *How to Solve It*, or Ekeland's *Mathematics and the Unexpected*. Such encounters with the meditative literature on mathematics become crucial biographical moments, able to light sparks in young minds. They succeed due to the openness of communicating mathematical ideas in accessible yet nontrivial ways. Although mathematicians find great company and friendship among themselves, over the twentieth century they rarely conveyed the beauty of their craft to outsiders in a sustained explanatory effort. This was an isolationist development, a retreat from the centuries-old tradition of extensive writing illustrated by great previous mathematicians. Over the last few decades, expository writing on mathematics gradually flourished again, while today the Internet is changing the nature of mathematical activity and the means of interacting between the mathematical community and the public at large. In this volume we include one text occasioned by such a shift (the article by Timothy Gowers and Michael Nielsen) and, in annual successors in this series, we will stay tuned for further developments and reactions.

Second, writing is crucial in *learning* and *teaching* mathematics. Most mathematicians rarely study mathematics only for personal pleasure. They also do it to instruct others, to transmit knowledge, and to stimulate curiosity, talent, and exploration. They do it with words, not just with symbols and figures. They *explain*, and the explanation is almost always discursive. In the common practice of teaching mathematics and publishing research on mathematics, the discursive part too often vanishes, wiped out by symbols. Most of our students—not to mention the nonspecialized public—are lost in the struggle to decipher a codelike message. Yet every student can learn mathematics better by reading and writing about it. If writing is included as a vital part of the mathematics instruction, it becomes an effective instrument for comprehending mathematics. But writing naturally follows reading; therefore in this book we are offering the first in an annual series of recent remarkable texts on mathematics.

Third, good writing on mathematics has a strategic role. Whether academics acknowledge it or not, their disciplines are engaged in a race for attracting interest and attention. In this competition, mathematics—a discipline requiring the mobilization of massive cognitive resources—is often shortchanged. A big part of the problem resides with the mathematics educators and instructors, who put too little effort into bridging their discipline to other learning domains. This reluctance is mirrored by nonmathematicians who use mathematics in their professions but fail to address the conceptual underpinnings of the mathematical tools they habitually employ. Good writing on mathematics can be instrumental in overcoming misperceptions between people trained strictly in sciences or strictly in humanities. Furthermore, good writing on mathematics requires not only an understanding of the subject matter but also the talent to persuade the surveyors of an eminently abstract subject that mathematics is more than an exercise in mental skills, that it is a powerful instrument humans use to represent and to study both simple and complex phenomena.

Overview of the Volume

I started to make the selection for this book by consulting all the 2009 periodicals and collective volumes pertaining to writing on mathematics that were accessible to me in the Cornell University Library or through its interlibrary services. I perused and read many more publications than those represented in the final content of this anthology, but constraints of space or related to copyright influenced the final selection. I had to leave aside excellent texts that were too long, or were slightly too technical for the intended profile of the book, or posed insurmountable copyright problems. Also, as a general rule, I did not consider for selection book reviews and interviews.

In a second phase I contacted competent people and asked for advice. This led to several deletions and additions; my debts for the latter are mentioned individually in the acknowledgments.

I purposefully covered all aspects of mathematics as a social enterprise, deliberately avoiding exclusivist compartmentalization. Mathematics is so vast that people working in some areas of it are ignorant of developments animating other areas. This volume is meant to facilitate a better awareness of the issues outstanding in various aspects of mathematics and to inform about reliable sources that can guide further research and interdisciplinary contacts. In doing so I made no attempt to sanitize the selection by avoiding controversial subjects. On the contrary, several of the texts in this collection are overt invitations to dissent.

For convenience, I divided the contents into six thematic sections, but the boundaries are not rigid, and there is considerable affinity among texts across themes.

MATHEMATICS ALIVE

In the first section of the volume I grouped several articles written from widely different perspectives, to illustrate the dynamism and the flexibility of the discourse on mathematics. These texts could have belonged just as well in later sections (or even in more than one), but in the present grouping they show that writing on mathematics, far from being arid or stultifying, is versatile, adaptive, and alive.

Chandler Davis argues that truth in mathematics has various strengths and that accepting this notion makes mathematics pertinent and relevant. He notes that mathematical theories worth their name are more than collections of truths; they concern the nature of conceptual relatedness, not only its existence.

Melvyn Nathanson discusses several difficulties in finding perfect, beautiful, insightful proofs for mathematical results and gives reasons for valuing these qualities over deductive proofs exclusively concerned with formal validity.

Branko Grünbaum details several errors of enumeration in the geometry of the polyhedra. He shows that some reside in logical flaws and proved surprisingly enduring, despite extended scrutiny of this subject over many centuries.

Keith Devlin points out that past mathematicians often solved problems by trial and error, but such experimentation disappeared when they published the results. Nowadays, with the advent and ever more widespread use of computers, distinctions between formal mathematical proofs and proofs by experimentation are likely to become blurred.

Henryk Woźniakowski describes the context and the parameters of employing mathematics to study the complexity of continuous problems generated by systems with partially available information.

Answering yet another "What is . . . mathematics?" question, Tim Johnson locates the origins of financial mathematics in the development of probability as abstract measure theory by the Russian mathematician Andrey Kolmogorov in the first part of the twentieth century. Johnson continues by sketching a brief list of the virtues and the dangers of using mathematics in finance.

Finally, in a piece with an amusing title, David Wagner invites us to give credit to the forbidden language of mathematics. Wagner argues that by turning to taboos and examining them, we escape the conformity inherent in the normative ways of practicing and teaching mathematics.

MATHEMATICIANS AND THE PRACTICE OF MATHEMATICS

Myriad stereotypes circulate about mathematicians and the work they do—so many, in fact, that one wonders why and how mathematicians *do* what they do. I selected for this section a few contributions that challenge stereotypical views of mathematics.

Freeman Dyson draws on his long familiarity with the work of past mathematicians and recalls his personal acquaintance with mathematicians prominent during the twentieth century, to observe that some were better at identifying broad, unifying concepts in mathematics, while others savored solving particular problems. He considers both styles equally valuable and necessary.

Robert Thomas discovers ludic aspects of mathematics and their virtues, but also the limitations of conceiving of the mathematical activity as a game. He thinks that communicating well is an essential part of doing mathematics, since—unlike in playing games—mathematicians are allowed to disobey strict rules as long as they can make a compelling case for doing so.

Timothy Gowers and Michael Nielsen describe a novel experiment in mathematical research: opening it up to the world, making it available for everybody to see, engage, and enjoy. The Polymath Project they present proposed finding an elementary solution to a particular case of a combinatorial problem; the result was a more effective collaboration than the initiators had expected. As we look into the future of mathematics as an activity, this experiment might well constitute a consequential breakthrough.

The last two texts in this section are similar, yet different. Philip J. Davis and Alicia Dickenstein, in separate contributions, write about mathematical reverberations in the work of a poet (Paul Valéry) and a physicist (Albert Einstein). Davis, following the Valéry exegete Julie Robinson, explains that the ubiquity of mathematical notes in the poet's famous *Cahiers* is far from accidental; instead, they betray Valéry's notion that mathematics succeeds in weeding out the opacity, vagueness, and ambiguity of ordinary language. Dickenstein's note traces the story of Einstein's first page of the original printing of his seminal 1905 article on the theory of relativity. That page, apparently lost through the mishandling of a paper copy in the process of reprinting, happened to contain Einstein's unambiguous (and early) recognition of important mathematicians as predecessors in the formulation of the general theory of relativity. This recovery, partly based on archival material, is not only valuable in itself but supports a line of counterarguments to a sizable, controversial, and conspiratorial literature claiming that Einstein was indebted, without acknowledgment, to the mathematical methods perfected by some of his mathematical predecessors.

MATHEMATICS AND ITS APPLICATIONS

Applied mathematics permeates our lives, even when we do not notice it. Without it we would have no Internet, no scanner at store counters, no airplane flights or space exploration, no global warming measurements, and of course no baseball statistics.

Walter Willinger, David Alderson, and John C. Doyle reconsider the relevance of mathematical modeling as an explanatory means for understanding the Internet. In a compelling piece they caution that scale-free network models ignore the idiosyncrasies and ambiguities characteristic of domain-specific data collection.

Brian Hayes discusses alternatives for solving the difficulties encountered in manipulating the ever-increasing gigantic numbers that pop up in computing, finance, and astronomy. The currently used floating-point methods are inconvenient for several reasons, most problematic being the overflow of digits. Hayes ponders the advantages and disadvantages of three other possibilities: the tapered floating point, logarithm replacement, and level-index systems.

Theodore P. Hill looks at the problem of timing. *When* is the best time to act (or cease acting) in circumstances of variable uncertainties, from economic competition, to war, to marriage, to betting, and beyond. He describes optimal strategies for stopping in situations where only partial information is available. Hill also gives a few examples of stopping problems that are simple to formulate and to understand, yet they remain unsolved.

Barry A. Cipra notes that applied mathematics increasingly uses topological ideas in the form of homology, a mathematical "mechanism" that makes sense globally of meaningless data gathered locally. Using homology, information recorded and transmitted by randomly distributed sensors can be analyzed with respect to matters concerning coverage, constraints, uncertainty, and optimization.

MATHEMATICS EDUCATION

Mathematics education is a rapidly expanding area of research that is riddled by controversial topics. In the United States the discontent is so widespread (touching many of the people who have a stake in mathematics instruction, including teachers, mathematicians, parents, students, and local and federal governments) that it seems to be the sole unifying issue among the participants in this social enterprise with wide-ranging consequences.

In an insightful essay, Anne Watson situates school mathematics in the broader context of the transition from adolescence to adulthood, underscoring both the tensions and the possibilities it holds. Her student-centered vision, enhanced by Vygotsky's ideas concerning close adult support in students' learning, echoes Dewey's century-old calls for transforming education into an institution that prepares youngsters for becoming informed and responsible members of communities.

Kathleen Ambruso Acker, Mary W. Gray, and Behzad Jalali present the legal, practical, and implicitly ethical ramifications of accommodating the learning space to the requirements of students with disabilities. This subject is rarely discussed with respect to mathematics instruction in higher education but is present with increasing frequency in court disputes involving colleges and universities.

David Pimm and Nathalie Sinclair discuss different styles of mathematical writing, another topic seldom thought about by many professional mathematicians and educators. The authors note that writing in mathematics is generally impersonal and contains implicit assumptions not always assumed by the reader or made plain to the learner.

In her solo text chosen for this volume, Nathalie Sinclair further elaborates on aesthetic aspects of mathematics. She examines the alternative elitist and frivolous perspectives on the axiological role of aesthetics in mathematics learning and relates them to pressing factors of concern in mathematics education, including student motivation, interest, creativity, and self confidence.

In a provocative article critical of the mathematics textbook industry, Ann Kajander and Miroslav Lovric review several sources of misconceptions that can be tracked to the sloppiness pervasive in some manuals. Among such sources they identify (by considering just the notion of "tangent") colloquial language, unwarranted generalizations, ignorance of the context, careless figures and diagrams, oversimplifications, and ill-defined concepts.

From the excellent, newly published yearbook of the National Council of Teachers of Mathematics I selected for this volume the instructive contribution by Howard Iseri on using paper models to explore the curvature of various surfaces. Iseri skillfully places the problem in historical context and suggests student-friendly ways for teaching the notion of curvature of both Euclidean and non-Euclidean surfaces.

Finally, in this section, Uri Leron and Orit Hazzan discern four different ways of distinguishing between intuitive and analytical thinking. They do it by placing mathematical thinking in a broader cognitive context that considers general theories of human learning.

HISTORY AND PHILOSOPHY OF MATHEMATICS

The history of mathematics and the philosophy of mathematics deserve separate sections but, for reasons of space, I grouped together several contributions from these two fields.

Judith V. Grabiner traces the unlikely circumstances of an eminent analyst, Joseph-Louis Lagrange, venturing into the fundaments of geometry with the illusion that he "proved" the parallel postulate. Grabiner shows that the incident and the paper that resulted from Lagrange's delusion, far from deserving laughs, allow reconstruction of the worldview held by the leading eighteenth-century scientists—one fundamentally different from ours.

Similarly, Harold M. Edwards recovers the algorithmic/constructivist meaning in Leopold Kronecker's mathematics, observing that not only for Kronecker but also for other distinguished past mathematicians (notably Evariste Galois and Niels Abel), mathematics meant something else than is generally implied today. Alluding to the latest advances in computational mathematics, Edwards asks whether a reconsideration of the definition of real numbers based on a logical foundation is in order.

Carlo Cellucci discusses Gian-Carlo Rota's place in the philosophy of mathematics as reflected by Rota's references to the questions surrounding the existence of mathematical objects, mathematical definitions, and proof in mathematics, as well as the relation between mathematics and the philosophy of mathematics.

Philip L. Bowers gives an engaging first-person account of the metamorphoses, over the last three decades, of problems related to circle coverings of a surface in various geometries and their interdependence on the changing role of topological considerations in geometry.

Mark Colyvan observes that inconsistent mathematical theories have worked for long periods of time, and still do. He then examines in detail this apparent oddity, contributing to the ongoing debate concerning the conditions that lead to the failure of mathematical models in applications.

Andrzej Pelc analyzes the role of informal proofs (which he simply calls proofs) and formal proofs (or logical derivations) in forming and consolidating confidence in mathematical results, as accepted by the mathematical community.

MATHEMATICS IN THE MEDIA

These days mathematical encounters await us wherever we turn—in architecture, arts, playgrounds, video games, movies, sports, politics, fashion, environment, the culinary arts, and countless other spheres of endeavor. This ubiquity is reflected in the media. Out of an enormous number of possible references, I selected a small sample of mathematical exploits that appeared in publications of general interest—not always in the usual places!

Erica Klarreich reports on the Kervaire conjecture, a result concerning the topological structure of manifolds of high dimension. The conjecture states that a large class of shapes in dimensions higher than 126 are fundamentally related to spheres. Although a few particular cases remain unsolved, the proof is important for combining topology with differential topology and for settling the quandary in an unexpected way.

In one of her periodical columns, Julie Rehmeyer traces the prehistory of the statistical Student's t-test, invented by William Gosset, to small-sample problems that preoccupied Darwin, long before Gosset.

Steven Strogatz plays with variations of Romeo and Juliet's love story to illustrate the explanatory power and limitations of mathematics in selecting a spouse. He shows that sophisticated mathematical notions, like that of differential equations, can be employed to describe a large variety of real-life phenomena.

Samuel Arbesman takes us to the movies to unscramble fine mathematical points in the twists of the plots: game theory strategies, the mathematics of networks, and mathematical epidemiology all come to help. Is watching movies enhanced by mathematical knowledge? With verve and conviction, Arbesman leaves no doubt it is.

Following countless anterior writers on the number harmony embedded in human creativity, Vijay Iyer discovers the joy of tracing proportions of consecutive numbers of the Fibonacci sequence in architecture and music.

Nick Paumgarten describes a living project, the idea of a mathematics museum in Manhattan. While taking the author on a midtown tour, the group of mathematicians who plan the museum offer a humorous account of the ubiquity of mathematics.

Other 2009 Writings on Mathematics

Inevitably, the content of this volume is unfairly parsimonious, leaving out many meritorious texts. Practical considerations—and, perhaps, personal bias—played an important role in selection. To alleviate some of the injustice done to the writings that did not make the final cut for various reasons, I offer a brief review of the 2009 literature on mathematics, with no pretension of completeness.

As runners-up for the book I considered numerous other articles, perhaps twice as many as the number of texts finally included. For instance, the publications of the Mathematical Association of America (both in print and online) are a treasure trove of good writing on mathematics. At various points during the selection I pondered whether to include one or another of the many articles published in *Math Horizons*, the MAA's *Focus*, or the monthly columns posted online—but most of them, at least for 2009, proved too technical for the book I had in mind, perhaps better suited for an anthology of remarkable mathematics writing (as distinct from one of writings on mathematics). As this annual series continues, all publications and authors consistently dedicated to mathematical subjects or engaged in debates concerning mathematics will be fairly represented.

There are many texts worth mentioning. For brevity, I will consider only books in the rest of this introduction—with the few exceptions of relatively new and yet little-known periodicals. Readers who wish to help improve my task in future volumes are encouraged to use the contact information provided at the end of this introduction.

Several beautifully illustrated volumes on mathematics appeared in 2009. The most notable is *Mathematicians: An Outer View of the Inner World*, a volume unique in content and graphical presentation. It contains 102 one-page autobiographical capsules written by some of the most renowned mathematicians alive, each essay facing the full-page portrait of its author—with photographs taken by Mariana Cook. A similar volume in graphic aspect, with each page of text facing the corresponding illustration, but organized historically by mathematical milestones, is *The Math Book* by Clifford Pickover. And a second edition of *Symmetry in Chaos*, by Michael Field and Martin Golubitsky, was also published in 2009.

Daina Taimina, who first crocheted a surface approximating a hyperbolic plane in the early 1990s (thus starting a phenomenon that has spread in many art galleries around the world as well as in mathematics classrooms), gives an excellent account of her craft in *Crocheting Adventures with Hyperbolic Planes*. In a similar vein, several articles published in the *Journal of Mathematics and the Arts* explore connections between handicrafts, mathematics, and the teaching of mathematics; most notable are the contributions authored respectively by Sarah-Marie Belcastro and Eva Knoll (see complete references at the end). The Bridges Organization (http://www.bridgesmathart .org/), dedicated to exploring the connections between mathematics and the arts, published its twelfth volume, edited by Craig Kaplan and Reza Sarhangi, with the title *Bridges Banff*. And Michele Emmer edited the sixth volume of the *Mathematics and Culture* series. Finally, I notice the publication of the new *Journal of Mathematics and Culture*.

The literature on mathematics and music has grown fast over the last decade, in periodical publications and in monographs. The year 2009 was particularly good. In *Mathematics and Music*, David Wright looks at the common foundations of the two subjects in a friendly, easy-to-follow format. Barry Parker takes a complementary perspective by analyzing the physics of music in *Good Vibrations*. A slightly more technical book exploring intimate connections not only between mathematics and music but also mathematical reverberations of natural phenomena is the massive *Mathematics of Harmony*, by Alexey Stakhov.

Mathematics education is now a vast field of research, rich in literature. I mention just a few of the many books and journals published in 2009. Caroline Baumann edited a collection of recent policy documents concerning mathematics education. Under the auspices of the National Council of Teachers of Mathematics, a group chaired by Gary Martin issued *Focus in High School Mathematics*. The International Commission for Mathematics Instruction published two new collective studies, *The Professional Education and Development of Teachers of Mathematics,* edited by Ruhama Even and Deborah Loewenberg Ball, and *Challenging Mathematics in and beyond the Classroom*, edited by Edward Barbeau and Peter Taylor. In the series Studies in Mathematical Thinking and Learning two volumes are notable, *Mathematics Teachers at Work*, edited by Janine Remillard and associates, and *Mathematical Literacy*, by Yvette Solomon. And another remarkable volume, in the Routledge Research in Education series, is *Mathematical Relationships in Education*, edited by Laura Black and associates. An extensive examination of contextual mathematics learning for small children is *Numeracy in Early Childhood*, by Agnes Macmillan. And a relatively new journal, *Investigations in Mathematics Learning*, is now in its second year of publication.

Several books on the history of mathematics appeared in 2009. Eleanor Robson and Jacqueline Stedall edited the massive *Oxford Handbook of the History of Mathematics*. William Adams published the second edition of his focused *Life and Times of the Central Limit Theorem*, while Ivor Grattan-Guinness authored *Routes of Learning*, a fascinating account of mathematical encounters throughout history and cultural history. An eclectic account of the origins of mathematical concepts and their various non-mathematical connotations over the centuries is Robert Tubbs's *What Is a Number?*

Bridging the history and the philosophy of mathematics is the volume *Development of Modern Logic*, edited by Leila Haaparanta. Similarly at the intersection of several thinking domains is the anthology of writings by Hermann Weyl, *Mind and Nature*, and the volume *The Big Questions*, by Steven Landsburg. In the Elsevier series Handbooks in the Philosophy of Science, Andrew Irvine edited the comprehensive *Philosophy of Mathematics*. And David Bostock published his one-author introduction with the same title, *Philosophy of Mathematics*.

Among other remarkable books on mathematics published in 2009 are *The Calculus of Friendship*, by Steven Strogatz, a moving account of the decades-long correspondence between the author and his high school calculus teacher; *A Mathematical Nature Walk*, by John Adams, an excellent compendium of questions and answers about mathematical facts found in nature; *Professor Stewart's Hoard of Mathematical Treasures*, a new collection of mathematical curiosities by the prolific Ian Stewart; and *Homage to a Pied Puzzler*, a collected volume edited by Ed Pegg Jr. and associates and dedicated to the foremost writer on entertaining (yet serious) mathematics, Martin Gardner. Also, in *Mythematics*, Michael Huber connects in an intriguing way some of the better-known myths of the Greek mythology to mathematical problems.

Most books on applied mathematics are highly technical, addressing professionals specialized in the respective disciplines and thus beyond the scope of this brief literature review. Somewhat more accessible is *Mathematical*

Methods in Counterterrorism, a volume edited by Nasrullah Memon and collaborators. Given the widespread use of statistics in sports, a useful book is Wayne Winston's *Mathletics: How Gamblers, Managers, and Sports Enthusiasts Use Mathematics in Baseball, Basketball, and Football*. We commonly use mathematics to solve practical problems, including physics problems, but in *The Mathematical Mechanic*, Mark Levi takes the opposite approach, solving mathematical problems by employing physical thinking. Also mentionable as an excellent example of applying the elementary logic of set theory to moral theory is Gustaf Arrhenius's contribution in the volume *Intergenerational Justice*.

Several newspapers of large circulation publish periodically pieces on mathematics, hosted by regular columnists or by invited contributors. This is an encouraging development. For instance, among many other authors, Carl Bialik publishes articles in the *Wall Street Journal*, Steven Strogatz posts insightful pieces for the *New York Times* online edition, and Masha Gessen sends excellent comments to a number of publications.

Mathematics entered cyberspace during the 1990s; by now the trend is maturing, with increasingly sophisticated software enriching the visual experience. The most convenient addresses for keeping up with mathematics in the media are the lists of occurrences updated daily on the homepages hosted by the American Mathematical Society (http://www.ams.org/mathmedia/) and the Mathematical Association of America (http://mathdl.maa.org/mathDL?pa=mathNews&sa=viewArchive). Mathematical Web sites, either sites with plain mathematical content or daily blogs maintained by professional mathematicians, are ever more numerous. They contribute to the fast circulation of ideas and original contributions, methods of research, and even contentious problems. They are so diverse that it is impossible to rank them or even to mention "the best" out there. In the last part of this introduction I commend the reader's attention to several Internet pages with unfailingly high presentation standards. No hierarchical ordering is intended. Further suggestions from the readers of this book are welcome.

John Baez, a mathematical physicist at the University of California, Riverside, maintains an excellent Web page focused on mathematical applications (http://math.ucr.edu/home/baez/TWF.html). Terence Tao of the University of California, Los Angeles, writes a research blog highly popular among mathematicians (http://terrytao.wordpress.com/); he includes wise advice on writing. A group of Berkeley mathematics PhD students maintain the Secret Blogging Seminar (http://sbseminar.wordpress.com/). A comprehensive blog on mathematics education is maintained by Reidar Mosvold, associate professor of mathematics education at the University of Stavanger in Norway (http://mathedresearch.blogspot.com/). The reader who is looking for a more complete list of sources can

consult the following URL address: http://www.ncatlab.org/nlab/show/
Online+Resources.

⚭

This cursory enumeration of remarkable writings on mathematics intelligible to a large readership and published during 2009 gives a measure of the lively scene animating the literature on mathematics and suggests the multifaceted character of mathematics as a mode of thought.

I hope you, the reader, find the same value and excitement in reading the texts in this volume as I found while searching, reading, and selecting them. For comments on this book and to suggest materials for consideration in future volumes, I encourage you to send correspondence to: Mircea Pitici, P.O. Box 4671, Ithaca, NY 14852, or to write electronically (bestmathwriting@ gmail.com).

Works Cited

Adam, John A. *A Mathematical Nature Walk*. Princeton, NJ: Princeton University Press, 2009.

Adams, William J. *The Life and Times of the Central Limit Theorem*. Providence, RI: American Mathematical Society, 2009.

Arrhenius, Gustaf. "Egalitarianism and Population Change." In *Intergenerational Justice*, ed. Alex Gosseries and Lukas H. Meyer, 323–46. Oxford: Oxford University Press, 2009.

Barbeau, Edward J., and Peter J. Taylor, eds. *Challenging Mathematics in and beyond the Classroom*. New York: Springer Science + Business Media, 2009.

Baumann, Caroline, ed. *Success in Mathematics Education*. New York: Nova Science Publishers Inc., 2009.

Belcastro, Sarah-Marie. "Every Topological Surface Can Be Knit: A Proof." *Journal of Mathematics and the Arts* 3, no. 2 (2009): 67–83.

Black, Laura, Heather Mendick, and Yvette Solomon, eds. *Mathematical Relationships in Education*. Abingdon, UK: Routledge, 2009.

Bostock, David. *Philosophy of Mathematics: An Introduction*. Malden, MA: Wiley-Blackwell, 2009

Cook, Mariana, ed. *Mathematicians: An Outer View of the Inner World*. Princeton, NJ: Princeton University Press, 2006.

Emmer, Michele, ed. *Mathematics and Culture VI*. Milan, Italy: Springer-Verlag, 2009.

Even, Ruhama, and Deborah Loewenberg Ball, eds. *The Professional Education and Development of Teachers of Mathematics*. New York: Springer Science + Business Media, 2009.

Field, Michael, and Martin Golubitsky. *Symmetry in Chaos: A Search for Pattern in Mathematics, Art, and Nature*. Philadelphia, PA: Society for Industrial and Applied Mathematics, 2009.

Grattan-Guinness, Ivor. *Routes of Learning: Highways, Pathways, and Byways in the History of Mathematics*. Baltimore, MD: Johns Hopkins University Press, 2009.

Haaparanta, Leila, ed. *The Development of Modern Logic*. Oxford: Oxford University Press, 2009.

Huber, Michael. *Mythematics: Solving the Twelve Labors of Hercules*. Princeton, NJ: Princeton University Press, 2009.

Irvine, Andrew, ed. *Philosophy of Mathematics*. Burlington, MA: Elsevier, 2009.

Kaplan, Craig S., and Reza Sarghangi, eds. *Bridges of Banff: Mathematics, Music, Art, Architecture, Culture*. Banff, AB: Banff Centre, 2009.

Knoll, Eva. "Pattern Transference: Making a 'Nova Scotia Tartan' Bracelet Using the Peyote Stitch." *Journal of Mathematics and the Arts* 3, no. 4 (2009): 185–94.

Landsburg, Steven E. *The Big Questions: Tackling the Problems of Philosophy with Ideas from Mathematics, Economics, and Physics*. New York: Free Press, 2009.

Levi, Mark. *The Mathematical Mechanic: Using Physical Reasoning to Solve Problems*. Princeton, NJ: Princeton University Press, 2009.

Macmillan, Agnes. *Numeracy in Early Childhood: Shared Contexts for Teaching and Learning*. Oxford: Oxford University Press, 2009.

Martin, Gary, at al. *Focus in High School Mathematics: Reasoning and Sense Making*. Reston, VA: National Council of Teachers of Mathematics, 2009.

Memon, Nassurllah, et al., eds. *Mathematical Methods in Counterterrorism*. New York: Springer-Verlag, 2009.

Parker, Barry. *Good Vibrations: The Physics of Music*. Baltimore, MD: Johns Hopkins University Press, 2009.

Pegg, Ed, Jr., Alan H. Schoen, and Tom Rodgers, eds. *Homage to a Pied Puzzler*. Wellesley, MA: A. K. Peters, 2009.

Pickover, Clifford. *The Math Book: From Pythagoras to the 57th Dimension, 250 Milestones in the History of Mathematics*. New York: Sterling Publishing, 2009.

Remillard, Janine T., Beth A. Herbel-Eisenmann, and Gwendolyn Lloyd, eds. *Mathematics Teachers at Work: Connecting Curriculum Materials and Classroom Instruction*. New York: Routledge, 2009.

Robson, Eleanor, and Jacqueline Stedall, eds. *The Oxford Handbook of the History of Mathematics*. Oxford: Oxford University Press, 2009.

Solomon, Yvette. *Mathematical Literacy*. New York: Routledge, 2009.

Stakhov, Alexey. *The Mathematics of Harmony: From Euclid to Contemporary Mathematics and Computer Science*. Singapore: World Scientific, 2009.

Stewart, Ian. *Professor Stewart's Hoard of Mathematical Treasures*. London: Profile Books, 2009.

Strogatz, Steven. *The Calculus of Friendship: What a Teacher and a Student Learned about Life While Corresponding about Math*. Princeton, NJ: Princeton University Press, 2009.

Taimina, Daina. *Crocheting Adventures with Hyperbolic Planes*. Wellesley, MA: A. K. Peters, 2009.

Tubbs, Robert. *What Is a Number? Mathematical Concepts and Their Origins*. Baltimore, MD: Johns Hopkins University Press, 2009.

Weyl, Hermann. *Mind and Nature: Selected Writings on Philosophy, Mathematics, and Physics*. Princeton, NJ: Princeton University Press, 2009.

Winston, Wayne. *How Gamblers, Managers, and Sports Enthusiasts Use Mathematics in Baseball, Basketball, and Football*. Princeton, NJ: Princeton University Press, 2009.

Wright, David. *Mathematics and Music*. Providence, RI: American Mathematical Society, 2009.

Mathematics Alive

The Role of the Untrue in Mathematics

CHANDLER DAVIS

We obtain perspective on any human activity by standing outside it.[*] If mathematics were really concerned mostly with truth, or entirely with truth, then we may imagine that in order to appreciate it fully we might be obliged to position ourselves squarely in a world of fallacy: get a little distance on it. I am not proposing anything quite that quirky. I will speak just as unparadoxically as my subjects permit, but even so, there will still be slippery borderlines. There are plenty of ways in which untrue assertions demand our respect.

Mathematical reasoning can be applied to untrue assertions in the same way as to true ones. We may say to an 18th-century geometer, "Let us assume that through any point not on line l there is more than one line which fails to intersect l," and our interlocutor, no matter how absurd this assumption seems, will be able to scrutinize our deductions in the same way as if we had made less preposterous assumptions. It is necessary for us to be able to reserve judgment in this way—for consider this example: we may say, "Suppose if possible that p/q is a fraction in lowest terms equal to the ratio of a square's diagonal to its side," and we may want to establish that that supposition is not admissible. Then it is important that our interlocutor agree as to what reasoning is valid. If the rules changed, if there were one way accepted for reasoning about true assertions and a different way for all others, then there would be no way to prove anything by contradiction.

Let me nail this observation down a little more snugly. It is not an observation about *tertium non datur*. Maybe our interlocutor is skeptical about the notion that every meaningful proposition must be either true or false; that's all right: even if we allow that truth status need not be a binary alternative, still when we want to argue that a proposition fails to have some truth

[*]This article is the text of a talk presented to the Joint Mathematics Meetings, Washington, D.C., USA, in 2009.

status, we may have to use methods of argument that do not depend on that truth status.

Lawyers are clearer about this, perhaps. They frequently say, "Supposing, arguendo, that . . ." and proceed, arguing temporarily as though they were conceding a premise that they are not at all willing to concede. I like that notion of arguendo. Mathematicians used to use it more than they do today. In particular, the whole magnificent edifice of classical continuum mechanics seems to me to be a case of supposing arguendo that continuous media obey laws of particle mechanics which, however, the Bernoullis and Euler did not really expect them to obey: a dollop of matter has mass as though it were localized at a point, and Newton's laws are invoked even in problems where the idealization to point masses would be nonsensical. Maybe I'm on safer ground if I cite a different example: the development of topology of manifolds in the 20th century. It was plain that certain aspects of manifolds deserved study, but it was not clear what they applied to—whether to certain chain complexes, or to certain abstract topological spaces, or what. The study proceeded arguendo by deductions as reliable as they could be made under the circumstances, and discrepancies between different entries into the subject were tidied up as well as might be. (The forging ahead and the tidying up are seen together in a book such as Raymond Wilder's.) A chain of reasoning belonging to such an intellectual domain may turn out in the future to relate two chapters of truth, or it may turn out to be part of a great *reductio ad absurdum*; we deal with the deductive chain arguendo, independent of its ultimate fate; the tests of its soundness are the same either way.

Even more persuasive for my purposes today is another centuries-old habit of mathematicians: to find what value of the variable makes a function zero, one pulls a guess out of the air and substitutes it into the function, finding of course that the function fails to be zero there; then one extracts information about the problem from the failed guess. Such calculations by *regula falsi* were used off and on over the centuries to become systematized and exceedingly fecund from the 16th century on. Their naïve motivation must have been, back then, like reasoning arguendo, and this doesn't seem far off to me even in retrospect. One would be justified in 1100 (or in 1600) in trying 1.5 to see whether its square was about 2, even if one did not have an algorithm of root-finding and therefore did not know how taking this stab at $\sqrt{2}$ would lead one to a better guess. One would anticipate that working with the blind guess would teach something. And I observe in this context too, of course, that in order to hope to be taught anything, one would surely commit to reasoning the same for a wrong guess as for a correct one.

The freakish notion that mathematics deals always with statements that are perfectly true would disallow any validity for this example, or for most discussion of approximation, for that matter. It would insist that "$\pi = 22/7$" be banned from mathematics as utterly as "$\pi = 59$"; half of our subject would be ruled out. There is no danger that mathematicians of this or any other age would really try to live by this freakish doctrine, but it persists in everyday discourse about mathematics. I began with discussion unrelated to approximate answers in order to emphasize that restriction to true statements would be crippling to even the most finite and discrete branches.

As we begin to examine the useful roles of less-than-true statements in our field, we have at once these two:

- Falsehood is something to avoid. We find it useful to reason by contradiction.
- Statements teach us something by their behaviour in reasoning arguendo, regardless of their truth value.

Then to continue the examination of the subject, we must recognize and defy the tradition that mathematics is truth, the whole truth, and nothing but the truth. Permit me to call it "truth-fetishism," though I accord it more respect than the playful label suggests. This tradition has taken many forms, and you may not agree with me in lumping them together.

By the 19th century, it had become clear that some true statements are contingent whereas others are essential to the cogency of human reasoning. (Thus it is merely a matter of observation that the South Pole is not in an ocean, and we can talk about it being in an ocean, even ask whether it once was; but we can not talk in any cogent way about the South Pole being on the Equator.) Truth-fetishism applied mostly to truths which were not contingent. In the decades after George Boole's "Investigation of the Laws of Thought," it became conventional to hold that, at least in philosophical and mathematical discourse, all true statements were equivalent, so that any true statement implied every true statement, and a false statement implied every statement whatever. If you have ever tried to get a freshman class to swallow this, you have probably appreciated the trouble Bertrand Russell had in his day.

Yes, I find it irresistible to retell the Russell anecdote: replying to the lay listener who objected that surely "$2 = 1$" does not imply that you are the Pope, Russell's put-down was, "You will agree that the Pope and I are two; then if $2 = 1$ it follows that the Pope and I are one." Now his verbal cleverness is charming, but it is off the point of the listener's objection, as the listener surely saw and we may hope Russell did as well. He was insisting that the only way to deal with truth and validity was the truth-fetishistic way, and that the only way to understand implication was material implication: that saying "A implies B" must be understood as saying "either B or not-A".

Boole's ambitious project of finding the laws of thought deserves the admiration it got. What kind of law should we hope for? We don't really want a prescriptive law ("thou shalt think in thy father's way") or a normative law ("here is the better way to think"), we want an empirical law, one that refers to thinking that is actually done. On the other hand, we can't insist that the laws of thought encompass our occasional pathology and our frequent simple blundering (to do that would be a formidable, never-ending task); so there is some normative selection; let us ask, however, for laws that apply to thinking as well as may be done. That doesn't mean surrendering to the truth-fetishists. Both before Boole's time and since, when given propositions *A* and *B* that have nothing to do with each other, a thinker does not set about inferring *B* from *A*.

It was natural, then, that even while truth-fetishism was extending its dominion, various resistance movements sprang up. Strict implication was distinguished from material implication, in the following sense. To say that *A* implies *B* could be regarded as a contingent statement even within logic, and there was a stronger statement that sometimes might hold: one distinguished the statement that *A* implies *B* from the statement that *A* must imply *B*, then one tried to elaborate rules of symbolic manipulation appropriate to thinking where both kinds of implication came into consideration. This "modal logic" of Langford and Lewis seemed to be a realistic strengthening of the vocabulary. Indeed, because there may be various bases for regarding an implication as necessitated, I even thought it worthwhile to allow for various strict implications within the same system. But in retrospect this program does not look like much of a success. None of the various algebraizations of strict implication seem to deepen one's understanding of thought.

At the same time there was some attention paid to allowing truth values intermediate between true and false, as by Jan Łukasiewicz. This is at least a start on embodying the notions expressed in everyday language by "sort of true" or "yes and no." It is a limitation to insist that the intermediate truth values be totally ordered—a limitation that could be overcome, and by the way, the corresponding limitation is not suffered by modal logic with multiple modal operators. In the later invention of "fuzzy logic" by Lotfi Zadeh, it is claimed that still greater flexibility is obtained.

In short, the 20th century brought us to an acknowledgement that truth may be of various strengths. The Gödel incompleteness theorems suggested that this was even unavoidable, that no matter how faithfully one hewed to the line that truth was the goal, there could never be a notion of truth that would sort all possible mathematical statements into an army of true ones and an army of others (their negations). If every axiom system leaves undecided propositions, then it seemed that every mathematician on the corridor might make a different choice of what arithmetic facts were facts.

Yet my deep discomfort with truth-fetishism is not addressed by making truth relative to a choice of axioms for set theory and arithmetic. We can agree that asserting "*A*" is distinct from asserting "*A* is provable (in some specified axiom system)" and distinct again from asserting "*A* is provable (in such-and-such other system)"; certainly all are distinct from asserting "*A* is nine-tenths true" or "*A* is sort of true"; and the list of options can be extended, as I will presently maintain. None of the options takes care of the big issue: logic based only on truth values is an impoverished logic, in that it sets aside intrinsic relations between concepts. I stipulate, in case it is not already plain, that by "relations" here I do not mean subsets of some direct product, as in many elementary developments of mathematics. I mean substantive relations.

Let me turn to some other quarrels I have with truth-fetishism.

Many spokesmen may say, since Boole, that all true theorems are equivalent and every true theorem is a tautology. As hyperbole, I understand this and endorse it; but oh, what it leaves out! First, it renounces any distinction between hard and easy theorems; second, it renounces any distinction based on what the theorems are about.

Similarly, many spokesmen may say that our aim in mathematics is to simplify every proof to self-evidence—that the ultimately desired proof of a statement's truth not only is necessarily tautological but also is plainly so. (I recently saw this thesis attributed to Gian-Carlo Rota, but many before him subscribed to it, and he, to my reading, did not.) Again, this ignores too much. Granted, we try to clear away the extraneous, and the Book proof never goes off on an unnecessary detour; but sometimes a theorem is valued for bringing together pieces from different conceptual sources, and this pay-off may be reduced if we simplify to allow quick attainment of a conclusion.

A third, grotesque example: some metamathematicians would have it that a mathematical theory is the set of propositions that are true in it. I have ranted against this interpretation elsewhere, maybe I don't need to belabour it here, but please indulge me while I do. A theory that is any good says something, opens some door. It has high points and central points; it has beginnings (I don't mean only axioms) and endings. Most likely it has avenues to other theories. The collection of all propositions that are true in it, on the other hand (if we could ever apprehend such a monster, which I doubt), consists mostly of banalities, so there can be no order, no revelation or insight. Such is not our science.

We must try to get realistic about deduction, the deduction we actually do at our best. What may one say in practice about a statement? On just the dimension of truth value—one may assert it; or going farther, one may claim to be able to prove it; or going less far, one may say one tends to believe it. If this is within mathematics, however, one probably says something with more structure. Let's take a realistic possibility. One may say, typically, "I think I can prove it with an additional hypothesis." But stop right there! The truth-fetishist calls this vacuous; for of course the additional hypothesis could be the conclusion, and then of course it can be proved. We know this is irrelevant; we want to protest, "This is beside the point," just as Bertrand Russell's listener did; we feel the need for a sense to ascribe to the property "provable with an additional hypothesis." But needs such as this have not been described by modal logic or many-valued logic or fuzzy logic, and I suggest that enlarging the lexicon of truth-values is not the way to go to describe them. Truth—even truth understood in some new sophisticated way—is not the point.

The point is pertinence. The point is relevance.

In the last half-century, serious efforts have gone into analyzing relevance, but they commonly rely on deliberately ignoring the content. This approach is admittedly a sidetrack from the direction of my quest, but I can't brush it off. Today one may marshal computer power to discover which of a large population (of people; of factors in a plan; of propositions) are most related, but one often is looking just for the existence of some strong relation rather than for its nature. Only connect, as E. M. Forster said! This may be done in a search engine by looking for the singular values of a very sparse matrix of very large order. It is the few non-zero entries that guide us. At the intersection of Bacon and Shakespeare in the matrix appears a rather large number; people who deny that Bacon wrote the Shakespeare plays are there, right alongside those who affirm it, indiscriminately; and the weighting is upped by any discussion of the issue, including the present one, whether or not any new insight is achieved. In short, the relations are in the form of a weighted graph with positive integer weights. I can even give an example much closer to home: suppose two words are connected if they are often used in the same utterance; then "knife" and "bandage" will both have connections to "wound" although one causes the wound and the other is a response to it. This is an approach that suppresses syntax and even the distinction between yes and no.

As psychologists speak of the impact of *mere exposure* to a stimulus regardless of positive or negative reinforcement, this way of boiling down intricate data uses *mere association* regardless of the nature of the association.

The approach seemed wrongheaded to me at first, I confess. Architects said they broke down their design task by drawing a graph of which considerations were related and then analyzing the graph computationally to find subgraphs—sub-tasks—which could be carried out by separate teams. This was said to lead to efficient sharing of design effort; I was skeptical. But mere association can be a precious bit of knowledge these days, and I will pause to acknowledge it.

What leads people to apply such blunt tools is the extremely large number of variables one may be trying to handle in many an application today. Take again minimization of an objective function. If there are fifty thousand independent variables, inevitably most of them will be without effect on the function's value, and in such a fix, the finding of one that does have an effect is a big part of the solution, even if one doesn't find out at once what value one should best assign to it. What's more, it may be that not one of the variables affects the value of the objective function enough to rise above roundoff error: only by a better choice of coordinate system, perhaps, can directions having noticeable effect be chosen. There are many contexts that impose hugely many interacting variables, but I want to mention one where this so-called "combinatorial explosion" sneaks up on us. In behavioural evolutionary theory, the traits whose selection one seeks to reconstruct are not life histories but *strategies*, that is, complete repertories of responses to life's predicaments; thus any serious attempt must run up against the game-theoretic feature that the number of strategies grows with the size of the game faster than polynomially.

Perceiving pertinence may be undertaken, then, by means having a family resemblance to the extensional characterization of properties, by methods in which the nature of relations between two things is banished from consideration.

I do not cease to feel that pertinence should be respected as having structure, that what we employ, whether in reasoning or in observational science, should be not mere association but the structure of the association. Even if we call on high technology to explore a graph of connections between items, it will be natural to refine it to be a directed graph, a colored graph, and surely much more. Only connect!—but there is such a wealth of ways that two nodes may be connected.

Between two propositions there may subsist (aside from their truth or falsehood, as those may be in doubt) the relation that one entails the other, or the converse, or both, or neither. Though this may be all the truth-fetishist recognizes, we see every day that the relations possible between propositions are much more diverse. Simple illustrations will make my point. Let me begin with a mantra of 20th-century math education: "'but' means 'and.'" We all know that this makes partial sense: namely, if one says "John is poor but happy" one is asserting both "John is poor" and "John is happy". Nevertheless "but" is a major component in the structure of thought (like "nevertheless"), and the version having "but" as the connective is not the same as the conjunction of the two simple assertions. Many English speakers would find "John is poor but happy" cogent but not "John is rich but happy." You will easily find more and subtler everyday examples. Examples within mathematics are subtler, inexhaustible, but more elusive; I will content myself with the one I already gave.

Of course I do not maintain that natural languages contain all the precision we seek for our logic. On the contrary, their ambiguities are sometimes just concessions to imprecision. If one says "John is rich so he is happy," it is not clear whether one means to assert that every rich person is sure to be happy; it is clear only that something is being said beyond the conjunction of "John is rich" with "John is happy." Similarly for the connective "aussi" in French. I do maintain that syntax of natural languages and our experience with reasoning can yield a great enrichment of our logical conceptual resources. The reason proofs are expressed in natural language is not only our deplorable lack of facility in reading formulas (however large a factor that may be), it is also the great power of nuance in natural language. The proper continuation of Boole's program is to do as much for relations as he did for truth-versus-falsehood. There is gold in those hills. My prospector's hunch is that the most promising underexploited lode is prepositions.

By now I have surely advanced enough dubious doctrines for one afternoon, but if I stopped now you would feel the lack of any mention of probability theory. It must fit into my talk somehow, right? Just so: it is a part of mathematics, it deals throughout with propositions which may turn out untrue, and I do have some dubious things to say about it. I was just saving it for last.

I have been discussing mostly the 19th and 20th centuries, but we must glance back now to the 16th. At its inception, was probability regarded as a competing notion of truth? There is no doubt that the idea of probability was close to the idea of truth at that early stage—etymologically, "probable" is "provable," and even today, "probity" means utter reliability—and the emerging notion of something having positive probability had to be disentangled from the different notion of appearing credible. This fascinating story has been closely studied in recent years, especially by Ian Hacking and Lorraine Daston, and I have nothing to add to their work. I pick up the story with the incorporation of probability into physics in the 19th century and its reconciliation with mathematics in the 20th.

The first development, the creation of statistical mechanics by Ludwig Boltzmann and others, and its success as a part of physics, has a consequence for the idea of truth in mathematics. Some statements about physical systems are definitively shown untenable if they are shown to hold with probability 0—or just with probability extremely close to 0. An applied mathematician has an obligation to accept the conclusion that the ice cube melting in your glass of water is not going to separate out again into ice, because the molecular theory that assigns to that outcome a prohibitively low probability is successful. The theory also says that the sequence followed by the molecules during melting is reversible; the reverse process is nevertheless ruled out, and the argument uses probability. If G. H. Hardy or some other deplorer of applying mathematics wants to retain a notion of possibility for the ice cube to be reborn, fine; all I am saying is that a different notion of possibility and impossibility has emerged in statistical physics. That it happens to involve probability extremely close to 0 instead of probability 0 is just an aggravation of the antithesis. We have a hierarchy: something may be known untrue; or more generally it may be known to have probability 0; or still more generally it may be known to have so low a probability as to be ruled out. Let me emphasize that the consequences for relating predicted behavior to observed behavior are the same for all three. A physical theory which disproves a phenomenon we observe is refuted; but a physical theory which assigns a prohibitively low probability to a phenomenon we observe is refuted just as thoroughly. Statistical physics has extended its sway in the last hundred years and we must live with it. Its criterion of truth deserves our respect.

Finally, a look at probability as a mathematical theory. With A. N. Kolmogorov's wonderful little book (1933), probability seemed to have earned a place at the table of mathematics. Its special notions had been put in correspondence with notions of analysis and measure theory which were as clear as the rest. Aside from its application to gambling, insurance, and statistical physics, probability was now welcomed as a tool within mathematics. To speak of an event having probability 0 was exactly to speak of a subset having measure 0. A striking string of theorems came forth over the years. One did not conclude that a phenomenon was certain to happen by proving its probability was 1; but if one could prove its probability was 1, or merely that it had probability greater than 0, one could conclude that it was *capable* of happening: a set of measure greater than 0 had to have some elements in it. The first striking achievement of this sort long predated Kolmogorov, actually (and it was expressed in the probabilistic terms): Emile Borel proved that a sequence of decimal digits chosen at random represents a normal number with probability 1, and this provided the first proof that it is even possible for a number to be normal. (Many of you know the definition: a decimal expansion is normal provided that all sequences of k digits occur as subsequences of it with the same asymptotic frequency, and that for every k. Correspondingly, other bases than ten can be brought in.) Now I comment first on how sharply the relation of probability to mathematical truth here contrasts with what we saw in physics. The number-theorist contentedly deals with sets of measure 0, and moreover values the positive measure of a set primarily for its guarantee that the set is not empty.

While we were talking physics, there was perhaps a temptation to live on the intermediate level of the hierarchy I mentioned: to disbelieve in events of probability 0. That doesn't work, of course. Yet there is a serious catch in the number-theorist's usage too: to say that measure greater than 0 ensures that a set has members is to defy intuition. The intuitionist responds, "The set has members? Really? Show me one." Today, after a century of debate, this catch is clarified, but, far from going away, it appears insuperable. The constructible numbers (by any appropriate definition) are a set of measure 0; yet they are the only numbers that might be shown. In the conventional terminology of 20th-century analysis, almost all real numbers are not constructible; in our experience, every real number that can be specified is constructible.

Now Kolmogorov knew all of this, he understood it better than the rest of us do, yet it seems not to have bothered him. Shall I assume that he had confidence that we would be able to straighten things out after his death? That's kind of him. By all means let us try.

Desperately Seeking Mathematical Proof

MELVYN B. NATHANSON

How do we decide if a proof is a proof? Why is a random sequence of true statements not a proof of the Riemann hypothesis? That is, how do we know if a purported proof of a true theorem is a proof?

Let's start with remarks on language. The phrase "true theorem" is redundant, and I won't use it again, since the definition of "theorem" is "true mathematical statement," and the expression "false theorem," like "false truth," is contradictory. I shall write "theorem" in quotation marks to denote a mathematical statement that is asserted to be true, that is, that someone claims is a theorem. It is important to remember that a theorem is not false until it has a proof; it is only unproven until it has a proof. Similarly, a false mathematical statement is not untrue until it has been shown that a counterexample exists or that it contradicts a theorem; it is just a "statement" until its untruth is demonstrated.

The history of mathematics is full of philosophically and ethically troubling reports about bad proofs of theorems. For example, the fundamental theorem of algebra states that every polynomial of degree n with complex coefficients has exactly n complex roots. D'Alembert published a proof in 1746, and the theorem became known as "D'Alembert's theorem," but the proof was wrong. Gauss published his first proof of the fundamental theorem in 1799, but this, too, had gaps. Gauss's subsequent proofs, in 1816 and 1849, were okay. It seems to have been difficult to determine if a proof of the fundamental theorem of algebra was correct. Why?

Poincaré was awarded a prize from King Oscar II of Sweden and Norway for a paper on the three-body problem, and his paper was published in *Acta Mathematica* in 1890. But the published paper was not the prize-winning paper. The paper that won the prize contained serious mistakes, and Poincaré and other mathematicians, most importantly, Mittag-Leffler, engaged in a conspiracy to suppress the truth and to replace the erroneous paper with an extensively altered and corrected one.

There are simple ways to show that a purported proof of a false mathematical statement is wrong. For example, one might find a mistake in the proof, that is, a line in the proof that is false. Or one might construct a counterexample to the "theorem." One might also be able to prove that the purported theorem is inconsistent with a known theorem. I assume, of course, that mathematics is consistent.

To find a flaw in the proof of a theorem is more complex, since no counterexample will exist, nor will the theorem contradict any other theorem. A proof typically consists of a series of assertions, each leading more or less to the next, and concluding in the statement of the theorem. How one gets from one assertion to the next can be complicated, since there are usually gaps. We have to interpolate the missing arguments, or at least believe that a good graduate student or an expert in the field can perform the interpolation. Often the gaps are explicit. A typical formulation (Massey [2, p. 88]) is:

> By the methods used in Chapter III, we can prove that the group $\Pi(X)$ is characterized up to isomorphism by this theorem. We leave the precise statement and proof of this fact to the reader.

There is nothing improper about gaps in proofs, nor is there any reason to doubt that most gaps can be filled by a competent reader, exactly as the author intends. The point is simply to emphasize that proofs have gaps and are, therefore, inherently incomplete and sometimes wrong. We frequently find statements such as the following (Washington [5, p. 321]):

> The Kronecker-Weber theorem asserts that every abelian extension of the rationals is contained in a cyclotomic field. It was first stated by Kronecker in 1853, but his proof was incomplete. . . . The first proof was given by Weber in 1886 (there was still a gap . . .).

There is a lovely but probably apocryphal anecdote about Norbert Wiener. Teaching a class at MIT, he wrote something on the blackboard and said it was "obvious." One student had the temerity to ask for a proof. Wiener started pacing back and forth, staring at what he had written on the board, saying nothing. Finally, he left the room, walked to his office, closed the door, and worked. After a long absence he returned to the classroom. "It *is* obvious," he told the class, and continued his lecture.

There is another reason why proofs are difficult to verify: Humans err. We make mistakes and others do not necessarily notice our mistakes. Hume [1, Part IV, Section I] expressed this beautifully in 1739:

> There is no Algebraist nor Mathematician so expert in his science, as to place entire confidence in any truth immediately upon his discovery of it, or regard it as any thing, but a mere probability. Every time he runs over his proofs, his confidence increases; but still more by the

approbation of his friends; and is raised to its utmost perfection by the universal assent and applauses of the learned world.

This suggests an important reason why "more elementary" proofs are better than "less elementary" proofs: The more elementary the proof, the easier it is to check and the more reliable is its verification. We are less likely to err. "Elementary" in this context does not mean elementary in the sense of elementary number theory, in which one tries to find proofs that do not use contour integrals and other tools of analytic function theory. On elementary versus analytic proofs in number theory, I once wrote [3, p. ix],

> In mathematics, when we want to prove a theorem, we may use any method. The rule is "no holds barred." It is OK to use complex variables, algebraic geometry, cohomology theory, and the kitchen sink to obtain a proof. But once a theorem is proved, once we know that it is true, particularly if it is a simply stated and easily understood fact about the natural numbers, then we may want to find another proof, one that uses only "elementary arguments" from number theory. Elementary proofs are not better than other proofs, . . .

I've changed my mind. In this paper I argue that elementary (at least, in the sense of easy to check) proofs really are better.

Many mathematicians have the opposite opinion; they do not or cannot distinguish the beauty or importance of a theorem from its proof. A theorem that is first published with a long and difficult proof is highly regarded. Someone who, preferably many years later, finds a short proof is "brilliant." But if the short proof had been obtained in the beginning, the theorem might have been disparaged as an "easy result."

Erdős was a genius at finding brilliantly simple proofs of deep results, but, until recently, much of his work was ignored by the mathematical establishment. Erdős often talked about "proofs from the Book." The "Book" would contain a perfect proof for every theorem, where a perfect proof was short, beautiful, insightful, and made the theorem instantaneously and obviously true. We already know the "Book proofs" of many results. I would argue that we do not, in fact, fully understand a theorem until we have a proof that belongs in the Book. It is impossible, of course, to know that every theorem has a "Book proof," but I certainly have the quasi-religious hope that all theorems do have such proofs.

There are other reasons for the persistence of bad proofs of theorems. Social pressure often hides mistakes in proofs. In a seminar lecture, for example, when a mathematician is proving a theorem, it is technically possible to interrupt the speaker in order to ask for more explanation of the argument. Sometimes the details will be forthcoming. Other times the response will be that it's "obvious" or "clear" or "follows easily from previous results." Occasionally speakers respond to a question from the audience with a look that conveys the message that the questioner is an idiot. That's why most mathematicians sit quietly through seminars, understanding very little after the introductory remarks, and applauding politely at the end of a mostly wasted hour.

Gel'fand's famous weekly seminar, in Moscow and at Rutgers, operated in marked contrast with the usual mathematics seminar or colloquium. One of the joys of Gel'fand's seminar was that he would constantly interrupt the speaker to ask simple questions and give elementary examples. Usually the speaker would not get to the end of his planned talk, but the audience would actually learn some mathematics.

The philosophical underpinning to this discussion is the belief that "mathematical" objects exist in the real world, and that mathematics is the science of discovering and describing their properties, just as "physical" objects exist in the real world and physics is the science of discovering and describing their properties. This is in contrast to an occasionally fashionable notion that mathematics is the logical game of deducing conclusions from interesting but arbitrarily chosen finite sets of axioms and rules of inference. If the mathematical world is real, then it is unlikely that it can be encapsulated in any finite system. There are, of course, masterpieces of mathematical exposition that develop a deep subject from a small set of axioms. Two examples of books that are perfect in this sense are Weil's *Number Theory for Beginners*, which is, unfortunately, out of print, and Artin's *Galois Theory*. Mathematics can be done scrupulously.

Different theorems can be proven from different assumptions. The compendium of mathematical knowledge, that is, the collection of theorems, becomes a social system with various substructures, analogous to clans and kinship systems, and a newly discovered theorem has to find its place in this social network. To the extent that a new discovery fits into an established community of mathematical truths, we believe it and tend to accept its proof. A theorem that is an "outsider"—a kind of social outlaw—requires more rigorous proof, and finds acceptance more difficult.

Wittgenstein [6, p. 401] wrote,

> If a contradiction were now actually found in arithmetic—that would only prove that an arithmetic with *such* a contradiction in it could render very good service; and it would be better for us to modify our concept of the certainty required, than to say it would really not yet have been a proper arithmetic.

This passage (still controversial in the philosophy of mathematics) evidences a pragmatic approach to mathematics that describes how mathematicians behave in the privacy of their offices, in contrast to our more pietistic public pronouncements.

Perhaps we should discard the myth that mathematics is a rigorously deductive enterprise. It may be more deductive than other sciences, but hand-waving is intrinsic. We try to minimize it and we can sometimes escape it, but not always, if we want to discover new theorems.

References

[1] David Hume, *A Treatise of Human Nature*, Barnes and Noble, New York, 2005.

[2] William S. Massey, *A Basic Course in Algebraic Topology*, Graduate Texts in Mathematics, vol. 127, Springer-Verlag, New York, 1991.

[3] Melvyn B. Nathanson, *Elementary Methods in Number Theory*, Graduate Texts in Mathematics, vol. 195, Springer-Verlag, New York, 2000.

[4] Melvyn B. Nathanson, *Desperately seeking mathematical truth*, Notices Amer. Math. Soc. 55:7 (2008), 773.

[5] Lawrence C. Washington, *Introduction to Cyclotomic Fields*, 2 ed., Graduate Texts in Mathematics, vol. 83, Springer-Verlag, New York, 1996.

[6] Ludwig Wittgenstein, *Remarks on the Foundations of Mathematics*, MIT Press, Cambridge, 1983.

An Enduring Error

Branko Grünbaum

Introduction

Mathematical truths are immutable, but mathematicians do make errors, especially when carrying out non-trivial enumerations. Some of the errors are "innocent"—plain mistakes that get corrected as soon as an independent enumeration is carried out. For example, Daublebsky [14] in 1895 found that there are precisely 228 types of configurations (123), that is, collections of 12 lines and 12 points, each incident with three of the others. In fact, as found by Gropp [19] in 1990, the correct number is 229. Another example is provided by the enumeration of the uniform tilings of 3-dimensional space by Andreini [1] in 1905; he claimed that there are precisely 25 types. However, as shown [20] in 1994, the correct number is 28. Andreini listed some tilings that should not have been included, and missed several others—but again, these are simple errors easily corrected.

Much more insidious are errors that arise by replacing enumeration of one kind of object by enumeration of some other objects—only to disregard the logical and mathematical distinctions between the two enumerations. It is surprising how errors of this type escape detection for a long time, even though there is frequent mention of the results. One example is provided by the enumeration of 4-dimensional simple polytopes with eight facets, by Brückner [7] in 1909. He replaced this enumeration by that of 3-dimensional "diagrams" that he interpreted as Schlegel diagrams of convex 4-polytopes, and claimed that the enumeration of these objects is equivalent to that of the polytopes. However, aside from several "innocent" mistakes in his enumeration, there is a fundamental error: While to all 4-polytopes correspond 3-dimensional diagrams, there is no reason to assume that every diagram arises from a polytope. At the time of Brückner's paper, even the corresponding fact about 3-polyhedra and 2-dimensional diagrams had not yet been established—this followed only from Steinitz's characterization of complexes that determine convex polyhedra [45], [46]. In fact, in the case considered by Brückner, the assumption is not only unjustified, but actually wrong: One of Brückner's polytopes does not exist, see [25]. Other examples of a similar nature involve the enumeration of types of isohedral or isogonal tilings of the plane. In many works, the tilings in question were replaced—for purposes of enumeration—by labeled or marked tilings, or by pairs consisting of a tiling and a group of symmetries. However, the results were erroneously claimed to represent classifications of the tilings proper. The literature is too numerous to be adequately quoted here; the reader should consult Chapters 6 to 9 of [24].

This brings us to the actual topic of this paper. Polyhedra have been studied since antiquity. It is, therefore, rather surprising that even concerning some of the polyhedra known since that time there is a lot of confusion, regarding both terminology and essence. But even more unexpected is the fact that many expositions of this topic commit serious mathematical and logical errors. Moreover, this happened not once or twice, but many times over the centuries, and continues to this day in many printed and electronic publications; the most recent case is in the second issue for 2008 of this journal. I will justify this harsh statement soon, after setting up the necessary background. We need first to clarify the enduring confusion in terminology, and then discuss the actual enduring errors.

Archimedean and Uniform Polyhedra

Several kinds of polyhedra have made their first appearance in antiquity, and continued to be investigated throughout the ages. Probably best known among these are the five *regular polyhedra*, also known as *Platonic solids*. Representatives of these five kinds of *convex* polyhedra are beautifully illustrated in countless publications, in print and electronic. Over the centuries, there have been many different definitions, all leading to the same set of five polyhedra—although in some cases a sizable grain of salt has to be supplied in order to reach that goal. One old and widely accepted definition is that a *convex* polyhedron is regular provided all its faces are congruent regular polygons, meeting in the same number at every vertex. A more recent definition stipulates that a convex polyhedron is regular provided the set of isometric symmetries of the polyhedron acts transitively on the family of all flags. (A *flag* is a triplet consisting of a face, an edge, and a vertex, all mutually incident). Although the former definition relies on strictly *local* conditions and the latter one on *global* properties, the remarkable fact is that they determine the same five polyhedra; many other definitions do the same.

Convex *Archimedean polyhedra* form another well-known family, also illustrated in many venues. They are frequently defined by the following requirement, similar to the first one for Platonic solids:

Local criterion: All faces are regular polygons, and the cyclic arrangement of the faces around each vertex is the same.

In this context "same" is understood to allow mirror images, and "around"
to include only faces that are incident with the vertex in question.

In contrast to this "local" definition stands the following "global" one:

Global criterion: All faces are regular polygons, and all vertices form
one orbit under isometric symmetries of the polyhedron.

Both definitions obviously include Platonic polyhedra, as well as regular-
faced prisms and antiprisms of arbitrarily many sides. However, many writers specify that the polyhedra just mentioned should (by fiat, and because of
tradition) be excluded and that by "polyhedra satisfying the local (or the
global) criterion" we should understand only those that are neither regular,
nor prisms or antiprisms. For simplicity of exposition, in what follows we
shall accede to this view even though it lacks any logical basis.

A lot of confusion surrounding the topic is the result of inconsistent
terminology. Many writers call **Archimedean** those polyhedra that satisfy
the local criterion, and many call **uniform** or **semiregular** the ones that
satisfy the global criterion. However, others give the name Archimedean
polyhedra to those satisfying the global definition. Still other writers (such
as Walsh [50] or Villarino [49]) consider "Archimedean" and "semiregular"
as denoting the same polyhedra, as specified by the local definition. Now,
since there are two differing definitions it is reasonable to give the two
classes of polyhedra different names. If this is accepted, then the polyhedra
satisfying the local criterion should be called Archimedean, since it is a
stretch to impute to Archimedes an approach via groups of symmetries—as
a matter of historical fact, before the nineteenth century nobody was thinking of groups, least of all in geometry. The polyhedra satisfying the global
criterion should be called uniform or semiregular.

The lack of standardization of terminology would be only a matter of
pedantic hairsplitting if it were not for the following two facts. First, "Archimedean" is the term used most frequently, even though many writers
using it do not specify what they mean by "Archimedean polyhedra" (or by
polyhedra designated by any of the other names). In the present paper we
shall consistently use Archimedean to denote polyhedra satisfying the local
criterion.

The second fact is much more important. One might expect that—in
analogy to the situation concerning regular polyhedra—the local and global
definitions yield the same polyhedra. If this were so, there would not be

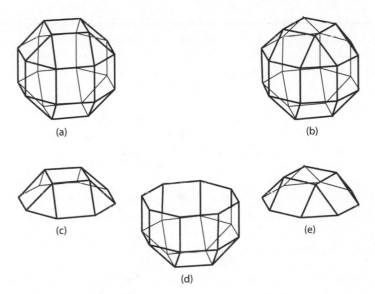

(a) (b)

(c) (e)

(d)

FIGURE 1. (a) The Archimedean (and uniform) rhombicuboctahedron (3.4.4.4), and (b) the Archimedean but non-uniform pseudorhombicuboctahedron. In (d) is shown the common bottom part of the two polyhedra, while (c) and (e) show how they differ in the top part—by a twist of $\pi/4 = 45°$. The top four vertices (and the bottom four) of the pseudorhombicuboctahedron are not equivalent by symmetries of the polyhedron to the middle eight vertices.

much point in insisting on different names; no confusion arises in the case of the regular polyhedra. However, **this coincidence does not occur**. With our understandings and exclusions, there are **fourteen** "Archimedean" convex polyhedra (that is, those that satisfy the local criterion), but only **thirteen** "uniform" (or "semiregular") that satisfy the global criterion. Representatives of the thirteen uniform convex polyhedra are shown in the sources mentioned above, while the fourteenth polyhedron is illustrated in Fig. 1. It satisfies the local criterion but not the global one, and therefore is—in our terminology—Archimedean but not uniform.

The history of the realization that the local criterion leads to fourteen polyhedra will be discussed in the next section; it is remarkable that this development occurred only in the 20th century. This implies that prior to the twentieth century *all enumerations of the polyhedra satisfying the local criterion were mistaken*. Unfortunately, many later enumerations make the same error.

It may seem that the confusion of terminology is a matter of little importance. Unfortunately, it is mirrored in surprisingly many writings in which the definitions of Archimedean and uniform polyhedra are conflated, by adopting the local criterion as definition but claiming that global symmetry results. In the sequel, this will be referred to as the **enduring error**. An unexpected aspect of this **mathematical error** is that it has been committed by many well-known mathematicians and other scientists. It gives me no pleasure to cite names, but obviously a claim as serious as the one I just made has to be documented. This occurs in Section 4. I should also mention that I have found no indication, in any of the three reviewing journals (*Jahrbuch über die Fortschritte der Mathematik, Zentralblatt für Mathematik, Mathematical Reviews*), that a reviewer of any of the papers in question was even aware of the problem. A search of "Uniform polyhedra" on Google™ in 2005 yielded "about 13,600" results. I checked the first three, namely [37], [51], [27], and found that **all three** committed this error. A search repeated in June 2008 yielded "about 90,100" items. The first four included [37], [51] and [27], but second on the list was [52]—which did not make the error. Details will be given in the section on the enduring error. The printed literature contains many books and articles that do not commit the enduring error, and the same is true for the Web. But my point is that in both kinds of publications, many are guilty of it. It should also be mentioned that many of the same sources contain a lot of other information, often quite valuable.

In the interest of full disclosure I should mention that a version of this paper was submitted some time ago to a different journal. The editor rejected the manuscript, because a referee stated that there are many correct expositions on the Internet. Following this peculiar logic you should not worry about the counterfeit banknote you found in your wallet since there are many genuine ones in circulation.

It is possible—maybe even probable—that the problem of incorrectly enumerating Archimedean polyhedra started with Archimedes. His writings have not survived, and we have only Pappus' word [48] that Archimedes found "the thirteen polyhedra." (The reference to Heron in [31, p. 327] appears to be in error.) But it is clear that if Archimedes used the local definition—as has been believed throughout history—then he should have found "the fourteen polyhedra." The first available enumeration, close to two millennia after Archimedes', was that of Kepler [32], who independently discovered and described thirteen polyhedra which have since been repeatedly rediscovered. Kepler used the local definition as well, hence committed the "enduring error." To Kepler's possible credit it should be said that on one occasion [33, p. 11] he stated without explanation that there are fourteen Archimedean polyhedra; see also Coxeter [10], Malkevitch [38, p. 85]. As far as is known, Kepler never publicly reconciled this statement with his detailed enumeration.

The Pseudorhombicuboctahedron

The first appearance—beyond the fleeting glimpse that Kepler may have had—of the fourteenth polyhedron, usually called the **pseudorhombicuboctahedron**, happened in a paper by Sommerville [43] in 1905. However, on the face of it this paper deals only with maps in 2-dimensional elliptic, Euclidean and hyperbolic spaces. It has been mentioned in several places (see, for example, [29]) that Sommerville at best has a map that can be interpreted as a Schlegel diagram of the pseudorhombicuboctahedron. However, a more careful reading of his paper (in particular, middle of p. 725), shows that he did actually have polyhedra in mind when discussing the diagrams.

Unfortunately, Sommerville's paper appears to have been completely forgotten for more than a quarter century. The next explicit mention of the pseudorhombicuboctahedron, discovered by J.C.P. Miller, is in [9, p. 336] in 1930. It has at times been called "Miller's mistake," because Miller allegedly intended only to make a model of the rhombicuboctahedron. Miller's polyhedron received wider exposure in Ball and Coxeter [4] and in Fejes Tóth [16, p. 111]; in both books the distinction between local and global properties is stressed.

Independently of Miller, the pseudorhombicuboctahedron was discovered by V. G. Ashkinuse in 1957 [2]. This was mentioned in Ashkinuse [3], and in Lyusternik [35]. A very strange presentation of Ashkinuse's polyhedron is in Roman [40]. While in Chapter 5 of [40] a detailed proof is presented of the "fact" that there are precisely thirteen polyhedra that satisfy the local criterion, in Chapter 6 is given a description of both the rhombicuboctahedron and the pseudorhombicuboctahedron. Roman correctly stresses their differences in regard to symmetries but apparently believes that they are isomorphic and should not be counted separately.

The Enduring Error

Skipping many early enumerations, I will mention only a few instances from more recent times where the enduring error has been committed in one form or another.

Badoreau [5] follows the global approach, and justly criticizes a well-known work of Catalan [8] for various errors; however, he does not observe that Catalan's local approach is incomplete and is not equivalent to his own global one.

Lines [34] devotes Chapter 12 to Archimedean polyhedra (as defined by the local criterion), and "proves" that there are precisely thirteen of them;

this is the source on which Cundly and Rollett [13] base the presentation in their well-known book. In his deservedly popular book, Wenninger [53, p. 2] states "Archimedean or semi-regular solids . . . have regular polygons as faces and all vertices equal but admit a variety of such polygons in one solid. There are thirteen such solids . . . ." Wenninger then gives as reference the book by Lines [34] just mentioned—even though it is clear that Wenninger is really concerned with polyhedra that satisfy the global criterion. Wenninger's book is quoted as the source in some "college geometry" texts (for example, Sibley [41, p. 55]). Lines [34] is also mentioned as a source for the definition of "Archimedean or semiregular polyhedra" by Villarino [49].

Fejes Tóth [15, p. 18] lists the thirteen Archimedean polyhedra as defined by the local criteria, quoting Brückner [6] as authority. This refers to Sections 99 and 106, pp. 121–122 and 132–133, of Brückner's book, where references to several earlier enumerations can be found. It should be noted that while Brückner *seems* to have had uniform polyhedra in mind, his definitions are local, although different from ours. He never mentions the necessity of investigating whether any of the combinatorial possibilities leads to more than a single polyhedron. (It is strange that the second edition of [15] repeats the error of the first edition mentioned above, although the situation is correctly presented in [16].) Williams [55] and Field [17] assert that local and global definitions yield the same polyhedra. Gerretsen and Verdenduin [18, p. 277] find only thirteen Archimedean polyhedra, although their definitions (different from the one accepted here) in fact allow *more* than fourteen. Peterson's statements in [39] concerning regular and Archimedean polyhedra are confused and incorrect in several respects.

Maeder [36], [37] states "*Uniform polyhedra* consist of regular faces and congruent vertices. Allowing for non-convex faces and vertex figures, there are 75 such polyhedra, as well as 2 infinite families of prisms and antiprisms." This is a typical "enduring error," since in the context of more general polyhedra considered by Maeder the correct number for polyhedra satisfying the local criteria he uses is at least 77. The second non-uniform one (first described in [30]; see also [29]) is shown in Fig. 2(c). In part (a) is shown the **quasirhombicuboctahedron** (see p. 132 of [53]) and in (b), (d) and (e) a certain partition of the set of its faces. At each vertex there are three squares and one triangle, as in the rhombicuboctahedron shown in Fig. 1, but their disposition in space is different. We note that with the global definition, 75 uniform polyhedra (other than prisms and antiprisms) are described in [11], with no claim that the enumeration is complete. (These 75 include the five Platonic and thirteen convex uniform polyhedra, as well as the four Kepler-Poinsot regular non-convex ones. The omission of the five Platonic polyhedra from Theorem 1 of [49] is in contradiction to

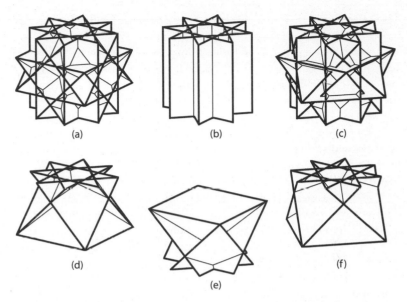

(a)   (b)   (c)

(d)   (f)

(e)

FIGURE 2. The Archimedean (and uniform) quasirhombicuboctahedron (3.4.4.4) is shown in (a), and the Archimedean but non-uniform pseudoquasirhombicuboctahedron is shown in (c). Part (b) shows the common octagrammatic mantle, and (e) the common part adjacent to the bottom octagram. Parts (d) and (f) are the top parts of the two polyhedra; as is visible, they differ by a rotation of $\pi/4 = 45°$.

the definitions given—although this may be deemed an "innocent" error.) All 75 are illustrated in [11], [53], and [26] and in many websites, such as [27], [37], [52]. The completeness of this enumeration was established independently by [44], [42] and [47].

Weisstein [51] in the MathWorld encyclopedia makes a different error, by stating: "The uniform polyhedra are polyhedra with identical polyhedron vertices." This clearly does not imply that the faces are regular polygons—any rectangular box satisfies this definition. Even if "identical" is understood as "equivalent under symmetries of the polyhedron" (and not as the more natural and more general interpretation that the vertices have congruent neighborhoods), then this is precisely the definition of "isogonal polyhedra." It is well-known that isogonal polyhedra exist in a virtually inexhaustible variety of types (see [21]), and not just 75 as claimed. There are many other errors on this site.

In the companion article [51] Weisstein writes: "The 13 Archimedean solids are the convex polyhedra that have a similar arrangement of non-intersecting regular convex polygons of two or more different types arranged in the same way about each vertex with all sides the same length."

This is precisely the "enduring error"; the final part "all sides the same length" is clearly superfluous.

Another recent example of the "enduring error" occurs in the book *Euler's Gem* by David S. Richeson (Princeton University Press 2008), pages 49 and 57. Richeson defines polyhedra he calls Archimedean or semiregular by the local criterion, and asserts that there are precisely thirteen of them (besides the regular polyhedra, prisms, and antiprisms).

A special mention should be made of several sources which are knowledgeable and authoritative in the matters under discussion, but contribute to the confusion by a different route, that is misleading and constitutes a **logical** error. They define the polyhedra they consider by the local criterion, state that their enumeration leads to the thirteen uniform ones, and then—later—observe that this is actually not so, but that the additional polyhedron can be excluded by some supplementary requirement.

Cromwell [12] describes the pseudorhombicuboctahedron, but does not include it among the Archimedean ones (which he defines by the local criterion, see [12, p. 79]). Cromwell's peculiar argument is that to deserve the name "Archimedean" polyhedra must possess properties of global symmetry, that is, be uniform. To quote (from p. 91 of [12]):

> . . . some writers have suggested that this polyhedron [the pseudo-rhombicuboctahedron] should be counted as a fourteenth Archimedean solid. This, however, misses the point. The true Archimedean solids, like the Platonic solids, have an aesthetic quality which Miller's solid does not possess. This attractiveness comes from their high degree of symmetry—a property that is easily appreciated and understood on an intuitive level. It is not the congruence of the solid angles that is the important characteristic but rather the fact that the solid angles are all indistinguishable from one another.

The present paper is not an appropriate place to debate the aesthetics of polyhedra. One important point to be made here is that the definition as accepted (but later disavowed) yields fourteen polyhedra; the decision to exclude one on other grounds should not be made after claiming that only thirteen polyhedra meet the original requirements. It may also be noted that even if one were to accept Cromwell's restricted concept of Archimedean polyhedra, the many authors who missed the pseudorhombicuboctahedron were not enumerating aesthetically appealing polyhedra. They were looking for polyhedra that satisfy the local condition of "congruence of solids angles"—and made an indisputable error in that enumeration. A moot point is the apparent claim that in the pseudorhombicuboctahedron the solid angles are **not** indistinguishable; how can they be distinguished seeing that they are congruent?

The presentation in Hart [27] is similar to Cromwell's, but at even greater remove. Uniform polyhedra are defined by the local criterion, and the "enduring error" is evident by the statement that there are 75 of them, although in the present context the number is at least 77. There is a link to [28], which deals with convex polyhedra and contains an explicit statement that there are just thirteen satisfying the local criteria. The polyhedra are listed by symbols and names, with links to impressive illustrations. However, after this, there is an "Exercise" stating:

> Just saying that the same regular polygons appear in the same sequence at each vertex is not a sufficient definition of these polyhedra. The Archimedean solid (shown at right) in which three squares and a triangle meet at each vertex is the *rhombicuboctahedron*. Look at it, and then imagine another, similar, convex solid with three squares and an equilateral triangle at each vertex.

The answer gives a link to [29], where the pseudorhombicuboctahedron is shown and its history is discussed. Following this link, Hart [28] states:

> A more precise definition of these Archimedean solids would be that [they] are convex polyhedra composed of regular polygons such that every vertex is equivalent.

For this replacement of the local criterion by the global one Hart [29] gives the same argument as Cromwell, by saying:

> The pseudo-rhombicuboctahedron is not classified as a semi-regular polyhedron because the essence (and beauty) of the semi-regular polyhedra is not about local properties of each vertex, but the symmetry operations under which the entire object appear unchanged.

Which is a correct argument for not including the pseudorhombicuboctahedron among uniform (or semiregular) polyhedra, but does not excuse its exclusion from the "Archimedean" polyhedra *as defined* in [28] by the local criterion.

Similar in spirit is the discussion in Kappraff [31, Ch. 9]. The Archimedean polyhedra (called there "semiregular") are defined by the local criterion, and it is stated that Archimedes discovered thirteen polyhedra of this kind. The existence of a fourteenth "semiregular" polyhedron is mentioned later, but it is not counted as an Archimedean polyhedron because it fails to have the following property [31, p. 328]:

> Archimedes' original 13 polyhedra can be inscribed in a regular tetrahedron so that four appropriate faces share the faces of a regular tetrahedron . . . This distinguishes them from prisms and antiprisms . . . and from . . . the pseudorhombicuboctahedron. . . .

Although this property of Archimedean solids is interesting, there is no indication in the literature that Archimedes had any such property in mind—
just as there is no indication he considered symmetry groups as operating
on the polyhedra in question.

The reason for the error of missing one of the Archimedean polyhedra
by all the authors that were actually enumerating them (and not just quoting other writers) is due to an error in the logic of enumeration. How
could such an error arise, and be perpetuated by being repeated over and
over again? Only through the neglect of rules to which we all profess to
adhere, but in practice often fail to follow.

To see this, let us recall the general procedure of determining all convex
Archimedean polyhedra. One first draws up a list of cycles of faces around a
vertex that are possible candidates for cycles of faces of Archimedean polyhedra. The precise steps of compiling such lists differ from author to author;
they rely either on the fact that the sum of angles of all faces incident to a
vertex of a convex polyhedron is less than $2\pi = 360°$, or on Euler's theorem, or on some other considerations. Using various arguments, the starting
list is pared down to one that consists of precisely thirteen different cycles
(besides those that correspond to regular polyhedra, to prisms or to antiprisms). Showing that each of these thirteen cycles actually corresponds to a
convex polyhedron with regular faces then completes the enumeration.

However, as anybody who tries to enumerate any sort of objects knows,
if you wish to get **all** the objects it is not enough to get a list of candidates
and determine which among them are actually realizable as objects of the
desired kind. You also have to find out if any candidate can be realized by
more than one object. Unfortunately, over the centuries, none of the geometers that dealt with Archimedean polyhedra bothered with this last task. As
shown in Fig. 1, in one of the cases a cycle corresponds to two distinct polyhedra, raising the number of Archimedean polyhedra to fourteen.

What is the moral of this story? Actually, there are several. First, define
precisely the objects you wish to consider, and stick consistently with the
definition. Second, when carrying out enumerations, be sure you do not
miss any of the objects. (Be also sure you do not count any twice!) Third,
and possibly most importantly, when quoting some results from the literature, apply common-sense precautions: Make sure that you understand the
definitions, verify the claims, and check the deductions.

In reality, if some "fact" is "well-known," one is often inclined to let one's
guard down. Put differently, when the result is a foregone conclusion, logical niceties get the short shrift in what amounts to "wishful seeing." Unfortunately, that is how many errors are propagated in the literature. The "fact"
that there are precisely thirteen Archimedean polyhedra is a prime example
of such a failure of our critical processes. Another illustration of the same

kind of error—turning a blind eye to facts that do not support the evidence or contradict it—appears in the famous and often reprinted work of Weyl [54]. Since he *knows* that any net (map) on a sphere with only meshes (faces) of at most six sides cannot have only hexagonal meshes, in his desire to bolster this claim Weyl produces several examples (Figs. 51, 52 and 55 of [54]) of spherical meshes. After each, he points out the existence of pentagons. However, his claims are logically invalid, since each of these nets contains heptagons. It is not relevant to the critique that, in fact, the presence of heptagons increases the number of pentagons that must be present.

With more general definitions of polygons and polyhedra (such as studied in [22] and [23]) it is easy to find many additional examples of polyhedra that satisfy the local—but not the global—criteria. However, one interesting question remains unsolved:

Are there any additional Archimedean but not uniform polyhedra in the class of polyhedra admitted by the definition in [11]?

It may be conjectured that the answer is negative, but a proof of this is probably quite complicated.

References

[1] Andreini, A.: Sulle reti di poliedri regolari e semiregolari e sulle corrispondenti reti correlative. *Mem. Società Italiana della Scienze* 14 (1905) 3, 75–129.

[2] Ashkinuse, V.G.: On the number of semiregular polyhedra. [In Russian] *Matemat. Prosveshch.* 1957, no. 1, 107–118.

[3] Ashkinuse, V.G.: Polygons and polyhedra. [In Russian] In: *Enciklopediya elementarnoi matematki.* Vol. 4. Gostehizdat, Moscow 1963. German translation: Aschkinuse, W.G.: Vielecke und Vielflache, 391–456. In: Alexandroff, P.S.; Markuschewitsch, A.I.; Chintschin, A.J. (eds.): *Enzyklopädie der Elementarmathematik.* Vol. 4 (Geometrie). VEB Deutscher Verlag der Wissenschaften, Berlin 1969.

[4] Ball, W.W.R.; Coxeter, H.S.M.: *Mathematical Recreations and Essays.* 11th ed. Macmillan, London 1939. 12th ed. Univ. of Toronto Press, 1973. 13th ed. Dover, New York 1987.

[5] Badoureau, A.: Mémoire sur les figures isoscèles. *J. École Polytech.* 30 (1881), 47–172.

[6] Brückner, M.: *Vielecke und Vielflache. Theorie und Geschichte.* Teubner, Leipzig 1900.

[7] Brückner, M.: Ueber die Ableitung der allgemeinen Polytope und die nach Isomorphismus verschiedenen Typen der allgemeinen Achtzelle (Oktatope). Verh. Nederl. Akad. Wetensch. Afd. Natuurk. Sect. 1, V. 10, No. 1 (1909), 27 pp. + 2 plates.

[8] Catalan, E.C.: Mémoire sur la théorie des polyèdres. *J. École Polytech.* 24 (1865), 1–71.

[9] Coxeter, H.S.M.: The polytopes with regular-prismatic vertex figures I. *Phil. Trans. Roy. Soc. London,* Ser. A, 229 (1930), 329–425.

[10] Coxeter, H.S.M.: Kepler and mathematics. In: Beer, A.; Beer, P. (eds.): *Four Hundred Years.* Proc. Conference held in Honour of Johannes Kepler. Vistas in Astronomy, Vol. 18. Pergamon Press, Oxford 1974, 661–670.

[11] Coxeter, H.S.M.; Longuet-Higgins, M.S.; Miller, J.C.P.: Uniform polyhedra. *Philos. Trans. Roy. Soc. London,* Ser. A, 246 (1953/54), 401–450.

[12] Cromwell, P.R.: *Polyhedra*. Cambridge Univ. Press 1997.

[13] Cundy, H.M.; Rollett, A.P.: *Mathematical Models*. Clarendon Press, Oxford 1952, 2nd ed. 1961.

[14] Daublebsky von Sterneck, R.: Die Configurationen 123. *Monatsh. Math. Phys.* 6 (1895), 223–255 + 2 plates.

[15] Fejes Tóth, L.: *Lagerungen in der Ebene, auf der Kugel und im Raum*. Springer, Berlin 1953. 2nd ed. 1972.

[16] Fejes Tóth, L.: *Reguläre Figuren*. Akadémiai Kiadó, Budapest 1965. English translation: *Regular Figures*. Macmillan, New York 1964.

[17] Field, J.V.: Rediscovering the Archimedean polyhedra: Piero della Francesca, Luca Pacioli, Leonardo da Vinci, Albrecht Dürer, Daniele Barbaro and Johannes Kepler. *Arch. Hist. Exact Sci.* 50 (1997), 241–289.

[18] Gerretsen, J.; Verdenduin, P.: Polygons and polyhedra. In: Behnke, H. et al. (eds.): *Fundamentals of Mathematics*. Vol. 2 (Geometry). Chapter 8. MIT Press, Cambridge, MA 1974. 2nd ed. 1983. Translation of the German original *Grundzüge der Mathematik*. Vandenhoeck & Ruprecht, Göttingen 1967.

[19] Gropp, H.: On the existence and nonexistence of configurations n_k. *J. Combin. Inform. System Sci.* 15 (1990), 34–48.

[20] Grünbaum, B.: Uniform tilings of 3-space. *Geombinatorics* 4 (1994), 49–56.

[21] Grünbaum, B.: Isogonal prismatoids. *Discrete Comput. Geom.* 18 (1997), 13–52.

[22] Grünbaum, B.: Are your polyhedra the same as my polyhedra? In: Aronov, B.; Basu, S.; Pach, J.; Sharir, M. (eds.): *Discrete and Computational Geometry: The Goodman-Pollack Festschrift*. Springer, New York 2003, 461–488.

[23] Grünbaum, B.: "New" uniform polyhedra. Discrete Geometry: In Honor of W. Kuperberg's 60th Birthday. Monographs and Textbooks in Pure and Applied Mathematics, Vol. 253. Marcel Dekker, New York 2003, 331–350.

[24] Grünbaum, B.; Shephard, G.C.: *Tilings and Patterns*. Freeman, New York 1987.

[25] Grünbaum, B.; Sreedharan, V.P.: An enumeration of simplicial 4-polytopes with 8 vertices. *J. Combin. Theory* 2 (1967), 437–465.

[26] Har'El, Z.: Uniform solutions for uniform polyhedra. *Geom. Dedicata* 47 (1993), 57–110.

[27] Hart, G.W.: Uniform polyhedra. http://www.georgehart.com/virtual-polyhedra/uniform-info.html.

[28] Hart, G.W.: Archimedean polyhedra. http://www.georgehart.com/virtual-polyhedra/archimedean-info.html.

[29] Hart, G.W.: Pseudorhombicuboctahedra. http://www.georgehart.com/virtual-polyhedra/pseudo-rhombicuboctahedra.html.

[30] Hughes Jones, R.: The pseudo-great rhombicuboctahedron. *Math. Sci.* 19 (1994) 1, 60–63.

[31] Kappraff, J.: *Connections. The Geometric Bridge Between Art and Science*. McGraw-Hill, New York 1991, 2nd ed. World Scientific, Singapore 2001.

[32] Kepler, J.: The Harmony of the World. *Mem. Amer. Philos. Soc.* 209 (1997). Translation of the Latin original published in 1619.

[33] Kepler, J.: *The Six-Cornered Snowflake*. Clarendon Press, Oxford 1966. Translation of the Latin original published in 1611.

[34] Lines, L.: *Solid Geometry*. Dover, New York 1965.

[35] Lyusternik, L.A.: *Convex Figures and Polyhedra*. [In Russian] English translation by T.J. Smith, Dover, New York 1963, and by D.L. Barnett, Heath, Boston 1966.

[36] Maeder, R.E.: Uniform polyhedra. *The Mathematica Journal* 3 (1993) 4, 48–57.

[37] Maeder, R.E.: Uniform Polyhedra. Excerpt of Chapter 9 of R. Maeder's book *The Mathematica Programmer II*. http://www.mathconsult.ch/showroom/unipoly/.

[38] Malkevitch, J.: Milestones in the history of polyhedra. In: Senechal, M.; Fleck, G. (eds.): *Shaping Space: A Polyhedral Approach*. Birkhäuser, Boston 1988, 80–92.

[39] Peterson, I.: Polyhedron man: Spreading the word about the wonders of crystal-like geometric forms. *Science News* 160 (2001), 396–398.

[40] Roman, T.: *Reguläre und halbreguläre Polyeder*. VEB Deutscher Verlag der Wissenschaften, Berlin 1968.

[41] Sibley, T.Q.: *The Geometric Viewpoint. A Survey of Geometries*. Addison-Wesley, Reading, MA 1998.

[42] Skilling, J.: The complete set of uniform polyhedra. *Philos. Trans. Roy. Soc. London*, Ser. A, 278 (1975), 111–135.

[43] Sommerville, D.M.Y.: Semi-regular networks of the plane in absolute geometry. *Trans. Roy. Soc. Edinburgh* 41 (1905), 725–747, with 12 plates.

[44] Sopov, S.P.: Proof of the completeness of the enumeration of uniform polyhedra. [In Russian] *Ukrain. Geom. Sb.* 8 (1970), 139–156.

[45] Steinitz, E.: *Polyeder und Raumeinteilungen*. Enzykl. Math. Wiss., Vol. 3 (Geometrie) Part 3AB12, 1922, 1–139.

[46] Steinitz, E.; Rademacher, H.: *Vorlesungen über die Theorie der Polyeder*. Springer, Berlin 1934.

[47] Szepesvári, I.: On the number of uniform polyhedra. I, II. [In Hungarian] *Mat. Lapok* 29 (1977/81), 273–328.

[48] Ver Eecke, P.: *Pappus d'Alexandrie: La collection mathématique*. Desclée, de Brouwer, Paris 1933.

[49] Villarino, M.B.: On the Archimedean or semiregular polyhedra. *Elem. Math.* 63 (2008), 76–87.

[50] Walsh, T.R.S.: Characterizing the vertex neighbourhoods of semi-regular polyhedra. *Geom. Dedicata* 1 (1972), 117–123.

[51] Weisstein, E.W.: *Uniform Polyhedron*. From *MathWorld*—A Wolfram Web Resource. http://mathworld.wolfram.com/UniformPolyhedron.html.

[52] Wikipedia: *Uniform polyhedron*. http://en.wikipedia.org/wiki/Uniform-polyhedron.

[53] Wenninger, M.J.: *Polyhedron Models*. Cambridge Univ. Press 1971.

[54] Weyl, H.: *Symmetry*. Princeton Univ. Press 1952. Reprinted 1982, 1989.

[55] Williams, R.: *Natural Structure*. Eudæmon Press, Moorpark, CA, 1972. Reprint: *The Geometrical Foundation of Natural Structures*. Dover, New York 1979.

What Is Experimental Mathematics?

KEITH DEVLIN

In my last column I gave some examples of mathematical hypotheses that, while supported by a mass of numerical evidence, nevertheless turn out to be false. Mathematicians know full well that numerical evidence, even billions of cases, does not amount to conclusive proof. No matter how many zeros of the Riemann Zeta function are computed and observed to have real-part equal to $1/2$, the Riemann Hypothesis will not be regarded as established until an analytic proof has been produced.

But there is more to mathematics than proof. Indeed, the vast majority of people who earn their living "doing math" are not engaged in finding proofs at all; their goal is to solve problems to whatever degree of accuracy or certainty is required. While proof remains the ultimate, "gold standard" for mathematical truth, conclusions reached on the basis of assessing the available evidence have always been a valid part of the mathematical enterprise. For most of the history of the subject, there were significant limitations to the amount of evidence that could be gathered, but that changed with the advent of the computer age.

For instance, the first published calculation of zeros of the Riemann Zeta function dates back to 1903, when J.P. Gram computed the first 15 zeros (with imaginary part less than 50). Today, we know that the Riemann Hypothesis is true for the first ten trillion zeros. While these computations do not prove the hypothesis, they constitute information about it. In particular, they give us a measure of confidence in results proved under the assumption of RH.

Experimental mathematics is the name generally given to the use of a computer to run computations—sometimes no more than trial-and-error tests—to look for patterns, to identify particular numbers and sequences, to gather evidence in support of specific mathematical assertions, that may themselves arise by computational means, including search.

Had the ancient Greeks (and the other early civilizations who started the mathematics bandwagon) had access to computers, it is likely that the word "experimental" in the phrase "experimental mathematics" would be superfluous; the kinds of activities or processes that make a particular mathematical activity "experimental" would be viewed simply as mathematics. On what basis do I make this assertion? Just this: if you remove from my above description the requirement that a computer be used, what would be left accurately describes what most, if not all, professional mathematicians have always spent much of their time doing!

Many readers, who studied mathematics at high school or university but did not go on to be professional mathematicians, will find that last remark surprising. For that is not the (carefully crafted) image of mathematics they were presented with. But take a look at the private notebooks of practically any of the mathematical greats and you will find page after page of trial-and-error experimentation (symbolic or numeric), exploratory calculations, guesses formulated, hypotheses examined, etc.

The reason this view of mathematics is not common is that you have to look at the private, unpublished (during their career) work of the greats in order to find this stuff (by the bucketful). What you will discover in their published work are precise statements of true facts, established by logical proofs, based upon axioms (which may be, but more often are not, stated in the work).

Because mathematics is almost universally regarded, and commonly portrayed, as the search for pure, eternal (mathematical) truth, it is easy to understand how the published work of the greats could come to be regarded as constitutive of what mathematics actually is. But to make such an identification is to overlook that key phrase "the search for." Mathematics is not, and never has been, merely the end product of the search; the process of discovery is, and always has been, an integral part of the subject. As the great German mathematician Carl Friedrich Gauss wrote to his colleague Janos Bolyai in 1808, "It is not knowledge, but the act of learning, not possession but the act of getting there, which grants the greatest enjoyment."

In fact, Gauss was very clearly an "experimental mathematician" of the first order. For example, his analysis—while still a child—of the density of prime numbers, led him to formulate what is now known as the Prime Number Theorem, a result not proved conclusively until 1896, more than 100 years after the young genius made his experimental discovery.

For most of the history of mathematics, the confusion of the activity of mathematics with its final product was understandable: after all, both activities were done by the same individual, using what to an outside observer were essentially the same activities—staring at a sheet of paper, thinking hard, and scribbling on that paper. But as soon as mathematicians started using computers to carry out the exploratory work, the distinction became obvious, especially when the mathematician simply hit the ENTER key to initiate the experimental work, and then went out to eat while the computer did its thing. In some cases, the output that awaited the mathematician on his or her return was a new "result" that no one had hitherto suspected and might have no inkling how to prove.

What makes modern experimental mathematics different (as an enterprise) from the classical conception and practice of mathematics is that the experimental process is regarded not as a precursor to a proof, to be relegated to private notebooks and perhaps studied for historical purposes only after a proof has been obtained. Rather, experimentation is viewed as a significant part of mathematics in its own right, to be published, considered by others, and (of particular importance) contributing to our overall mathematical knowledge. In particular, this gives an epistemological status to assertions that, while supported by a considerable body of experimental results, have not yet been formally proved, and in some cases may never be proved. (It may also happen that an experimental process itself yields a formal proof. For example, if a computation determines that a certain parameter p, known to be an integer, lies between 2.5 and 3.784, that amounts to a rigorous proof that $p = 3$.)

When experimental methods (using computers) began to creep into mathematical practice in the 1970s, some mathematicians cried foul, saying that such processes should not be viewed as genuine mathematics—that the one true goal should be formal proof. Oddly enough, such a reaction would not have occurred a century or more earlier, when the likes of Fermat, Gauss, Euler, and Riemann spent many hours of their lives carrying out (mental) calculations in order to ascertain "possible truths" (many but not all of which they subsequently went on to prove). The ascendancy of the notion of proof as the sole goal of mathematics came about in the late nineteenth and early twentieth centuries, when attempts to understand the infinitesimal calculus led to a realization that the intuitive concepts of such basic concepts as function, continuity, and differentiability were highly problematic, in some cases leading to seeming contradictions. Faced with the uncomfortable reality that their intuitions could be inadequate or just plain misleading, mathematicians began to insist that value judgments were hitherto to be banished to off-duty chat in the university mathematics common room and nothing would be accepted as legitimate until it had been formally proved.

What swung the pendulum back toward (openly) including experimental methods, was in part pragmatic and part philosophical. (Note that word "including." The inclusion of experimental processes in no way eliminates proofs.)

The pragmatic factor behind the acknowledgment of experimental techniques was the growth in the sheer power of computers, to search for patterns and to amass vast amounts of information in support of a hypothesis.

At the same time that the increasing availability of ever cheaper, faster, and more powerful computers proved irresistible for some mathematicians, there was a significant, though gradual, shift in the way mathematicians viewed their discipline. The Platonistic philosophy that abstract mathematical objects have a definite existence in some realm outside of Mankind, with the task of the mathematician being to uncover or discover eternal, immutable truths about those objects, gave way to an acceptance that the subject is the product of Mankind, the result of a particular kind of human thinking.

The shift from Platonism to viewing mathematics as just another kind of human thinking brought the discipline much closer to the natural sciences, where the object is not to establish "truth" in some absolute sense, but to analyze, to formulate hypotheses, and to obtain evidence that either supports or negates a particular hypothesis.

In fact, as the Hungarian philosopher Imre Lakatos made clear in his 1976 book *Proofs and Refutations,* published two years after his death, the distinction between mathematics and natural science—as practiced—was always more apparent than real, resulting from the fashion among mathematicians to suppress the exploratory work that generally precedes formal proof. By the mid-1990s, it was becoming common to "define" mathematics as a science—"the science of patterns."

The final nail in the coffin of what we might call "hard-core Platonism" was driven in by the emergence of computer proofs, the first really major example being the 1974 proof of the famous Four Color Theorem, a statement that to this day is accepted as a theorem solely on the basis of an argument (actually, today at least two different such arguments) of which a significant portion is of necessity carried out by a computer.

The degree to which mathematics has come to resemble the natural sciences can be illustrated using the example I have already cited: the Riemann Hypothesis. As I mentioned, the hypothesis has been verified compuationally for the ten trillion zeros closest to the origin. But every mathematician will agree that this does not amount to a conclusive proof. Now suppose that, next week, a mathematician posts on the Internet a five-hundred page argument that she or he claims is a proof of the hypothesis. The argument is very dense and contains several new and very deep ideas. Several years go by, during which many mathematicians around the world pore over the proof in every detail, and although they discover (and continue to discover) errors, in each case they or someone else (including the original author) is able to find a correction. At what point does the mathematical community as a whole declare that the hypothesis has indeed been proved? And even then, which do you find more convincing, the fact that there is an argument—which you have never read, and have no intention of reading—for which none of the hundred or so errors found so far have proved to be fatal, or the fact that the hypothesis has been verified computationally (and, we shall assume, with total certainty) for 10 trillion cases? Different mathematicians will give differing answers to this question, but their responses are mere opinions.

With a substantial number of mathematicians these days accepting the use of computational and experimental methods, mathematics has indeed grown to resemble much more the natural sciences. Some would argue that it simply is a natural science. If so, it does however remain, and I believe ardently will always remain, the most secure and precise of the sciences. The physicist or the chemist must rely ultimately on observation, measurement, and experiment to determine what is to be accepted as "true," and there is always the possibility of a more accurate (or different) observation, a more precise (or different) measurement, or a new experiment (that modifies or overturns the previously accepted "truths"). The mathematician, however, has that bedrock notion of proof as the final arbitrator. Yes, that method is not (in practice) perfect, particularly when long and complicated proofs are involved, but it provides a degree of certainty that the natural sciences rarely come close to.

So what kinds of things does an experimental mathematician do? (More precisely, what kinds of activity does a mathematician do that classify, or can be classified, as "experimental mathematics"?) Here are a few:

- Symbolic computation using a computer algebra system such as Mathematica or Maple
- Data visualization methods
- Integer-relation methods, such as the PSLQ algorithm
- High-precision integer and floating-point arithmetic
- High-precision numerical evaluation of integrals and summation of infinite series
- Iterative approximations to continuous functions
- Identification of functions based on graph characteristics.

References

Want to know more? As a mathematician who has not actively worked in an experimental fashion (apart from the familiar trial-and-error playing with ideas that are part and parcel of any mathematical investigation), I did, and I recently had an opportunity to learn more by collaborating with one of the leading figures in the area, the Canadian mathematician Jonathan Borwein, on an introductory-level book about the subject. The result was published recently by A.K. Peters: *The Computer as Crucible: An Introduction to Experimental Mathematics*. This month's column is abridged from that book.

What Is Information-Based Complexity?

Henryk Woźniakowski

The purpose of this short note is to informally introduce information-based complexity (IBC). We describe the basic notions of IBC for the approximate solution of continuous mathematically posed problems.

IBC is the branch of computational complexity that studies continuous mathematical problems. Typically, such problems are defined on spaces of functions of d variables and often d is huge. Since one cannot enter a function of real or complex variables into a digital computer, the available information is usually given by finitely many function values at some prescribed sample points.[1] The sample points can be chosen adaptively, that is, the choice of the jth point may be a function of the already computed function values at the $j - 1$ previously used points.

Such information is

partial, contaminated and *priced.*

It is *partial* since knowing finitely many function values, we cannot in general recover the function exactly and we are unable to find the exact solution of a continuous problem. It is *contaminated* since the function values are computed with model, experimental, and/or rounding errors. It is also *priced* since we are charged for each experiment leading to a function value or for each computation needed to obtain a function value. Often, it is expensive to obtain a function value. For example, some functions occurring in computational practice require thousands or millions of arithmetic operations to compute one function value.

Continuous problems for which partial, contaminated and priced information is available arise in many areas including numerical analysis, statistics, physics, chemistry and many computational sciences. Such problems can only be solved approximately to within some error threshold ε.

The goal of IBC is to create a theory of computational complexity for such problems. Intuitively, complexity is defined as the minimal cost of all possible algorithms that compute the solution of a continuous problem to within error at most ε. For many problems, the minimal cost is determined by the minimal number of function values needed for computing the solution to within ε.

Depending on precisely how the error and the cost are defined we have various settings. In the *worst case* setting, the error and cost of algorithms are defined by their worst case performance. That is, the error is the supremum of the distance between the exact solution and the approximation computed by the algorithm for all functions from a given set. The distance may be defined by a norm, or by a metric, and by a specific error criterion. For example, we may have the absolute, relative or normalized error criterion. Similarly, the cost is defined as the supremum of the costs of the algorithm for all functions from the same set of functions. The cost for a single function is equal to the sum of *information* and *combinatory* costs. The information cost is the number of function values times the cost of computing one function value. The combinatory cost is the number of all arithmetic operations needed to combine the already computed function values. Here we assume for simplicity that the cost of one arithmetic operation is taken as unity, and by arithmetic operations we mean such operations as addition, multiplication, subtraction, division and comparison of numbers. The set of permissible arithmetic operations can be extended by permitting the computation of special functions such as logarithms, exponential and trigonometric functions. The sets of functions studied in IBC are usually unit balls or whole spaces, depending on the error criterion. Typical spaces are Hilbert or Banach spaces of infinite dimension.

In the *average case* setting, the error and the cost of algorithms are defined by their average performance. That is, we assume a probability measure on a given set of functions and we take the expectation of the errors and costs of the algorithm for all functions from the given set with respect to this probability measure. The probability measures studied in IBC in the average case setting are usually Gaussian or truncated Gaussian measures defined on infinitely dimensional spaces.

In *the probabilistic* setting, the error and the cost of algorithms are defined as in the worst case setting by taking the supremum over a given set of functions modulo a subset of small measure. That is, we agree that the algorithms behave properly on a set of measure at least, say, $1 - \delta$, and we do not control their behavior on a set of measure at most δ.

In the worst case, average case and probabilistic settings discussed so far, we consider only *deterministic* algorithms. In the *randomized* setting, we also permit randomization. That is, we use *randomized* algorithms that can compute function values at randomized points, and can combine these values using also randomization. The randomized error and the randomized cost are defined analogously as before by taking the expectation with respect to a random element of the algorithm and then the worst case, average case or probabilistic performance with respect to functions from a given set.

Quite recently, one more setting of IBC has been added. This is the *quantum setting* where computations are performed on a (still) hypothetical quantum computer. This leads to different definitions of the error and the cost of algorithms, and the quantum complexity is defined as the minimal cost needed to compute the solution of a continuous problem to within error at most ε with a given probability.

The model of computation used in IBC is the *real number model* which is consistent with the fact that most continuous problems are solved today using floating point arithmetic, usually with a fixed precision. The cost of operations in floating point arithmetic does not depend on the size of the input and modulo rounding errors is equivalent to the real number model. We simplify the IBC analysis by not considering rounding errors. Surprisingly enough, for most algorithms whose cost in the real number model is close to the complexity of a continuous problem we can find numerically stable implementation of such algorithms. Then we obtain essentially the same results in floating point arithmetic as in the real number model when we assume that the problem is not too ill-conditioned and that ε is related to the relative precision of floating point arithmetic.

Today we know rather tight bounds on the complexity of many continuous problems, and this holds in all settings we mentioned above. There are many multivariate problems whose complexity in the worst case, average case and randomized settings is exponential in d. These IBC results may be contrasted with discrete problems, such as factorization, where the exponential complexity is conjectured and not yet proved. Of course, for discrete problems we have complete information and we cannot use powerful proof techniques for partial information that essentially allow us to obtain lower bounds and prove intractability.

We now briefly explain how lower and upper bounds are obtained in IBC. For simplicity, we assume that function values can be computed exactly. We start with lower bounds since they are usually harder to obtain.

Lower bounds are possible to obtain by using so-called *adversary* arguments. That is, we want to identify two functions that are indistinguishable with respect to finitely many function values used by the algorithm, and with the most widely separated solutions. That is, they have the same function values used by the algorithm but with the maximal distance between solutions we are trying to approximate. Clearly, the algorithm cannot distinguish between these two functions and the best it can do is to take the mean of their two solutions. So no matter how the algorithm is defined, there is no way to beat half of the distance between these two solutions.

Hence, the maximal distance between the solutions for indistinguishable functions gives us a lower bound on the error. This is usually expressed as a function of n, where n is the number of function values used by algorithms. We stress that in many cases, it is quite hard to find this function of n, although we may use a whole arsenal of mathematical tools to help us to find this function. Today, there are many techniques for finding lower bounds. Just to name a few, they are based on n-widths, especially Gelfand and Kolmogorov n-widths, on decomposable reproducing kernels, on optimality of linear algorithms for linear problems, on adaption versus non-adaption techniques etc.

Upper bounds can be obtained, for example, by using so-called *interpolatory* algorithms. Namely, when we have already computed, say n function values $f(x_j)$ for $j = 1, 2, \ldots, n$, we want to find a function g belonging to the same set of functions as f, and sharing the same function values as f. That is, $g(x_j) = f(x_j)$ for all $j = 1, 2, \ldots, n$, so we interpolate the data. Then the interpolatory algorithm takes the exact solution for g as the approximate solution for f. For given information by n function values at x_j, the error of the interpolatory algorithm is almost minimal since it can differ from the lower bound only by a factor of at most 2. Obviously, we still need to find optimal information, i.e., points x_j for which the error is minimal, and it is usually a hard nonlinear problem.

The cost of the interpolatory algorithm is, in general, more tricky. For some spaces and solutions, it turns out that *splines* are interpolatory and we can use vast knowledge about splines to compute them efficiently. For some spaces or solutions, it may, however, happen that the cost of an interpolatory algorithm is large. For some cases, we can use different algorithms. For example, for many linear problems, it has been proved that the error is minimized by linear algorithms. This is a vast simplification that helps enormously for the search of easy to implement algorithms with almost minimal error. The first such result for general linear functionals defined over balanced and convex sets of functions was proved by S. Smolyak in 1965. There are many generalizations of this result for linear operators but this is beyond this note.

We want to mention one more technique of obtaining upper bounds which uses a *randomization* argument, although the original problem is studied, say, in the worst case setting. It turns out that for Hilbert spaces and linear functionals, we can explicitly compute the worst case error of any linear algorithm. This worst case error obviously depends on the sample points used by the algorithm. So assume for a moment that the sample points are independent randomly chosen points. We then compute the expected worst case error with respect to some distribution of these points. It turns out that for many cases, the expectation is small. By the mean value theorem, we know that there must be sample points for which the worst case error is at least as small as the expectation. Furthermore, using Chebyshev's inequality, we know that the measure of sample points with error exceeding the expectation by a factor larger than one is large. This proves non-constructively that good linear algorithms exist. Obviously, we now face the problem of how to construct them. There are a number of options but we want to mention only one of them for multivariate integration of d-variate functions with d in the hundreds or thousands. There is a beautiful algorithm, called the CBC algorithm, that permits the construction of n sample points component by component (so the name CBC) by using the fast Fourier transform (FFT) in time of order $n \ln(n) d$. The CBC algorithm was designed by the Australian school of I. Sloan, S. Joe, F. Kuo and J. Dick starting from 2001, and its fast implementation was proposed by R. Cools and D. Nuyens in 2006. In this way, today we approximate integrals of functions with even 9125 variables.

For many multivariate problems defined on spaces of d-variate functions, we know that the worst case complexity of computing the solution to within ε is $\Theta(\varepsilon^{-p_d})$. That is, lower and upper bounds are proportional to ε^{-p_d} with factors in the Θ notation independent of ε^{-1} but possibly dependent on d. Sometimes such estimates hold modulo a power of $\ln \varepsilon^{-1}$ which we omit for simplicity. The exponent p_d usually depends on the smoothness of the set of functions. If they are r times differentiable then usually

$$p_d = \frac{d}{r}.$$

Note that for fixed r and varying d, we have an arbitrarily large power of ε^{-1}. In general, we cannot, however, claim that the complexity is exponential in d since it also depends on how the factor in the upper bound in the Θ notation depends on d. For Lipschitz functions we have $r = 1$, and for multivariate integration it is known due to A. Sukharev from 1979, who used the proof technique and results of S. Bakhvalov mostly from 1959, that the complexity in the worst case setting is roughly

$$\frac{1}{e}\left(\frac{1}{2\varepsilon}\right) \quad \text{for} \quad \varepsilon < \frac{1}{2},$$

with $e = \exp(1)$. Hence, it depends exponentially on d. If the complexity of a multivariate problem is exponential in d, we say that the problem is *intractable*, or suffers from the *curse of dimensionality* following Bellman who coined this phrase in 1957. One central issue of IBC research today is to determine for which multivariate problems and for which settings we have tractability, that is, when the complexity is *not* exponential in ε^{-1} and d. Depending on how we measure the lack of exponential dependence, we have various notions of tractability such as polynomial, strong polynomial and weak tractability. There is a huge literature on the complexity of multivariate problems. However, most of these papers and books have results which are sharp with respect to ε^{-1} but have unfortunately unknown dependence on d. To prove tractability we must establish also sharp dependence on d. Therefore tractability requires new proof techniques to obtain sharp bounds also on d. The second book in the list below is devoted to tractability of multivariate problems.

We end this note with a list of IBC books, where the reader can find more information, results and proofs on the complexity of continuous problems.

Notes

1. Sometimes we may use more general information consisting of finitely many linear functionals but, for simplicity, we restrict ourselves in this note only to function values as available information.

Further Reading

- E. Novak, *Deterministic and stochastic error bounds in numerical analysis*. Lecture Notes in Math. 1349, Springer-Verlag, Berlin, 1988.
- E. Novak and H. Woźniakowski, *Tractability of multivariate problems*. Volume I: Linear information, EMS Tracts Math. 6, European Mathematical Society Publishing House, Zürich, 2008.
- L. Plaskota, *Noisy information and computational complexity*. Cambridge University Press, Cambridge, UK, 1996.
- K. Ritter, *Average-case analysis of numerical problems*. Lecture Notes in Math. 1733, Springer-Verlag, Berlin, 2000.
- K. Sikorski, *Optimal solution of nonlinear equations*. Oxford University Press, 2000.
- J. F. Traub and H. Woźniakowski, *A general theory of optimal algorithms*. Academic Press, New York, 1980.
- J. F. Traub, G. W. Wasilkowski, and H. Woźniakowski, *Information, uncertainty, complexity*. Addison-Wesley, Reading, MA, 1983.
- J. F. Traub, G. W. Wasilkowski, and H. Woźniakowski, *Information-based complexity*. Comput. Sci. Sci. Comput, Academic Press, New York, 1988.
- J. F. Traub and A. G. Werschulz, *Complexity and information*. Lezioni Lincee, Cambridge University Press, Cambridge, UK, 1998.
- A. G. Werschulz, *The computational complexity of differential and integral equations: An information-based approach*. Oxford Math. Monogr., Oxford University Press, Oxford, 1991.

What Is Financial Mathematics?

TIM JOHNSON

If I tell someone I am a financial mathematician, they often think I am an accountant with pretensions. Since accountants do not like using negative numbers, one of the older mathematical technologies, I find this irritating.

I was drawn into financial maths not because I was interested in finance, but because I was interested in making good decisions in the face of uncertainty. Mathematicians have been interested in the topic of decision-making since Girolamo Cardano explored the ethics of gambling in his *Liber de Ludo Aleae* of 1564, which contains the first discussion of the idea of mathematical probability. Cardano famously commented that knowing that the chance of a fair dice coming up with a six is one in six is of no use to the gambler since probability does not predict the future. But it is of interest if you are trying to establish whether a gamble is fair or not; it helps in making good decisions.

With the exception of Pascal's wager (essentially, you've got nothing to lose by betting that God exists), the early development of probability, from Cardano, through Galileo and Fermat and Pascal up to Daniel Bernoulli in the 1720s, was driven by considering gambling problems. These ideas about probability were collected by Jacob Bernoulli (Daniel's uncle), in his work *Ars Conjectandi*. Jacob introduced the *law of large numbers*, proving that if you repeat the same experiment (say rolling a dice) a large number of times, then the observed mean (the average of the scores you have rolled) will converge to the expected mean. (For a fair dice each of the six scores is equally likely, so the expected mean is $(1 + 2 + 3 + 4 + 5 + 6)/6 = 3.5$.)

Building on Jacob Bernoulli's work, probability theory, in conjunction with statistics, became an essential tool of the scientist and was developed by the likes of Laplace in the eighteenth century and Fisher, Neyman and Pearson in the twentieth. For the first third of the twentieth century, probability was associated with inferring results, such as the life expectancy of a person, from observed data. But as an inductive science (i.e. the results were inspired by experimental observations, rather than the *deductive* nature of mathematics that builds on axioms), probability was not fully integrated into maths until 1933 when Andrey Kolmogorov identified probability with measure theory, defining probability to be any measure on a collection of events—not necessarily based on the frequency of events.

This idea is counter-intuitive if you have been taught to calculate probabilities by counting events, but can be explained with a simple example. If I want to measure the value of a painting, I can do this by measuring the area that the painting occupies, base it on the price an auctioneer gives it or base it on my own subjective assessment. For Kolmogorov, these are all acceptable measures which could be transformed into probability measures. The measure you choose to help you make decisions will depend on the problem you are addressing, if you want to work out how to cover a wall with pictures, the area measure would be best, if you are speculating, the auctioneer's would be better.

Kolmogorov formulated the axioms of probability that we now take for granted. Firstly, that the probability of an event happening is a non-negative real number ($P(E) \geq 0$). Secondly, that you know all the possible outcomes, and the probability of one of these outcomes occurring is 1 (e.g. for a six-sided dice, the probability of rolling a 1, 2, 3, 4, 5, or 6 is $P(1,2,3,4,5,6) = 1$). And finally, that you can sum the probability of mutually exclusive events (e.g. the probability of rolling an even number is $P(2,4,6) = P(2) + P(4) + P(6) = 1/2$).

Why is the measure theoretic approach so important in finance? Financial mathematicians investigate markets on the basis of a simple premise; when you price an asset it should be impossible to make money without the risk of losing money, and by symmetry, it should be impossible to lose money without the chance of making money. If you stop and think about this premise you should quickly realise it has little to do with the practicalities of business, where the objective is to make money without the risk of losing it, which is called an arbitrage, and financial institutions invest millions in technology that helps them identify arbitrage opportunities.

Based on the idea that the price of an asset should be such as to prevent arbitrages, financial mathematicians realised that an asset's price can be represented as an expectation under a special probability measure, a risk-neutral measure, which bears no direct relation to the 'natural' probability of the asset price rising or falling based on past observations. The basic arguments are pretty straightforward and can be simply explained using that result from algebra that tells us if we have two equations we can find two unknowns.

However, as with much of probability, what seems simple can be very subtle. A no-arbitrage price is not simply an expectation using a special probability; it is only a price if it is 'risk neutral' and will not result in the possibility of making or losing money. You have to undertake an investment strategy, known as hedging, that removes these possibilities. In the real world, which involves awkward things like taxes and transaction costs, it is impossible to find a unique risk-neutral measure that will ensure all these risks can be hedged away. One of the key objectives of financial maths is to understand how to construct the best investment strategies that minimises risks in the real world.

Financial mathematics is interesting because it synthesizes a highly technical and abstract branch of maths, measure theoretic probability, with practical applications that affect people's everyday lives. Financial mathematics is exciting because, by employing advanced mathematics, we are developing the theoretical foundations of finance and economics. To appreciate the impact of this work, we need to realise that much of modern financial theory, including Nobel prize winning work, is based on assumptions that are imposed, not because they reflect observed phenomena but because they enable mathematical tractability. Just as physics has motivated new maths, financial mathematicians are now developing new maths to model observed economic, rather than physical, phenomena.

Financial innovation currently has a poor reputation and some might feel that mathematicians should think twice before becoming involved with 'filthy lucre'. However, Aristotle tells us the Thales, the father of western science, became rich by applying his scientific knowledge to speculation, Galileo left the University of Padua to work for Cosimo II de Medici, and wrote *On the discoveries of dice*, becoming the first "quant". Around a hundred years after Galileo left Padua, Sir Isaac Newton left Cambridge to become warden of the Royal Mint, and lost the modern equivalent of £2,000,000 in the South Sea Bubble. Personally, what was good enough for Newton is good enough for me.

Moreover, interesting things happen when maths meets finance. In 1202 Fibonacci wrote a book, the *Liber abaci*, in which he introduced his series and the Hindu-Arabic numbers we use today, in order to help merchants conduct business. At the time merchants had to deal with dozens of different currencies and conduct risky trading expeditions that might last years, to keep on top of this they needed mathematics. There is an argument that the reason western science adopted maths, uniquely, around 1500–1700 was because people realised the importance of maths in commerce. The concept of conditional probability, behind Bayes's Theorem and the solution to the Monty Hall Problem, emerged when Christian Huygens considered a problem in life-insurance, while the number e was identified when Jacob Bernoulli thought about charging interest on a bank loan. Today, looking at the 23 DARPA Challenges for Mathematics, the first three, the mathematics of the brain, the dynamics of networks and capturing and harnessing stochasticity in nature, and the eighth, beyond convex optimization, are all highly relevant to finance.

The Credit Crisis did not affect all banks in the same way, some banks engaged with mathematics and made good decisions, like J.P. Morgan and described in Gillian Tett's book *Fools' Gold*, while others did not, and caused mayhem. Since Cardano, financial maths has been about understanding how humans make decisions in the face of uncertainty and then establishing how to make good decisions. Making, or at least not losing, money is simply a by-product of this knowledge. As Xunyu Zhou, who is developing the rigorous mathematical basis for behavioural economics at Oxford, recently commented:

> *financial mathematics needs to tell not only what people ought to do, but also what people actually do. This gives rise to a whole new horizon for mathematical finance research: can we model and analyse (what we are most capable of) the consistency and predictability in human flaws so that such flaws can be explained, avoided or even exploited for profit?*

This is the theory. In practice, in the words of one investment banker, banks

> *need high level maths skills because that is how the bank makes money.*

If Mathematics Is a Language,
How Do You Swear in It?

DAVID WAGNER

Swears are words that are considered rude or offensive. Like most other words, they are arbitrary symbols that index meaning: there is nothing inherently wrong with the letters that spell a swear word, but strung together they conjure strong meaning. This reminds us that language has power. This is true in mathematics classrooms too, where language practices structure the way participants understand mathematics and where teachers and students can use language powerfully to shape their own mathematical experience and the experiences of others.

When people swear they are either ignoring cultural norms or tromping on them for some kind of effect. In any language and culture there are ways of speaking and acting that are considered unacceptable. Though there is a need for classroom norms, there are some good reasons for encouraging alternatives to normal behavior and communication. In this sense, I want my mathematics students to swear regularly, creatively and with gusto. To illustrate, I give four responses to the question: If mathematics is a language, how do you swear in it?

Response #1: To Swear Is to Say Something Non-permissible

I've asked the question about swearing in various discussions amongst mathematics teachers. The first time I did this, we thought together about what swearing is and agreed that it is the expression of the forbidden or taboo. With this in mind, someone wrote $\sqrt{-1}$ on the whiteboard and giggled with delight. Another teacher reveled in the sinful pleasure of scrawling $\frac{\pi}{0}$ as if it were graffiti. These were mathematical swears.

Though I can recall myself as a teacher repeating "we can't have a negative radicand" and "we can't have a zero denominator," considering the possibility of such things helps me understand real numbers and expressions. For example, when the radicand in the quadratic formula is negative ($b^2 - 4ac < 0$), I know the quadratic has no roots. And when graphing rational expressions, I even sketch in the non-permissible values to help me sketch the actual curve. Considering the forbidden has even more value than this.

Though it is usually forbidden to have a negative radicand or a zero denominator, significant mathematics has emerged when mathematicians have *challenged* the forbidden. Imaginary numbers opened up significant real-world applications, and calculus rests on imagining denominators that approach zero. This history ought to remind us to listen to students who say things that we think are wrong, and to listen to students who say things in *ways* we think are wrong (which relates to response #3 in this chapter). We can ask them to explain their reasoning or to explain why they are representing ideas in a unique way.

Knowing what mathematical expressions are not permitted helps us understand the ones that are permitted. Furthermore, pursuing the non-permissible opens up new realities.

Response #2: Wait a Minute. Let's Look at Our Assumptions. Is Mathematics Really a Language?

Good mathematics remains cognizant of the assumptions behind any generalization or exploration. Thus, in this exploration of mathematical swearing, it is worth questioning how mathematics is a language, if it is at all.

It is often said that mathematics is the universal language. For example, Keith Devlin has written a wonderful book called "The Language of Mathematics," which is a history of mathematics that draws attention to the prevalence of pattern in the natural world. The book is not about language in the sense that it is about words and the way people use them. Its connection to language is more implicit. Humans across cultures can understand each other's mathematics because we share common experiences of patterns in the world and of trying to make sense of these patterns. We can understand each other. Understanding is an aspect of language. There are other ways in which mathematics can be taken as a language, and there can be value in treating it as a language, as demonstrated by Usiskin (1996).

However, it would not be so easy to find a linguist who calls mathematics a language. Linguists use the expression 'mathematics register' (e.g. Halliday, 1978) to describe the peculiarities of a mainstream language used in a mathematical context. David Pimm (1987) writes extensively about aspects of this register. It is still English, but a special kind of English. For example, a 'radical expression' in mathematics (e.g. "$3\sqrt{2} + \sqrt{5}$") is different from a 'radical expression' over coffee (e.g. "To achieve security, we have to make ourselves vulnerable") because they appear in different contexts, different registers.

Multiple meanings for the same word in different contexts are not uncommon. Another example significantly related to this article is the word 'discourse', which has emerged as a buzzword in mathematics teaching circles since reforms led by the National Council of Teachers of Mathematics in North America. The word is often used as a synonym for 'talking' (the practice of language in any situation) and also to describe the structure and history of mathematics classroom communication (the discipline of mathematics in general), which, of course, has a powerful influence on the practice of language in the classroom. Both meanings have validity, so it is up to the people in a conversation to find out what their conversation partners are thinking about when they use the word 'discourse'. It is the same for the word 'language'.

Who has the right to say mathematics is a language, or mathematics is not a language? Language belongs to all the people who use it. Dictionaries *describe* meanings typically associated with words more than they *prescribe* meaning. By contrast, students in school often learn definitions and prescribed meanings—especially in mathematics classes. This is significantly different from the way children learn language for fluency.

When we are doing our own mathematics—noticing patterns, describing our observations, making and justifying conjectures—language is alive and we use it creatively. When we make real contributions to a conversation it is often a struggle to represent our ideas and to find words and diagrams that will work for our audience. For example, I have shown some excerpts from students' mathematical explorations in Wagner (2003). The students who worked on the given task developed some new expressions to refer to their new ideas and in the article I adopted some of these forms, calling squares '5-squares' and '45-squares' (expressions that have no conventional meaning). When we are doing our own mathematics we try various words to shape meaning. By contrast, mathematical exercises—doing someone else's mathematics repeatedly—are an exercise in conformity and rigidity.

One role of a mathematics teacher is to engage students in solving real problems that require mathematical ingenuity, which also requires ingenuity in communication because students have to communicate ideas that are new to them. Once the students have had a chance to explore mathematically, the teacher has another role—to draw their attention to each other's mathematics. When students compare their mathematical ideas to those of their peers and to historical or conventional mathematical practices, there is a need to standardize word-choice so people can understand each other's ideas. In this sense, the mathematics register is a significant language phenomenon worth attending to. However, there is also value in deviating from it with awareness. Teachers who resist the strong tradition of pre-reform mathematics teaching are swearing, in a way, by deviating from tradition.

Response #3: Swear Words Remind Us of the Relationship between Language and Action

There are connections between inappropriate words (swearing) and inappropriate actions. For example, it is inappropriate to use swear words publicly to refer to our bodies' private parts, but it is even less appropriate to show these private parts in public. It is taboo.

This connection between action and words exists for appropriate as well as inappropriate action. Yackel and Cobb (1996) describe the routines of mathematics class communication as 'sociomathematical norms.' These norms significantly influence students' understanding of what mathematics is. Because teachers use language and gesture to guide the development of these norms, this language practice relates to conceptions of what mathematics is and does. Thus, I suggest that there is value in drawing students' attention to the way words are used in mathematics class, to help them understand the nature of their mathematical action. This goes beyond the common and necessary practice of helping students mimic the conventions of the mathematics register. Students can be encouraged to investigate some of the peculiarities of the register, and to find a range of ways to participate in this register.

For example, we might note that our mathematics textbook does not use the personal pronouns 'I' and 'we' and then ask students whether (or when) they should use these pronouns in mathematics class. When I asked this question of a class I was co-teaching for a research project, most of the students said personal pronouns were not appropriate because mathematics is supposed to be independent of personal particularities, yet these same students continued to use personal pronouns when they were constructing their new mathematical ideas. A student who said, "Personally, I think you shouldn't use 'I', 'you', or 'we' or 'me' or whatever" also said later "I'm always thinking in the 'I' form when I'm doing my math. I don't know why. It's just, I've always thought that way. Because I'm always doing something." (The research that this is part of is elaborated in Wagner, 2007.) The tension between students' personal agency in mathematical action and their sense of how mathematics ought to appear is central to what mathematics is.

Mathematical writing tends to obscure the decisions of the people doing the mathematics. Students are accustomed to word problems like this: "The given equation represents the height of a football in relation to time . . . ." The reality that equations come from people acting in particular contexts is glossed over by the structure of the sentence. Where did the equation come from? The perennial student question, "Why are we doing this?," may seem like a swear itself as it seems to challenge the authority of classroom practice. However, it is the most important question students can ask because even their so-called applications of mathematics typically suggest that equations exist without human involvement.

Though I find it somewhat disturbing when mathematicians and others ignore human particularities, it is important to recognize that this loss is central to the nature of mathematics. Generalization and abstraction are features of mathematical thinking, and they have their place in thoughtful human problem solving. There is value in asking what is always true regardless of context. There is also value in prompting mathematics students to realize how mathematics obscures context and to discuss the appropriateness of this obfuscation. Mathematics students should make unique contributions (using the word 'I') *and* find ways of generalizing (losing the word 'I').

This connection between agency-masking language form and mathematics' characteristic generalization, is merely one example of the way language and action are connected. Whenever we read research on discourse in mathematics classrooms we can consider the connections between mathematics and the aspects of discourse described in the research. As with the example given here, talking about these connections with students can help them understand both the nature of mathematics and the peculiarities of the mathematics register. A good way of starting such a conversation is to notice the times when students break the normal discourse rules—the times that they 'swear'. We can take their mathematical swears as an opportunity to discuss different possible ways of structuring mathematical conversations.

Response #4: I'm Not Sure How to Swear Mathematically, But I Know When I Swear in Mathematics Class!

The connection between human intention and mathematics reminds me of one student's work on the above-described investigation that had '5-squares' and '45-squares'. For the research, there was a tape recorder at each group's table. Ryan's group was working on the task, which is described in the same article (Wagner, 2003). Ryan had made a conjecture and was testing it with various cases. Listening later, I heard his quiet work punctuated with muffled grunts of affirmation for each example that verified his conjecture, until he exclaimed a loud and clear expletive, uttered when he proved his conjecture false.

Linguistic analysis of swearing practices shows how it marks a sense of attachment (Wajnryb, 2005). Ryan swore because he cared. He cared about his mathematics. He cared about his conjecture and wanted to know whether it was generalizable. His feverish work and his frustrated expletive made this clear. I want my students to have this kind of attachment to the tasks I give them, even if it gets them swearing in frustration or wonder (though I'd rather have them express their frustration and wonder in other ways). The root of their frustration is also behind their sense of satisfaction when they develop their own ways of understanding. As is often the case with refuted conjectures, finding a counterexample helped Ryan refine the conjecture into one he could justify.

To help my students develop a sense of attachment to their mathematics, I need to give them mathematical investigations that present them with real problems. They may swear in frustration but they will also find satisfaction and pleasure.

Reflection

Swearing is about bucking the norm. The history of mathematics is rich with examples of the value of people doing things that others say should not be done. Thus there is a tension facing mathematics teachers who want both a disciplined class and one that explores new ideas.

Though my own experiences as a mathematics student were strictly discipline-oriented, I try to provide for my students a different kind of discourse—a classroom that encourages creativity. I want my students to swear mathematically for at least four reasons. 1) Understanding the non-permissible helps us understand normal practice and to open up new forms of practice. 2) Creative expression casts them as participants in the long and diverse history of mathematical understanding, which is sometimes called the universal language of mathematics. 3) Attention to the relationship between language and action can help students understand both. 4) The student who swears cares: the student who chooses a unique path is showing engagement in the discipline.

References

Halliday, M. (1978). *Language as social semiotic: the social interpretation of language and meaning.* Baltimore, Maryland: University Park Press.

Pimm, D. (1987). *Speaking mathematically.* London and New York: Routledge and Kegan Paul.

Usiskin, Z. (1996). Mathematics as a language. In P. Elliott and M. Kenney (Eds.), *Communication in mathematics, K–12 and beyond* (pp. 231–243). Reston, VA: National Council of Teachers of Mathematics.

Wagner, D. (2007). Students' critical awareness of voice and agency in mathematics classroom discourse. *Mathematical Thinking and Learning,* 9 (1), 31–50.

Wagner, D. (2003). We have a problem here: 5 + 20 = 45? *Mathematics Teacher,* 96 (9), 612–616.

Wajnryb, R. (2005). *Expletive deleted: a good look at bad language.* New York: Free Press.

Yackel, E. and Cobb, P. (1996). Sociomathematical norms, argumentation, and autonomy in mathematics. *Journal for Research in Mathematics Education,* 27 (4), 458–477.

Mathematicians and the Practice of Mathematics

Birds and Frogs

FREEMAN DYSON

Some mathematicians are birds, others are frogs.* Birds fly high in the air and survey broad vistas of mathematics out to the far horizon. They delight in concepts that unify our thinking and bring together diverse problems from different parts of the landscape. Frogs live in the mud below and see only the flowers that grow nearby. They delight in the details of particular objects, and they solve problems one at a time. I happen to be a frog, but many of my best friends are birds. The main theme of my talk tonight is this. Mathematics needs both birds and frogs. Mathematics is rich and beautiful because birds give it broad visions and frogs give it intricate details. Mathematics is both great art and important science, because it combines generality of concepts with depth of structures. It is stupid to claim that birds are better than frogs because they see farther, or that frogs are better than birds because they see deeper. The world of mathematics is both broad and deep, and we need birds and frogs working together to explore it.

This talk is called the Einstein lecture, and I am grateful to the American Mathematical Society for inviting me to do honor to Albert Einstein. Einstein was not a mathematician, but a physicist who had mixed feelings about mathematics. On the one hand, he had enormous respect for the power of mathematics to describe the workings of nature, and he had an instinct for mathematical beauty which led him onto the right track to find nature's laws. On the other hand, he had no interest in pure mathematics, and he had no technical skill as a mathematician. In his later years he hired younger colleagues with the title of assistants to do mathematical calculations for him. His way of thinking was physical rather than mathematical. He was supreme among physicists as a bird who saw further than others. I will not talk about Einstein since I have nothing new to say.

---

*This article is a written version of Freeman Dyson's AMS Einstein Lecture.

Francis Bacon and René Descartes

At the beginning of the seventeenth century, two great philosophers, Francis Bacon in England and René Descartes in France, proclaimed the birth of modern science. Descartes was a bird, and Bacon was a frog. Each of them described his vision of the future. Their visions were very different. Bacon said, "All depends on keeping the eye steadily fixed on the facts of nature." Descartes said, "I think, therefore I am." According to Bacon, scientists should travel over the earth collecting facts, until the accumulated facts reveal how Nature works. The scientists will then induce from the facts the laws that Nature obeys. According to Descartes, scientists should stay at home and deduce the laws of Nature by pure thought. In order to deduce the laws correctly, the scientists will need only the rules of logic and knowledge of the existence of God. For four hundred years since Bacon and Descartes led the way, science has raced ahead by following both paths simultaneously. Neither Baconian empiricism nor Cartesian dogmatism has the power to elucidate Nature's secrets by itself, but both together have been amazingly successful. For four hundred years English scientists have tended to be Baconian and French scientists Cartesian. Faraday and Darwin and Rutherford were Baconians; Pascal and Laplace and Poincaré were Cartesians. Science was greatly enriched by the cross-fertilization of the two contrasting cultures. Both cultures were always at work in both countries. Newton was at heart a Cartesian, using pure thought as Descartes intended, and using it to demolish the Cartesian dogma of vortices. Marie Curie was at heart a Baconian, boiling tons of crude uranium ore to demolish the dogma of the indestructibility of atoms.

In the history of twentieth century mathematics, there were two decisive events, one belonging to the Baconian tradition and the other to the Cartesian tradition. The first was the International Congress of Mathematicians in Paris in 1900, at which Hilbert gave the keynote address, charting the course of mathematics for the coming century by propounding his famous list of twenty-three outstanding unsolved problems. Hilbert himself was a bird, flying high over the whole territory of mathematics, but he addressed his problems to the frogs who would solve them one at a time. The second decisive event was the formation of the Bourbaki group of mathematical birds in France in the 1930s, dedicated to publishing a series of textbooks that would establish a unifying framework for all of mathematics. The Hilbert problems were enormously successful in guiding mathematical research into fruitful directions. Some of them were solved and some remain unsolved, but almost all of them stimulated the growth of new ideas and new fields of mathematics. The Bourbaki project was equally influential. It changed the style of mathematics for the next fifty years, imposing a logical coherence that did not exist before, and moving the emphasis from concrete examples to abstract generalities. In the Bourbaki scheme of things, mathematics is the abstract structure included in the Bourbaki textbooks. What is not in the textbooks is not mathematics. Concrete examples, since they do not appear in the textbooks, are not mathematics. The Bourbaki program was the extreme expression of the Cartesian style. It narrowed the scope of mathematics by excluding the beautiful flowers that Baconian travelers might collect by the wayside.

Jokes of Nature

For me, as a Baconian, the main thing missing in the Bourbaki program is the element of surprise. The Bourbaki program tried to make mathematics logical. When I look at the history of mathematics, I see a succession of illogical jumps, improbable coincidences, jokes of nature. One of the most profound jokes of nature is the square root of minus one that the physicist Erwin Schrödinger put into his wave equation when he invented wave mechanics in 1926. Schrödinger was a bird who started from the idea of unifying mechanics with optics. A hundred years earlier, Hamilton had unified classical mechanics with ray optics, using the same mathematics to describe optical rays and classical particle trajectories. Schrödinger's idea was to extend this unification to wave optics and wave mechanics. Wave optics already existed, but wave mechanics did not. Schrödinger had to invent wave mechanics to complete the unification. Starting from wave optics as a model, he wrote down a differential equation for a mechanical particle, but the equation made no sense. The equation looked like the equation of conduction of heat in a continuous medium. Heat conduction has no visible relevance to particle mechanics. Schrödinger's idea seemed to be going nowhere. But then came the surprise. Schrödinger put the square root of minus one into the equation, and suddenly it made sense. Suddenly it became a wave equation instead of a heat conduction equation. And Schrödinger found to his delight that the equation has solutions corresponding to the quantized orbits in the Bohr model of the atom.

It turns out that the Schrödinger equation describes correctly everything we know about the behavior of atoms. It is the basis of all of chemistry and most of physics. And that square root of minus one means that nature works with complex numbers and not with real numbers. This discovery came as a complete surprise, to Schrödinger as well as to everybody else. According to Schrödinger, his fourteen-year-old girl friend Itha Junger said to him at the time, "Hey, you never even thought when you began that so much sensible stuff would come out of it." All through the nineteenth century, mathematicians from Abel to Riemann and Weierstrass had been creating a magnificent theory of functions of complex variables. They had discovered that the theory of functions became far deeper and more powerful when it was extended from real to complex numbers. But they always thought of complex numbers as an artificial construction, invented by human mathematicians as a useful and elegant abstraction from real life. It never entered their heads that this artificial number system that they had invented was in fact the ground on which atoms move. They never imagined that nature had got there first.

Another joke of nature is the precise linearity of quantum mechanics, the fact that the possible states of any physical object form a linear space. Before quantum mechanics was invented, classical physics was always nonlinear, and linear models were only approximately valid. After quantum mechanics, nature itself suddenly became linear. This had profound consequences for mathematics. During the nineteenth century Sophus Lie developed his elaborate theory of continuous groups, intended to clarify the behavior of classical dynamical systems. Lie groups were then of little interest either to mathematicians or to physicists. The nonlinear theory of Lie groups was too complicated for the mathematicians and too obscure for the physicists. Lie died a disappointed man. And then, fifty years later, it turned out that nature was precisely linear, and the theory of linear representations of Lie algebras was the natural language of particle physics. Lie groups and Lie algebras were reborn as one of the central themes of twentieth century mathematics.

A third joke of nature is the existence of quasi-crystals. In the nineteenth century the study of crystals led to a complete enumeration of possible discrete symmetry groups in Euclidean space. Theorems were proved, establishing the fact that in three-dimensional space discrete symmetry groups could contain only rotations of order three, four, or six. Then in 1984 quasi-crystals were discovered, real solid objects growing out of liquid metal alloys, showing the symmetry of the icosahedral group, which includes fivefold rotations. Meanwhile, the mathematician Roger Penrose discovered the Penrose tilings of the plane. These are arrangements of parallelograms that cover a plane with pentagonal long-range order. The alloy quasi-crystals are three-dimensional analogs of the two-dimensional Penrose tilings. After these discoveries, mathematicians had to enlarge the theory of crystallographic groups to include quasi-crystals. That is a major program of research which is still in progress.

A fourth joke of nature is a similarity in behavior between quasi-crystals and the zeros of the Riemann Zeta function. The zeros of the zeta-function are exciting to mathematicians because they are found to lie on a straight line and nobody understands why. The statement that with trivial exceptions they all lie on a straight line is the famous Riemann Hypothesis. To prove the Riemann Hypothesis has been the dream of young mathematicians for more than a hundred years. I am now making the outrageous suggestion that we might use quasi-crystals to prove the Riemann Hypothesis. Those of you who are mathematicians may consider the suggestion frivolous. Those who are not mathematicians may consider it uninteresting. Nevertheless I am putting it forward for your serious consideration. When the physicist Leo Szilard was young, he became dissatisfied with the ten commandments of Moses and wrote a new set of ten commandments to replace them. Szilard's second commandment says: "Let your acts be directed towards a worthy goal, but do not ask if they can reach it: they are to be models and examples, not means to an end." Szilard practiced what he preached. He was the first physicist to imagine nuclear weapons and the first to campaign actively against their use. His second commandment certainly applies here. The proof of the Riemann Hypothesis is a worthy goal, and it is not for us to ask whether we can reach it. I will give you some hints describing how it might be achieved. Here I will be giving voice to the mathematician that I was fifty years ago before I became a physicist. I will talk first about the Riemann Hypothesis and then about quasi-crystals.

There were until recently two supreme unsolved problems in the world of pure mathematics, the proof of Fermat's Last Theorem and the proof of the Riemann Hypothesis. Twelve years ago, my Princeton colleague Andrew Wiles polished off Fermat's Last Theorem, and only the Riemann Hypothesis remains. Wiles' proof of the Fermat Theorem was not just a technical stunt. It required the discovery and exploration of a new field of mathematical ideas, far wider and more consequential than the Fermat Theorem itself. It is likely that any proof of the Riemann Hypothesis will likewise lead to a deeper understanding of many diverse areas of mathematics and perhaps of physics too. Riemann's zeta-function, and other zeta-functions similar to it, appear ubiquitously in number theory, in the theory of dynamical systems, in geometry, in function theory, and in physics. The zeta-function stands at a junction where paths lead in many directions. A proof of the hypothesis will illuminate all the connections. Like every serious student of pure mathematics, when I was young I had dreams of proving the Riemann Hypothesis. I had some vague ideas that I thought might lead to a proof. In recent years, after the discovery of quasi-crystals, my ideas became a little less vague. I offer them here for the consideration of any young mathematician who has ambitions to win a Fields Medal.

Quasi-crystals can exist in spaces of one, two, or three dimensions. From the point of view of physics, the three-dimensional quasi-crystals are the most interesting, since they inhabit our three-dimensional world and can be studied experimentally. From the point of view of a mathematician, one-dimensional quasi-crystals are much more interesting than two-dimensional or three-dimensional quasi-crystals because they exist in far greater variety. The mathematical definition of a quasi-crystal is as follows. A quasi-crystal is a distribution of discrete point masses whose Fourier transform is a distribution of discrete point frequencies. Or to say it more briefly, a quasi-crystal is a pure point distribution that has a pure point spectrum. This definition includes as a special case the ordinary crystals, which are periodic distributions with periodic spectra.

Excluding the ordinary crystals, quasi-crystals in three dimensions come in very limited variety, all of them associated with the icosahedral group. The two-dimensional quasi-crystals are more numerous, roughly one distinct type associated with each regular polygon in a plane. The two-dimensional quasi-crystal with pentagonal symmetry is the famous Penrose tiling of the plane. Finally, the one-dimensional quasi-crystals have a far richer structure since they are not tied to any rotational symmetries. So far as I know, no complete enumeration of one-dimensional quasi-crystals exists. It is known that a unique quasi-crystal exists corresponding to every Pisot-Vijayaraghavan number or PV number. A PV number is a real algebraic integer, a root of a polynomial equation with integer coefficients, such that all the other roots have absolute value less than one, [1]. The set of all PV numbers is infinite and has a remarkable topological structure. The set of all one-dimensional quasi-crystals has a structure at least as rich as the set of all PV numbers and probably much richer. We do not know for sure, but it is likely that a huge universe of one-dimensional quasi-crystals not associated with PV numbers is waiting to be discovered.

Here comes the connection of the one-dimensional quasi-crystals with the Riemann hypothesis. If the Riemann hypothesis is true, then the zeros of the zeta-function form a one-dimensional quasi-crystal according to the definition. They constitute a distribution of point masses on a straight line, and their Fourier transform is likewise a distribution of point masses, one at each of the logarithms of ordinary prime numbers and prime-power numbers. My friend Andrew Odlyzko has published a beautiful computer calculation of the Fourier transform of the zeta-function zeros, [8]. The calculation shows precisely the expected structure of the Fourier transform, with a sharp discontinuity at every logarithm of a prime or prime-power number and nowhere else.

My suggestion is the following. Let us pretend that we do not know that the Riemann Hypothesis is true. Let us tackle the problem from the other end. Let us try to obtain a complete enumeration and classification of one-dimensional quasi-crystals. That is to say, we enumerate and classify all point distributions that have a discrete point spectrum. Collecting and classifying new species of objects is a quintessentially Baconian activity. It is an appropriate activity for mathematical frogs. We shall then find the well-known quasi-crystals associated with PV numbers, and also a whole universe of other quasi-crystals, known and unknown. Among the multitude of other quasi-crystals we search for one corresponding to the Riemann zeta-function and one corresponding to each of the other zeta-functions that resemble the Riemann zeta-function. Suppose that we find one of the quasi-crystals in our enumeration with properties that identify it with the zeros of the Riemann zeta-function. Then we have proved the Riemann Hypothesis and we can wait for the telephone call announcing the award of the Fields Medal.

These are of course idle dreams. The problem of classifying one-dimensional quasi-crystals is horrendously difficult, probably at least as difficult as the problems that Andrew Wiles took seven years to explore. But if we take a Baconian point of view, the history of mathematics is a history of horrendously difficult problems being solved by young people too ignorant to know that they were impossible. The classification of quasi-crystals is a worthy goal, and might even turn out to be achievable. Problems of that degree of difficulty will not be solved by old men like me. I leave this problem as an exercise for the young frogs in the audience.

Abram Besicovitch and Hermann Weyl

Let me now introduce you to some notable frogs and birds that I knew personally. I came to Cambridge University as a student in 1941 and had the tremendous luck to be given the Russian mathematician Abram Samoilovich Besicovitch as my supervisor. Since this was in the middle of World War Two, there were very few students in Cambridge, and almost no graduate students. Although I was only seventeen years old and Besicovitch was already a famous professor, he gave me a great deal of his time and attention, and we became life-long friends. He set the style in which I began to work and think about mathematics. He gave wonderful lectures on measure-theory and integration, smiling amiably when we laughed at his glorious abuse of the English language. I remember only one occasion when he was annoyed by our laughter. He remained silent for a while and then said, "Gentlemen. Fifty million English speak English you speak. Hundred and fifty million Russians speak English I speak."

Besicovitch was a frog, and he became famous when he was young by solving a problem in elementary plane geometry known as the Kakeya problem. The Kakeya problem was the following. A line segment of length one is allowed to move freely in a plane while rotating through an angle of 360 degrees. What is the smallest area of the plane that it can cover during its rotation? The problem was posed by the Japanese mathematician Kakeya in 1917 and remained a famous unsolved problem for ten years. George Birkhoff, the leading American mathematician at that time, publicly proclaimed that the Kakeya problem and the four-color problem were the outstanding unsolved problems of the day. It was widely believed that the minimum area was $\pi/8$, which is the area of a three-cusped hypocycloid. The three-cusped hypocycloid is a beautiful three-pointed curve. It is the curve traced out by a point on the circumference of a circle with radius one-quarter, when the circle rolls around the inside of a fixed circle with radius three-quarters. The line segment of length one can turn while always remaining tangent to the hypocycloid with its two ends also on the hypocycloid. This picture of the line turning while touching the inside of the hypocycloid at three points was so elegant that most people believed it must give the minimum area. Then Besicovitch surprised everyone by proving that the area covered by the line as it turns can be less than $\mathcal{E}$ for any positive $\mathcal{E}$.

Besicovitch had actually solved the problem in 1920 before it became famous, not even knowing that Kakeya had proposed it. In 1920 he published the solution in Russian in the *Journal of the Perm Physics and Mathematics Society*, a journal that was not widely read. The university of Perm, a city 1,100 kilometers east of Moscow, was briefly a refuge for many distinguished mathematicians after the Russian revolution. They published two volumes of their journal before it died amid the chaos of revolution and civil war. Outside Russia the journal was not only unknown but unobtainable. Besicovitch left Russia in 1925 and arrived at Copenhagen, where he learned about the famous Kakeya problem that he had solved five years earlier. He published the solution again, this time in English in the *Mathematische Zeitschrift*. The Kakeya problem as Kakeya proposed it was a typical frog problem, a concrete problem without much connection with the rest of mathematics. Besicovitch gave it an elegant and deep solution, which revealed a connection with general theorems about the structure of sets of points in a plane.

The Besicovitch style is seen at its finest in his three classic papers with the title, "On the fundamental geometric properties of linearly measurable plane sets of points", published in *Mathematische Annalen* in the years 1928, 1938, and 1939. In these papers he proved that every linearly measurable set in the plane is divisible into a regular and an irregular component, that the regular component has a tangent almost everywhere, and the irregular component has a projection of measure zero onto almost all directions. Roughly speaking, the regular component looks like a collection of continuous curves, while the irregular component looks nothing like a continuous curve. The existence and the properties of the irregular component are connected with the Besicovitch solution of the Kakeya problem. One of the problems that he gave me to work on was the division of measurable sets into regular and irregular components in spaces of higher dimensions. I got nowhere with the problem, but became permanently imprinted with the Besicovitch style. The Besicovitch style is architectural. He builds out of simple elements a delicate and complicated architectural structure, usually with a hierarchical plan, and then, when the building is finished, the completed structure leads by simple arguments to an unexpected conclusion. Every Besicovitch proof is a work of art, as carefully constructed as a Bach fugue.

A few years after my apprenticeship with Besicovitch, I came to Princeton and got to know Hermann Weyl. Weyl was a prototypical bird, just as Besicovitch was a prototypical frog. I was lucky to overlap with Weyl for one year at the Princeton Institute for Advanced Study before he retired from the Institute and moved back to his old home in Zürich. He liked me because during that year I published papers in the *Annals of Mathematics* about number theory and in the *Physical Review* about the quantum theory of radiation. He was one of the few people alive who was at home in both subjects. He welcomed me to the Institute, in the hope that I would be a bird like himself. He was disappointed. I remained obstinately a frog. Although I poked around in a variety of mud-holes, I always looked at them one at a time and did not look for connections between them. For me, number theory and quantum theory were separate worlds with separate beauties. I did not look at them as Weyl did, hoping to find clues to a grand design.

Weyl's great contribution to the quantum theory of radiation was his invention of gauge fields. The idea of gauge fields had a curious history. Weyl invented them in 1918 as classical fields in his unified theory of general relativity and electromagnetism, [10]. He called them "gauge fields" because they were concerned with the non-integrability of measurements of length. His unified theory was promptly and publicly rejected by Einstein. After this thunderbolt from on high, Weyl did not abandon his theory but moved on to other things. The theory had no experimental consequences that could be tested. Then wave mechanics was invented by Schrödinger, and three independent publications by Fock, Klein and Gordon in 1926 proposed a relativistic wave equation for a charged particle interacting with an electromagnetic field. Only Fock noticed that the wave equation was invariant under a group of transformations which he called "gradient transformations" [4]. The authoritative Russian text-book on Classical Field Theory [5] calls the invariance "gradient-invariance" and attributes its discovery to Fock. Meanwhile, F. London in 1927 and Weyl in 1928 observed that the gradient-invariance of quantum mechanics is closely related to the gauge-invariance of Weyl's version of general relativity. For a detailed account of this history see [9]. Weyl realized that his gauge fields fitted far better into the quantum world than they did into the classical world, [11]. All that he needed to do, to change a classical gauge into a quantum gauge, was to change real numbers into complex numbers. In quantum mechanics, every quantum of electric charge carries with it a complex wave function with a phase, and the gauge field is concerned with the non-integrability of measurements of phase. The gauge field could then be precisely identified with the electromagnetic potential, and the law of conservation of charge became a consequence of the local phase invariance of the theory.

Weyl died four years after he returned from Princeton to Zürich, and I wrote his obituary for the journal *Nature,* [3]. "Among all the mathematicians who began their working lives in the twentieth century," I wrote, "Hermann Weyl was the one who made major contributions in the greatest number of different fields. He alone could stand comparison with the last great universal mathematicians of the nineteenth century, Hilbert and Poincaré. So long as he was alive, he embodied a living contact between the main lines of advance in pure mathematics and in theoretical physics. Now he is dead, the contact is broken, and our hopes of comprehending the physical universe by a direct use of creative mathematical imagination are for the time being ended." I mourned his passing, but I had no desire to pursue his dream. I was happy to see pure mathematics and physics marching ahead in opposite directions.

The obituary ended with a sketch of Weyl as a human being: "Characteristic of Weyl was an aesthetic sense which dominated his thinking on all subjects. He once said to me, half joking, 'My work always tried to unite the true with the beautiful; but when I had to choose one or the other, I usually chose the beautiful'. This remark sums up his personality perfectly. It shows his profound faith in an ultimate harmony of Nature, in which the laws should inevitably express themselves in a mathematically beautiful form. It shows also his recognition of human frailty, and his humor, which always stopped him short of being pompous. His friends in Princeton will remember him as he was when I last saw him, at the Spring Dance of the Institute for Advanced Study last April: a big jovial man, enjoying himself splendidly, his cheerful frame and his light step giving no hint of his sixty-nine years."

The fifty years after Weyl's death were a golden age of experimental physics and observational astronomy, a golden age for Baconian travelers picking up facts, for frogs exploring small patches of the swamp in which we live. During these fifty years, the frogs accumulated a detailed knowledge of a large variety of cosmic structures and a large variety of particles and interactions. As the exploration of new territories continued, the universe became more complicated. Instead of a grand design displaying the simplicity and beauty of Weyl's mathematics, the explorers found weird objects such as quarks and gamma-ray bursts, weird concepts such as supersymmetry and multiple universes. Meanwhile, mathematics was also becoming more complicated, as exploration continued into the phenomena of chaos and many other new areas opened by electronic computers. The mathematicians discovered the central mystery of computability, the conjecture represented by the statement P is not equal to NP. The conjecture asserts that there exist mathematical problems which can be quickly solved in individual cases but cannot be solved by a quick algorithm applicable to all cases. The most famous example of such a problem is the traveling salesman problem, which is to find the shortest route for a salesman visiting a set of cities, knowing the distance between each pair. For technical reasons, we do not ask for the shortest route but for a route with length less than a given upper bound. Then the traveling salesman problem is conjectured to be NP but not P. But nobody has a glimmer of an idea how to prove it. This is a mystery that could not even have been formulated within the nineteenth-century mathematical universe of Hermann Weyl.

Frank Yang and Yuri Manin

The last fifty years have been a hard time for birds. Even in hard times, there is work for birds to do, and birds have appeared with the courage to tackle it. Soon after Weyl left Princeton, Frank Yang arrived from Chicago and moved into Weyl's old house. Yang took Weyl's place as the leading bird among my generation of physicists. While Weyl was still alive, Yang and his student Robert Mills discovered the Yang-Mills theory of non-Abelian gauge fields, a marvelously elegant extension of Weyl's idea of a gauge field, [14]. Weyl's gauge field was a classical quantity, satisfying the commutative law of multiplication. The Yang-Mills theory had a triplet of gauge fields which did not commute. They satisfied the commutation rules of the three components of a quantum mechanical spin, which are generators of the simplest non-Abelian Lie algebra A_2. The theory was later generalized so that the gauge fields could be generators of any finite-dimensional Lie algebra. With this generalization, the Yang-Mills gauge field theory provided the framework for a model of all the known particles and interactions, a model that is now known as the Standard Model of particle physics. Yang put the finishing touch to it by showing that Einstein's theory of gravitation fits into the same framework, with the Christoffel three-index symbol taking the role of gauge field, [13].

In an appendix to his 1918 paper, added in 1955 for the volume of selected papers published to celebrate his seventieth birthday, Weyl expressed his final thoughts about gauge field theories (my translation), [12]: "The strongest argument for my theory seemed to be this, that gauge invariance was related to conservation of electric charge in the same way as coordinate invariance was related to conservation of energy and momentum." Thirty years later Yang was in Zürich for the celebration of Weyl's hundredth birthday. In his speech, [15], Yang quoted this remark as evidence of Weyl's devotion to the idea of gauge invariance as a unifying principle for physics. Yang then went on, "Symmetry, Lie groups, and gauge invariance are now recognized, through theoretical and experimental developments, to play essential roles in determining the basic forces of the physical universe. I have called this the principle that symmetry dictates interaction." This idea, that symmetry dictates interaction, is Yang's generalization of Weyl's re-mark. Weyl observed that gauge invariance is intimately connected with physical conservation laws. Weyl could not go further than this, because he knew only the gauge invariance of commuting Abelian fields. Yang made the connection much stronger by introducing non-Abelian gauge fields. With non-Abelian gauge fields generating nontrivial Lie algebras, the possible forms of interaction between fields become unique, so that symmetry dic-tates interaction. This idea is Yang's greatest contribution to physics. It is the contribution of a bird, flying high over the rain forest of little problems in which most of us spend our lives.

Another bird for whom I have a deep respect is the Russian mathemati-cian Yuri Manin, who recently published a delightful book of essays with the title *Mathematics as Metaphor* [7]. The book was published in Moscow in Rus-sian, and by the American Mathematical Society in English. I wrote a pref-ace for the English version, and I give you here a short quote from my preface. *"Mathematics as Metaphor* is a good slogan for birds. It means that the deepest concepts in mathematics are those which link one world of ideas with another. In the seventeenth century Descartes linked the disparate worlds of algebra and geometry with his concept of coordinates, and New-ton linked the worlds of geometry and dynamics with his concept of flux-ions, nowadays called calculus. In the nineteenth century Boole linked the worlds of logic and algebra with his concept of symbolic logic, and Rie-mann linked the worlds of geometry and analysis with his concept of Rie-mann surfaces. Coordinates, fluxions, symbolic logic, and Riemann sur-faces are all metaphors, extending the meanings of words from familiar to unfamiliar contexts. Manin sees the future of mathematics as an exploration of metaphors that are already visible but not yet understood. The deepest such metaphor is the similarity in structure between number theory and physics. In both fields he sees tantalizing glimpses of parallel concepts, sym-metries linking the continuous with the discrete. He looks forward to a unification which he calls the quantization of mathematics.

"Manin disagrees with the Baconian story, that Hilbert set the agenda for the mathematics of the twentieth century when he presented his famous list of twenty-three unsolved problems to the International Congress of Mathematicians in Paris in 1900. According to Manin, Hilbert's problems were a distraction from the central themes of mathematics. Manin sees the important advances in mathematics coming from programs, not from problems. Problems are usually solved by applying old ideas in new ways. Programs of research are the nurseries where new ideas are born. He sees the Bourbaki program, rewriting the whole of mathematics in a more abstract language, as the source of many of the new ideas of the twentieth century. He sees the Langlands program, unifying number theory with geometry, as a promising source of new ideas for the twenty-first. People who solve famous unsolved problems may win big prizes, but people who start new programs are the real pioneers."

The Russian version of *Mathematics as Metaphor* contains ten chapters that were omitted from the English version. The American Mathematical Society decided that these chapters would not be of interest to English language readers. The omissions are doubly unfortunate. First, readers of the English version see only a truncated view of Manin, who is perhaps unique among mathematicians in his broad range of interests extending far beyond mathematics. Second, we see a truncated view of Russian culture, which is less compartmentalized than English language culture, and brings mathematicians into closer contact with historians and artists and poets.

John von Neumann

Another important figure in twentieth century mathematics was John von Neumann. Von Neumann was a frog, applying his prodigious technical skill to solve problems in many branches of mathematics and physics. He began with the foundations of mathematics. He found the first satisfactory set of axioms for set-theory, avoiding the logical paradoxes that Cantor had encountered in his attempts to deal with infinite sets and infinite numbers. Von Neumann's axioms were used by his bird friend Kurt Gödel a few years later to prove the existence of undecidable propositions in mathematics. Gödel's theorems gave birds a new vision of mathematics. After Gödel, mathematics was no longer a single structure tied together with a unique concept of truth, but an archipelago of structures with diverse sets of axioms and diverse notions of truth. Gödel showed that mathematics is inexhaustible. No matter which set of axioms is chosen as the foundation, birds can always find questions that those axioms cannot answer.

Von Neumann went on from the foundations of mathematics to the foundations of quantum mechanics. To give quantum mechanics a firm mathematical foundation, he created a magnificent theory of rings of operators.

Every observable quantity is represented by a linear operator, and the peculiarities of quantum behavior are faithfully represented by the algebra of operators. Just as Newton invented calculus to describe classical dynamics, von Neumann invented rings of operators to describe quantum dynamics.

Von Neumann made fundamental contributions to several other fields, especially to game theory and to the design of digital computers. For the last ten years of his life, he was deeply involved with computers. He was so strongly interested in computers that he decided not only to study their design but to build one with real hardware and software and use it for doing science. I have vivid memories of the early days of von Neumann's computer project at the Institute for Advanced Study in Princeton. At that time he had two main scientific interests, hydrogen bombs and meteorology. He used his computer during the night for doing hydrogen bomb calculations and during the day for meteorology. Most of the people hanging around the computer building in daytime were meteorologists. Their leader was Jule Charney. Charney was a real meteorologist, properly humble in dealing with the inscrutable mysteries of the weather, and skeptical of the ability of the computer to solve the mysteries. John von Neumann was less humble and less skeptical. I heard von Neumann give a lecture about the aims of his project. He spoke, as he always did, with great confidence. He said, "The computer will enable us to divide the atmosphere at any moment into stable regions and unstable regions. Stable regions we can predict. Unstable regions we can control." Von Neumann believed that any unstable region could be pushed by a judiciously applied small perturbation so that it would move in any desired direction. The small perturbation would be applied by a fleet of airplanes carrying smoke generators, to absorb sunlight and raise or lower temperatures at places where the perturbation would be most effective. In particular, we could stop an incipient hurricane by identifying the position of an instability early enough, and then cooling that patch of air before it started to rise and form a vortex. Von Neumann, speaking in 1950, said it would take only ten years to build computers powerful enough to diagnose accurately the stable and unstable regions of the atmosphere. Then, once we had accurate diagnosis, it would take only a short time for us to have control. He expected that practical control of the weather would be a routine operation within the decade of the 1960s.

Von Neumann, of course, was wrong. He was wrong because he did not know about chaos. We now know that when the motion of the atmosphere is locally unstable, it is very often chaotic. The word "chaotic" means that motions that start close together diverge exponentially from each other as time goes on. When the motion is chaotic, it is unpredictable, and a small perturbation does not move it into a stable motion that can be predicted. A small perturbation will usually move it into another chaotic motion that is equally unpredictable. So von Neumann's strategy for controlling the weather fails. He was, after all, a great mathematician but a mediocre meteorologist.

Edward Lorenz discovered in 1963 that the solutions of the equations of meteorology are often chaotic. That was six years after von Neumann died. Lorenz was a meteorologist and is generally regarded as the discoverer of chaos. He discovered the phenomena of chaos in the meteorological context and gave them their modern names. But in fact I had heard the mathematician Mary Cartwright, who died in 1998 at the age of 97, describe the same phenomena in a lecture in Cambridge in 1943, twenty years before Lorenz discovered them. She called the phenomena by different names, but they were the same phenomena. She discovered them in the solutions of the van der Pol equation which describe the oscillations of a nonlinear amplifier, [2]. The van der Pol equation was important in World War II because nonlinear amplifiers fed power to the transmitters in early radar systems. The transmitters behaved erratically, and the Air Force blamed the manufacturers for making defective amplifiers. Mary Cartwright was asked to look into the problem. She showed that the manufacturers were not to blame. She showed that the van der Pol equation was to blame. The solutions of the van der Pol equation have precisely the chaotic behavior that the Air Force was complaining about. I heard all about chaos from Mary Cartwright seven years before I heard von Neumann talk about weather control, but I was not far-sighted enough to make the connection. It never entered my head that the erratic behavior of the van der Pol equation might have something to do with meteorology. If I had been a bird rather than a frog, I would probably have seen the connection, and I might have saved von Neumann a lot of trouble. If he had known about chaos in 1950, he would probably have thought about it deeply, and he would have had something important to say about it in 1954.

Von Neumann got into trouble at the end of his life because he was really a frog but everyone expected him to fly like a bird. In 1954 there was an International Congress of Mathematicians in Amsterdam. These congresses happen only once in four years and it is a great honor to be invited to speak at the opening session. The organizers of the Amsterdam congress invited von Neumann to give the keynote speech, expecting him to repeat the act that Hilbert had performed in Paris in 1900. Just as Hilbert had provided a list of unsolved problems to guide the development of mathematics for the first half of the twentieth century, von Neumann was invited to do the same for the second half of the century. The title of von Neumann's talk was announced in the program of the congress. It was "Unsolved Problems in Mathematics: Address by Invitation of the Organizing Committee". After the congress was over, the complete proceedings were published, with the texts of all the lectures except this one. In the proceedings there is a blank page with von Neumann's name and the title of his talk. Underneath, it says, "No manuscript of this lecture was available."

What happened? I know what happened, because I was there in the audience, at 3:00 p.m. on Thursday, September 2, 1954, in the Concertgebouw concert hall. The hall was packed with mathematicians, all expecting to hear a brilliant lecture worthy of such a historic occasion. The lecture was a huge disappointment. Von Neumann had probably agreed several years earlier to give a lecture about unsolved problems and had then forgotten about it. Being busy with many other things, he had neglected to prepare the lecture. Then, at the last moment, when he remembered that he had to travel to Amsterdam and say something about mathematics, he pulled an old lecture from the 1930s out of a drawer and dusted it off. The lecture was about rings of operators, a subject that was new and fashionable in the 1930s. Nothing about unsolved problems. Nothing about the future. Nothing about computers, the subject that we knew was dearest to von Neumann's heart. He might at least have had something new and exciting to say about computers. The audience in the concert hall became restless. Somebody said in a voice loud enough to be heard all over the hall, "Aufgewärmte Suppe", which is German for "warmed-up soup". In 1954 the great majority of mathematicians knew enough German to understand the joke. Von Neumann, deeply embarrassed, brought his lecture to a quick end and left the hall without waiting for questions.

Weak Chaos

If von Neumann had known about chaos when he spoke in Amsterdam, one of the unsolved problems that he might have talked about was weak chaos. The problem of weak chaos is still unsolved fifty years later. The problem is to understand why chaotic motions often remain bounded and do not cause any violent instability. A good example of weak chaos is the orbital motions of the planets and satellites in the solar system. It was discovered only recently that these motions are chaotic. This was a surprising discovery, upsetting the traditional picture of the solar system as the prime example of orderly stable motion. The mathematician Laplace two hundred years ago thought he had proved that the solar system is stable. It now turns out that Laplace was wrong. Accurate numerical integrations of the orbits show clearly that neighboring orbits diverge exponentially. It seems that chaos is almost universal in the world of classical dynamics.

Chaotic behavior was never suspected in the solar system before accurate long-term integrations were done, because the chaos is weak. Weak chaos means that neighboring trajectories diverge exponentially but never diverge far. The divergence begins with exponential growth but afterwards remains bounded. Because the chaos of the planetary motions is weak, the solar system can survive for four billion years. Although the motions are chaotic, the planets never wander far from their customary places, and the system as a whole does not fly apart. In spite of the prevalence of chaos, the Laplacian view of the solar system as a perfect piece of clockwork is not far from the truth.

We see the same phenomena of weak chaos in the domain of meteorology. Although the weather in New Jersey is painfully chaotic, the chaos has firm limits. Summers and winters are unpredictably mild or severe, but we can reliably predict that the temperature will never rise to 45 degrees Celsius or fall to minus 30, extremes that are often exceeded in India or in Minnesota. There is no conservation law of physics that forbids temperatures from rising as high in New Jersey as in India, or from falling as low in New Jersey as in Minnesota. The weakness of chaos has been essential to the long-term survival of life on this planet. Weak chaos gives us a challenging variety of weather while protecting us from fluctuations so severe as to endanger our existence. Chaos remains mercifully weak for reasons that we do not understand. That is another unsolved problem for young frogs in the audience to take home. I challenge you to understand the reasons why the chaos observed in a great diversity of dynamical systems is generally weak.

The subject of chaos is characterized by an abundance of quantitative data, an unending supply of beautiful pictures, and a shortage of rigorous theorems. Rigorous theorems are the best way to give a subject intellectual depth and precision. Until you can prove rigorous theorems, you do not fully understand the meaning of your concepts. In the field of chaos I know only one rigorous theorem, proved by Tien-Yien Li and Jim Yorke in 1975 and published in a short paper with the title, "Period Three Implies Chaos", [6]. The Li-Yorke paper is one of the immortal gems in the literature of mathematics. Their theorem concerns nonlinear maps of an interval onto itself. The successive positions of a point when the mapping is repeated can be considered as the orbit of a classical particle. An orbit has period N if the point returns to its original position after N mappings. An orbit is defined to be chaotic, in this context, if it diverges from all periodic orbits. The theorem says that if a single orbit with period three exists, then chaotic orbits also exist. The proof is simple and short. To my mind, this theorem and its proof throw more light than a thousand beautiful pictures on the basic nature of chaos. The theorem explains why chaos is prevalent in the world. It does not explain why chaos is so often weak. That remains a task for the future. I believe that weak chaos will not be understood in a fundamental way until we can prove rigorous theorems about it.

String Theorists

I would like to say a few words about string theory. Few words, because I know very little about string theory. I never took the trouble to learn the subject or to work on it myself. But when I am at home at the Institute for Advanced Study in Princeton, I am surrounded by string theorists, and I sometimes listen to their conversations. Occasionally I understand a little of what they are saying. Three things are clear. First, what they are doing is first-rate mathematics. The leading pure mathematicians, people like Michael Atiyah and Isadore Singer, love it. It has opened up a whole new branch of mathematics, with new ideas and new problems. Most remarkably, it gave the mathematicians new methods to solve old problems that were previously unsolvable. Second, the string theorists think of themselves as physicists rather than mathematicians. They believe that their theory describes something real in the physical world. And third, there is not yet any proof that the theory is relevant to physics. The theory is not yet testable by experiment. The theory remains in a world of its own, detached from the rest of physics. String theorists make strenuous efforts to deduce consequences of the theory that might be testable in the real world, so far without success.

My colleagues Ed Witten and Juan Maldacena and others who created string theory are birds, flying high and seeing grand visions of distant ranges of mountains. The thousands of humbler practitioners of string theory in universities around the world are frogs, exploring fine details of the mathematical structures that birds first saw on the horizon. My anxieties about string theory are sociological rather than scientific. It is a glorious thing to be one of the first thousand string theorists, discovering new connections and pioneering new methods. It is not so glorious to be one of the second thousand or one of the tenth thousand. There are now about ten thousand string theorists scattered around the world. This is a dangerous situation for the tenth thousand and perhaps also for the second thousand. It may happen unpredictably that the fashion changes and string theory becomes unfashionable. Then it could happen that nine thousand string theorists lose their jobs. They have been trained in a narrow specialty, and they may be unemployable in other fields of science.

Why are so many young people attracted to string theory? The attraction is partly intellectual. String theory is daring and mathematically elegant. But the attraction is also sociological. String theory is attractive because it offers jobs. And why are so many jobs offered in string theory? Because string theory is cheap. If you are the chairperson of a physics department in a remote place without much money, you cannot afford to build a modern laboratory to do experimental physics, but you can afford to hire a couple of string theorists. So you offer a couple of jobs in string theory, and you have a modern physics department. The temptations are strong for the chairperson to offer such jobs and for the young people to accept them. This is a hazardous situation for the young people and also for the future of science. I am not saying that we should discourage young people from working in string theory if they find it exciting. I am saying that we should offer them alternatives, so that they are not pushed into string theory by economic necessity.

Finally, I give you my own guess for the future of string theory. My guess is probably wrong. I have no illusion that I can predict the future. I tell you my guess, just to give you something to think about. I consider it unlikely that string theory will turn out to be either totally successful or totally useless. By totally successful I mean that it is a complete theory of physics, explaining all the details of particles and their interactions. By totally useless I mean that it remains a beautiful piece of pure mathematics. My guess is that string theory will end somewhere between complete success and failure. I guess that it will be like the theory of Lie groups, which Sophus Lie created in the nineteenth century as a mathematical framework for classical physics. So long as physics remained classical, Lie groups remained a failure. They were a solution looking for a problem. But then, fifty years later, the quantum revolution transformed physics, and Lie algebras found their proper place. They became the key to understanding the central role of symmetries in the quantum world. I expect that fifty or a hundred years from now another revolution in physics will happen, introducing new concepts of which we now have no inkling, and the new concepts will give string theory a new meaning. After that, string theory will suddenly find its proper place in the universe, making testable statements about the real world. I warn you that this guess about the future is probably wrong. It has the virtue of being falsifiable, which according to Karl Popper is the hallmark of a scientific statement. It may be demolished tomorrow by some discovery coming out of the Large Hadron Collider in Geneva.

Manin Again

To end this talk, I come back to Yuri Manin and his book *Mathematics as Metaphor.* The book is mainly about mathematics. It may come as a surprise to Western readers that he writes with equal eloquence about other subjects such as the collective unconscious, the origin of human language, the psychology of autism, and the role of the trickster in the mythology of many cultures. To his compatriots in Russia, such many-sided interests and expertise would come as no surprise. Russian intellectuals maintain the proud

tradition of the old Russian intelligentsia, with scientists and poets and artists and musicians belonging to a single community. They are still today, as
we see them in the plays of Chekhov, a group of idealists bound together by
their alienation from a superstitious society and a capricious government.
In Russia, mathematicians and composers and film-producers talk to one
another, walk together in the snow on winter nights, sit together over a
bottle of wine, and share each other's thoughts.

Manin is a bird whose vision extends far beyond the territory of mathematics into the wider landscape of human culture. One of his hobbies is
the theory of archetypes invented by the Swiss psychologist Carl Jung. An
archetype, according to Jung, is a mental image rooted in a collective unconscious that we all share. The intense emotions that archetypes carry with
them are relics of lost memories of collective joy and suffering. Manin is
saying that we do not need to accept Jung's theory as true in order to find
it illuminating.

More than thirty years ago, the singer Monique Morelli made a recording of songs with words by Pierre MacOrlan. One of the songs is *La Ville
Morte,* the dead city, with a haunting melody tuned to Morelli's deep con-
tralto, with an accordion singing counterpoint to the voice, and with verbal
images of extraordinary intensity. Printed on the page, the words are nothing special:

> *En pénétrant dans la ville morte,*
> *Je tenait Margot par le main. . .*
> *Nous marchions de la nécropole,*
> *Les pieds brisés et sans parole,*
> *Devant ces portes sans cadole,*
> *Devant ces trous indéfinis,*
> *Devant ces portes sans parole*
> *Et ces poubelles pleines de cris.*

"As we entered the dead city, I held Margot by the hand . . . We walked
from the graveyard on our bruised feet, without a word, passing by these
doors without locks, these vaguely glimpsed holes, these doors without a
word, these garbage cans full of screams."

I can never listen to that song without a disproportionate intensity of
feeling. I often ask myself why the simple words of the song seem to reso-
nate with some deep level of unconscious memory, as if the souls of the
departed are speaking through Morelli's music. And now unexpectedly in
Manin's book I find an answer to my question. In his chapter, "The Empty
City Archetype", Manin describes how the archetype of the dead city
appears again and again in the creations of architecture, literature, art
and film, from ancient to modern times, ever since human beings began to

congregate in cities, ever since other human beings began to congregate in armies to ravage and destroy them. The character who speaks to us in Mac-Orlan's song is an old soldier who has long ago been part of an army of occupation. After he has walked with his wife through the dust and ashes of the dead city, he hears once more:

Chansons de charme d'un clairon
Qui fleurissait une heure lointaine
Dans un rêve de garnison.

"The magic calls of a bugle that came to life for an hour in an old soldier's dream".

The words of MacOrlan and the voice of Morelli seem to be bringing to life a dream from our collective unconscious, a dream of an old soldier wandering through a dead city. The concept of the collective unconscious may be as mythical as the concept of the dead city. Manin's chapter describes the subtle light that these two possibly mythical concepts throw upon each other. He describes the collective unconscious as an irrational force that powerfully pulls us toward death and destruction. The archetype of the dead city is a distillation of the agonies of hundreds of real cities that have been destroyed since cities and marauding armies were invented. Our only way of escape from the insanity of the collective unconscious is a collective consciousness of sanity, based upon hope and reason. The great task that faces our contemporary civilization is to create such a collective consciousness.

References

[1] M. J. BERTIN ET AL., *Pisot and Salem Numbers*, Birkhäuser Verlag, Basel, 1992.

[2] M. L. CARTWRIGHT and J. E. LITTLEWOOD, On nonlinear differential equations of the second order, I, *Jour. London Math. Soc.* **20** (1945), 180–189.

[3] FREEMAN DYSON, Prof. Hermann Weyl, For.Mem. R.S., *Nature* **177** (1956), 457–458.

[4] V. A. FOCK, On the invariant form of the wave equation and of the equations of motion for a charged massive point, *Zeits. Phys.* 39(1926), 226–232.

[5] L. LANDAU and E. LIFSHITZ, *Teoria Polya*, GITTL. M.-L.(1941), Section 16.

[6] TIEN-YIEN LI and JAMES A. YORKE, Period three implies chaos, *Amer. Math. Monthly* **82** (1975), 985–992.

[7] YURI I. MANIN, *Mathematics as Metaphor: Selected Essays*, American Mathematical Society, Providence, Rhode Island, 2007. [The Russian version is: MANIN, YU. I., *Matematika kak Metafora*, Moskva, Izdatyelstvo MTsNMO, 2008.]

[8] ANDREW M. ODLYZKO, Primes, quantum chaos and computers, in *Number Theory, Proceedings of a Symposium*, National Research Council, Washington DC, 1990, pp. 35–46.

[9] L. B. OKUN, V. A. Fock and Gauge Symmetry, in *Quantum Theory, in honour of Vladimir A. Fock*, Proceedings of the VIII UNESCO International School of Physics, St. Petersburg, 1998, ed. Yuri Novozhilov and Victor Novozhilov, Part II, 13–17.

[10] HERMANN WEYL, Gravitation und elektrizität, *Sitz. König. Preuss. Akad. Wiss.* **26** (1918), 465–480.

[11] ———, Elektron und gravitation, *Zeits. Phys.* **56** (1929), 350–352.

[12] ———, *Selecta,* Birkhäuser Verlag, Basel, 1956, p. 192.

[13] CHEN NING YANG, Integral formalism for gauge fields, *Phys. Rev. Letters* **33** (1974), 445–447.

[14] CHEN NING YANG and ROBERT L. MILLS, Conservation of isotopic spin and isotopic gauge invariance, *Phys. Rev.* **96** (1954), 191–195.

[15] ———, Hermann Weyl's contribution to physics, in *Hermann Weyl, 1885–1985,* (K. Chandrasekharan, ed.), Springer-Verlag, Berlin, 1986, p. 19.

Mathematics Is Not a Game But . . .

Robert Thomas

As a mathematician I began to take an interest in philosophy of mathematics on account of my resentment at the incomprehensible notion I encountered that mathematics was

a game played with meaningless symbols on paper

—not a quotation to be attributed to anyone in particular, but a notion that was around before Hilbert [1]. Various elements of this notion are false and some are also offensive.

Mathematical effort, especially in recent decades—and the funding of it—indicate as clearly and concretely as is possible that mathematics is a serious scientific-type activity pursued by tens of thousands of persons at a professional level. While a few games may be pursued seriously by many and lucratively by a professional few, no one claims spectator sports are like mathematics. At the other end of the notion, paper is inessential, merely helpful to the memory. Communication (which is what the paper might hint at) *is* essential; our grip on the objectivity of mathematics depends on our being able to communicate our ideas effectively.

Turning to the more offensive aspects of the notion, we think often of competition when we think of games, and in mathematics one has no opponent. Such competitors as there are not opponents. Worst of all is the meaninglessness attributed to the paradigm of clear meaning; what could be clearer than $2 + 2 = 4$? Is this game idea not irredeemably outrageous?

Yes, it is outrageous, but there is within it a kernel of useful insight that is often obscured by outrage at the main notion, which is not often advocated presumably for that reason. I know of no one that claims that mathematics is a game or bunch of games. The main advocate of the idea that doing mathematics is *like* playing a game is David Wells [3]. It is the purpose of this essay to point to the obscured kernel of insight.

Mathematics Is Not a Game

Mathematics is not a collection of games, but perhaps it is somehow like games, as written mathematics is somehow like narrative. I became persuaded of the merit of some comparison with games in two stages, during one of which I noticed a further fault with the notion itself: there are no meaningless games. Meaningless activities such as tics and obsessions are not games, and no one mistakes them for games. Meanings in games are internal, not having to do with reference to things outside the game (as electrons, for example, in physics are supposed to refer to electrons in the world). The kings and queens of chess would not become outdated if all nations were republics.

The 'meaningless' aspect of the mathematics-as-game notion is self-contradictory; it might be interesting to know how it got into it and why it stayed so long.

Taking it as given then that games are meaningful to their players and often to spectators, how are mathematical activities like game-playing activities? The first stage of winning me over to a toleration of this comparison came in my study of the comparison with narrative [4].

One makes sense of narrative, whether fictional or factual, by a mental construction that is sometimes called the world of the story. Keeping in mind that the world of the story may be the real world at some other time or right now in some other place, one sees that this imaginative effort is a standard way of understanding things that people say; it need have nothing at all to do with an intentionally creative imagining like writing fiction. In order to understand connected speech about concrete things, one imagines them. This is as obvious as it is unclear how we do it. We often say that we pretend that we are in the world of the story. This pretence is one way—and a very effective way—of indicating how we imagine what one of the persons we are hearing about can see or hear under the circumstances of the story. If I want to have some idea what a person in certain circumstances can see, for example, I imagine myself in those circumstances and ask myself what I can see [5]. Pretending to be in those circumstances does not conflict with my certain knowledge that on the contrary I am listening to the news on my radio at home. This may make it a weak sort of pretence, but it is no less useful for that. The capacity to do this is of some importance. It encourages empathy, but it also allows one to do mathematics. One can pretend what one likes and consider the consequences at any length, entirely without commitment. This is often fun, and it is a form of playing with ideas. Some element of this pretence is needed, it seems to me, in changing one's response to 'what is $2 + 2$?' from '$2 + 2$ what?' to the less concrete 'four' [6].

This ludic aspect of mathematics is emphasized by Brian Rotman in his semiotic analysis of mathematics [7] and acknowledged by David Wells in his comparison of mathematics and games. Admitting this was the first stage of my coming to terms with games. The ludic aspect is something that undergraduates, many of whom have decided that mathematics is either a guessing game (a bad comparison of mathematics and games) or the execution of rigidly defined procedures, need to be encouraged to do when they are learning new ideas. They need to fool around with them to become familiar with them. Changing the parameters and seeing what a function looks like with that variety of parameter values is a good way to learn how the function behaves. And it is by no means only students that need to fool around with ideas in order to become familiar with them. Mathematical research involves a good deal of fooling around, which is part of why it is a pleasurable activity. This sort of play is the kind of play that Kendall Walton illustrates with the example of boys in woods not recently logged pretending that stumps are bears [8]. This is not competitive, just imaginative fooling around.

I do not think that this real and fairly widely acknowledged—at least never denied—aspect of mathematics has much to do with the canard with which I began. The canard is a reductionistic attack on mathematics, for it says it is 'nothing but' something it is not: the standard reductionist tactic. In my opinion, mathematics is an objective science, but a slightly strange one on account of its subtle subject matter; in some hands it is also an art [9]. Having discussed this recently at some length [10], I do not propose to say anything about what mathematics *is* here, but to continue with what mathematics is *like*; because such comparisons, like that with narrative, are instructive and sometimes philosophically interesting.

The serious comparison of mathematics with games is due in my experience to David Wells, who has summed up what he has been saying on the matter for twenty years in a strange document, draft zero of a book or two called *Mathematics and Abstract Games: An Intimate Connection* [3]. Wells is no reductionist and does not think that mathematics is any sort of a game, meaningless or otherwise. He confines himself to the comparison ('like a collection of abstract games'—p. 7, a section on differences—pp. 45–51), and I found this helpful in the second stage of my seeking insight in the comparison. But I did not find Wells's direct comparison as helpful as I hope to make my own, which builds on his with the intent of making it more comprehensible and attractive (*cf.* my opening sentence).

Doing Mathematics Is Not Like Playing a Game

Depending on when one thinks the activities of our intellectual ancestors began to include what we acknowledge as mathematics, one may or may not include as mathematics the thoughts lost forever of those persons with the cuneiform tablets on which they solved equations. The tablets them-selves indicate procedures for solving those particular equations. Just keep-ing track of quantities of all sorts of things obviously extended still farther back, to something we would not recognize as mathematics but which gave rise to arithmetic. Keeping track of some of the many things that one can-not count presumably gave rise to geometrical ideas. It does seem undeni-able that such procedural elements are the historical if not the logical basis of mathematics, and not only in the Near East but also in India and China. I do not see how mathematics could arise without such pre-existing proce-dures and reflections on them—probably written down, for it is so much easier to reflect on what is written down.

This consideration of procedures, and of course their raw material and results, is of great importance to my comparison of mathematics and games because *my comparison is not between playing games and doing mathematics,* I am taking mathematics to be the sophisticated activity that is the subject mat-ter of philosophy of mathematics and research in mathematics. I do not mean actions such as adding up columns of figures. Mathematics is not even those more complicated actions that we are happy to transfer to comput-ers. Mathematics is what we want to keep for ourselves. When playing games, we stick to the rules (or we are changing the game being played), but when doing serious mathematics (not executing algorithms) we make up the rules—definitions, axioms, and some of us even logics. As Wells points out in the section of his book on differences between games and mathemat-ics, in arithmetic we find prime numbers, which are a whole new 'game' in themselves (metaphorically speaking).

While mathematics requires reflection on pre-existing procedures, re-flection on procedures does not become recognizable as mathematics until the reflection has become sufficiently communicable to be convincing. Con-viction of something is a feeling, and so it can occur without communica-tion and without verbalizing or symbolizing. But to convince *someone else* of something, we need to communicate, and that does seem to be an essential feature of mathematics, whether anything is written down or not—*a forti-ori* whether anything is symbolic. And of course convincing argument is proof.

The analogy with games that I accept is based on the possibility of con-vincing argument about abstract games. Anyone knowing the rules of chess can be convinced that a move has certain consequences. Such argument does not follow the rules of chess or any other rules, but it is *based* on the rules of chess in a way different from the way it is based on the rules of logic that it might obey. To discuss the analogy of this with mathematics, I think it may be useful to call upon two ways of talking about mathematics, those of Philip Kitcher and of Brian Rotman.

Ideal Agents

In his book *The Nature of Mathematical Knowledge* [11], Kitcher introduced a theoretical device he called the ideal agent. 'We can conceive of the principles of [empirical] Arithmetic as implicit definitions of an ideal agent. An ideal agent is a being whose physical operations of segregation do satisfy the principles [that allow the deduction in physical terms of the theorems of elementary arithmetic].' (p. 117). No ontological commitment is given to the ideal agent; in this it is likened to an ideal gas. And for this reason we are able to 'specify the capacities of the ideal agent by abstracting from the incidental limitations on our own collective practice' (ibid.).

The agent can do what we can do but can do it for collections however large, as we cannot. Thus modality is introduced without regard to human physical limitations. 'Our geometrical statements can finally be understood as describing the performances of an ideal agent on ideal objects in an ideal space.' (p. 124). Kitcher also alludes to the 'double functioning of mathematical language—its use as a vehicle for the performance of mathematical operations as well as its reporting on those operations' (p. 130). 'To solve a problem is to discover a truth about mathematical operations, and to fiddle with the notation or to discern analogies in it is, on my account, to engage in those mathematical operations which one is attempting to characterize.' (p. 131).

Rotman is at pains to distinguish what he says from what Kitcher had written some years before the 1993 publication of *The Ghost in Turing's Machine* [7] because he developed his theory independently and with different aims, but we readers can regard his apparatus as a refinement of Kitcher's, for Rotman's cast of characters includes an Agent to do the bidding of the character called the Subject. The Subject is Rotman's idealization of the person that reads and writes mathematical text, and also the person that carries out some of the commands of the text. For example, it is the reader that obeys the command, 'Consider triangle *ABC*.' But it is the Agent (p. 73) that carries out such commands as, 'Drop a perpendicular from vertex *A* to the line *BC*,' provided that the command is within the Agent's capacities. We humans are well aware that we cannot draw straight lines; that is the work of the agents, Kitcher's and Rotman's. We reflect on the potential actions of these agents and address our reflections to other thinking Subjects. Rotman's discussion of this is rich with details like the tenselessness of the commands to the Agent, indeed the complete lack of all indexicality in such texts. The tenselessness is an indication of how the Subject is an idealization as the Agent is, despite not being blessed with the supernatural powers of the Agent. The Agent, Rotman says, is like the person in a dream, the Subject like the person dreaming the dream, whereas in our normal state we real folk are more like the dreamer awake, what Rotman calls the Person to complete his semiotic hierarchy.

Rotman then transfers the whole enterprise to the texts, so that mathematical statements are claims about what will result when certain operations are performed on signs (p. 77). We need, not follow him there to appreciate the serviceability of his semiotic distinctions.

The need for superhuman capacities was noted long ago in Frege's ridicule [2] of the thought that mathematics is about empty symbols:

([. . .] we would need an infinitely long blackboard, an infinite supply of chalk, and an infinite length of time—p. 199, § 124).

He also objected to a comparison to chess for Thomae's formal theory of numbers, while admitting that 'there can be theorems in a *theory* of chess' (p. 168, § 93, my emphasis). According to Frege,

The distinction between the game itself and its theory, not drawn by Thomae, makes an essential contribution towards our understanding of the matter. [. . .] in the theory of chess it is not the chess pieces which are actually investigated; it is a question of the rules and their consequences. (pp. 168–169, § 93)

The Analogies between Mathematics and Games

Having at our disposal the superhuman agents of Kitcher and Rotman, we are in a position to see what is analogous between mathematics and games.

It is not playing the game that is analogous to mathematics, but our reflection in the role of subject on the playing of the game, which is done by the agent. When a column of figures is added up, we do it, and sometimes when the product of two elements of a group is required, we calculate it; but mathematics in the sense I am using here is not such mechanical processes at all, but the investigation of their possibility, impossibility, and results. For that highly sophisticated reflective mathematical activity, the agent does the work because the agent can draw straight lines. Whether points are collinear depends on whether they are on the agent's straight lines, not on whether they appear on the line in our sketch. We can put them on or off the line at will; the agent's results are constrained by the rules of the system in which the agent is working. Typically we have to deduce whether the agent's line is through a point or not. The agent, 'playing the game' according to the rules, gets the line through the point or not, but we have to figure it out. We *can* figure it out; the agent just does it. The analogy to games is two-fold.

1. The agent's mathematical activity (not playing a game) is analogous to the activity of playing a game like chess where it is clear what is possible and what is impossible—the same for every player—often superhuman but bound by rules. (Games like tennis depend for what is possible on physical skill, which has no pertinence here.)

2. Our mathematical activity is analogous to (a) game invention and development, (b) the reflection on the playing of a game like chess that distinguishes expert play from novice play, or (c) consideration of matters of play for their intrinsic interest apart from playing any particular match—merely human but not bound by rules.

It is we that deduce; the agent just does what it is told, provided that it is within the rules we have chosen. Analogous to the hypotheses of our theorems, are chess positions, about which it is possible to reason as dependably as in mathematics because the structure is sufficiently precisely set out that everyone who knows the rules can see what statements about chess positions are legitimate and what are not. Chains of reasoning can be as long as we like without degenerating into the vagueness that plagues chains of reasoning about the real world. The ability to make and depend on such chains of reasoning in chess and other games is the ability that we need to make such chains in mathematics, as David Wells points out.

To obtain a useful analogy here, it is necessary to rise above the agent in the mathematics and the mere physical player in the game, but the useful analogy is dependent on the positions in the game and the relations in the mathematics. The reflection in the game is about positions more than the play, and the mathematics is about relations and their possibility more than drawing circles or taking compact closures. Certainly the physical pieces used in chess and the symbols on paper are some distance below what is importantly going on.

I hope that the previous discussion makes clear why some rules are necessary to the analogy despite the fact that we are not bound by those rules. The rules are essential because we could not do what we do without them, but it is the agent that is bound by them. We are talking about, as it were, what a particular choice of them does and does not allow. But our own activity is not bound by rules; we can say anything that conveys our meaning, anything that is convincing to others.

Here is objectivity without objects. Chess reasoning is not dependent upon chess boards and chess men; it is dependent on the relations of positions mandated by the rules of the game of chess. Mathematics is not dependent on symbols (although they are as handy as chess sets) but on the relations of whatever we imagine the agent to work on, specified and reasoned about. Our conclusions are right or wrong as plainly as if we were ideal agents loose in Plato's heaven, but they are right or wrong dependent on what the axioms, conventions, or procedures we have chosen dictate.

Outside mathematics, we reason routinely about what does not exist, most particularly about the future. As the novelist Jim Crace was quoted on page R10 of the February 6, 2007, Toronto *Globe and Mail,* 'As a good Darwinist, I know that what doesn't confer an advantage dies out. One advantage [of narrative (*Globe and Mail* addition)] is that it enables us to play out the bad things that might happen to us and to rehearse what we might do.' In order to tell our own stories, it is essential to project them hypothetically into the future based on observations and assumptions about the present. At its simplest and most certain, the skill involved is what allows one to note that if one moves this pawn forward one square the opponent's pawn can take it. It's about possibilities and of course impossibilities, all of them hypothetical. It is this fundamental skill that is used both in reflection on games and in mathematics to see what is necessary in their respective worlds.

I must make clear that David Wells thinks that entities in maths and abstract games have the same epistemological status but that doing mathematics is like (an expert's) playing a game in several crucial respects, no more; he disagrees with the usefulness of bringing in ideal agents, indeed opposes doing so, apparently not seeing the advantage of splitting the analogy into the two numbered aspects above. This section is my attempt to outline a different but acceptable game analogy—a game-analysis analogy.

Conclusion

Games such as bridge and backgammon, which certainly involve strategy, have a stochastic element that prevents long chains of reasoning from being as useful as they are in chess. Such chains are, after all, an important part of how computers play chess. The probabilistic mathematics advocated by Doron Zeilberger [12] is analogous to the analysis of such a stochastic game, and will be shunned by those uninterested in such analysis of something in which they see nothing stochastic. Classical (von Neumann) game theory, on the other hand, actually is the analysis of situations that are called games and do involve strategy. The game theory of that current Princeton genius, John Conway, is likewise the actual analysis of game situations [13]. Does the existence of such mathematical analysis count for or against the general analogy between mathematics and game analysis?

On my version of the analogy, to identify mathematics with games would be one of those part-for-whole mistakes (like 'all geometry is projective geometry' or 'arithmetic is just logic' from the nineteenth century); but identification is not the issue. It seems to me that my separation of game analysis from playing games tells in favour of the analogy of mathematics to analysis of games played by other—not *necessarily* superhuman—agents, and against the analogy of mathematics to the expert play of the game itself. This is not a question David Wells has discussed. For Wells, himself an expert at abstract games such as chess and go, play *is* expert play based unavoidably on analysis; analysis is just part of playing the game. Many are able to distinguish these activities, and not just hypothetically.

One occasionally hears the question, is mathematics invented or discovered?—or an answer. As David Wells points out, even his game analogy shows why both answers and the answer 'both' are appropriate. Once a game is invented, the consequences are discovered—genuinely discovered, as it would require a divine intelligence to know just from the rules how a complex game could best be played. When in practice rules are changed, one makes adjustments that will not alter the consequences too drastically. Analogously, axioms are usually only adjusted and the altered consequences discovered.

What use can one make of this analogy? One use that one cannot make of it is as a stick to beat philosophers into admitting that mathematics is not problematic. Like mathematicians, philosophers thrive on problems. Problems are the business of both mathematics and philosophy. Solving problems is the business of mathematics. If a philosopher came to regard the analogy as of some validity, then she would import into the hitherto unexamined territory of abstract games all of the philosophical problems concerning mathematics. Are chess positions real? How do we know about them? And so on; a new branch of philosophy would be invented.

What use then can mathematicians make of the analogy? We can use it as comparatively unproblematic material in discussing mathematics with those nonphilosophers desiring to understand mathematics better. I have tried to indicate above some of the ways in which the analogy is both apt and of sufficient complexity to be interesting; it is no simple metaphor but can stand some exploration. Some of this exploration has been carried out by David Wells, to whose work I need to refer the reader.

References

[1] Thomae, Johannes. *Elementare Theorie der analytischen Functionen einer complexen Veränderlichen*. 2nd ed. Halle: L. Nebert, 1898 (1st ed. 1880), ridiculed by Frege [2].

[2] Frege, Gottlob. 'Frege against the formalists' from *Grundgesetze der Arithmetik*, Jena: H. Pohle. Vol. 2, Sections 86–137, in *Translations from the philosophical writings of Gottlob Frege*. 3rd ed. Peter Geach and Max Black, eds. Oxford: Blackwell, 1980. I am grateful to a referee for pointing this out to me.

[3] Wells, David. *Mathematics and Abstract Games: An Intimate Connection*. London: Rain Press, 2007. (Address: 27 Cedar Lodge, Exeter Road, London NW2 3UL, U.K. Price: £10; $20 including surface postage.)

[4] Thomas, Robert. 'Mathematics and Narrative'. *The Mathematical Intelligencer* 24 (2002), 3, 43–46.

[5] O'Neill, Daniella K., and Rebecca M. Shultis. 'The emergence of the ability to track a character's mental perspective in narrative', *Developmental Psychology*, 43 (2007), 1032–1037.

[6] Donaldson, Margaret. *Human Minds*, London: Penguin, 1993.

[7] Rotman, Brian. *Ad Infinitum: The Ghost in Turing's Machine*. Stanford: Stanford University Press, 1993.

[8] Walton, Kendall L. *Mimesis as Make-Believe: On the Foundations of the Representational Arts*. Cambridge, Mass.: Harvard University Press, 1990. Also Currie, Gregory. *The Nature of Fiction*. Cambridge: Cambridge University Press, 1990.

[9] Davis, Chandler, and Erich W. Ellers. *The Coxeter Legacy: Reflections and Projections*. Providence, R.I.: American Mathematical Society, 2006.

[10] Thomas, Robert. 'Extreme Science: Mathematics as the Science of Relations as such,' in *Proof and Other Dilemmas: Mathematics and Philosophy*, Ed. Bonnie Gold and Roger Simons. Washington, D.C.: Mathematical Association of America, 2008.

[11] Kitcher, Philip. *The Nature of Mathematical Knowledge*. New York: Oxford University Press, 1984.

[12] Zellberger, Doron. 'Theorems for a price: Tomorrow's semi-rigorous mathematical culture'. *Notices of the Amer. Math. Soc.* 40 (1993), 978–981. Reprinted in *The Mathematical Intelligencer* 16 (1994), 4, 11–14.

[13] Conway, John H. *On Numbers and Games*. 2nd ed. Natick, Mass.: AK Peters, 2001. (1st ed. LMS Monographs; 6. London: Academic Press, 1976.) Also Berlekamp, E.R., J.H. Conway, and R.K. Guy. *Winning Ways for your Mathematical Plays*. 2 volumes. London: Academic Press, 1982.

Massively Collaborative Mathematics

TIMOTHY GOWERS AND MICHAEL NIELSEN

On 27 January 2009, one of us—Gowers—used his blog to announce an unusual experiment. The Polymath Project had a conventional scientific goal: to attack an unsolved problem in mathematics. But it also had the more ambitious goal of doing mathematical research in a new way. Inspired by open-source enterprises such as Linux and Wikipedia, it used blogs and a wiki to mediate a fully open collaboration. Anyone in the world could follow along and, if they wished, make a contribution. The blogs and wiki functioned as a collective short-term working memory, a conversational commons for the rapid fire exchange and improvement of ideas.

The collaboration achieved far more than Gowers expected, and showcases what we think will be a powerful force in scientific discovery—the collaboration of many minds through the Internet.

The specific aim of the Polymath Project was to find an elementary proof of a special case of the density Hales-Jewett theorem (DHJ), which is a central result of combinatorics, the branch of mathematics that studies discrete structures (see 'Multidimensional noughts and crosses'). This theorem was already known to be true, but for mathematicians, proofs are more than guarantees of truth: they are valued for their explanatory power, and a new proof of a theorem can provide crucial insights. There were two reasons to want a new proof of the DHJ theorem. First, it is one of a cluster of important related results, and although almost all the others have multiple proofs, DHJ had just one—a long and complicated proof that relied on heavy mathematical machinery. An elementary proof—one that starts from first principles instead of relying on advanced techniques—would require many new ideas. Second, DHJ implies another famous theorem, called Szemerédi's theorem, novel proofs of which have led to several breakthroughs over the past decade, so there was reason to expect that the same would happen with a new proof of the DHJ theorem.

The project began with Gowers posting a description of the problem, pointers to background materials and a preliminary list of rules for collaboration (see go.nature.com/DrCmnC). These rules helped to create a polite, respectful atmosphere, and encouraged people to share a single idea in each comment, even if the idea was not fully developed. This lowered the barrier to contribution and kept the conversation informal.

Building Momentum

When the collaborative discussion kicked off on 1 February, it started slowly: more than seven hours passed before Jozsef Solymosi, a mathematician at the University of British Columbia in Vancouver made the first comment. Fifteen minutes later a comment came in from Arizona-based high-school teacher Jason Dyer. Three minutes after that Terence Tao (winner of a Fields Medal, the highest honour in mathematics) at the University of California, Los Angeles, made a comment. Over the next 37 days, 27 people contributed approximately 800 substantive comments, containing 170,000 words. No one was specifically invited to participate: anybody, from graduate student to professional mathematician, could provide input on any aspect. Nielsen set up the wiki to distil notable insights from the blog discussions. The project received commentary on at least 16 blogs, reached the front page of the Slashdot technology-news aggregator, and spawned a closely related project on Tao's blog. Things went smoothly: neither Internet 'trolls'—persistent posters of malicious or purposefully distracting comments—nor well-intentioned but unhelpful comments were significant problems, although spam was an occasional issue on the wiki. Gowers acted as a moderator, but this involved little more than correcting a few typos.

Progress came far faster than anyone expected. On 10 March, Gowers announced that he was confident that the Polymath participants had found an elementary proof of the special case of DHJ, but also that, very surprisingly (in the light of experience with similar problems), the argument could be straightforwardly generalized to prove the full theorem. A paper describing this proof is being written up, along with a second paper describing related results. Also during the project, Tim Austin, a graduate student at the University of California, Los Angeles, announced another new (but non-elementary) proof of DHJ that made crucial use of ideas from the Polymath Project.

The working record of the Polymath Project is a remarkable resource for students of mathematics and for historians and philosophers of science. For the first time one can see on full display a complete account of how a serious mathematical result was discovered. It shows vividly how ideas grow, change, improve and are discarded, and how advances in understanding may come not in a single giant leap, but through the aggregation and refinement of many smaller insights. It shows the persistence required to solve a difficult problem, often in the face of considerable uncertainty, and how even the best mathematicians can make basic mistakes and pursue many failed ideas. There are ups, downs and real tension as the participants close in on a solution. Who would have guessed that the working record of a mathematical project would read like a thriller?

Broader Implications

The Polymath Project differed from traditional large-team collaborations in other parts of science and industry. In such collaborations, work is usually divided up in a static, hierarchical way. In the Polymath Project, everything was out in the open, so anybody could potentially contribute to any aspect. This allowed ideas to be explored from many different perspectives and allowed unanticipated connections to be made.

The process raises questions about who should count as an author: it is difficult to set a hard-and-fast bar for authorship without causing contention or discouraging participation. What credit should be given to contributors with just a single insightful contribution, or to a contributor who is prolific but not insightful? As a provisional solution, the project is signing papers with a group pseudonym, 'DHJ Polymath', and a link to the full working record. One advantage of Polymath-style collaborations is that because all contributions are out in the open, it is transparent what any given person contributed. If it is necessary to assess the achievements of a Polymath contributor, then this may be done primarily through letters of recommendation, as is done already in particle physics, where papers can have hundreds of authors.

The project also raises questions about preservation. The main working record of the Polymath Project is spread across two blogs and a wiki, leaving it vulnerable should any of those sites disappear. In 2007, the US Library of Congress implemented a programme to preserve blogs by people in the legal profession; a similar but broader programme is needed to preserve research blogs and wikis.

New projects now under way will help to explore how collaborative mathematics works best (see go.nature.com/4Zfldc). One question of particular interest is whether the process can be scaled up to involve more contributors. Although DHJ Polymath was large compared with most mathematical collaborations, it fell short of being the mass collaboration initially envisaged. Those involved agreed that scaling up much further would require changes to the process. A significant barrier to entry was the linear narrative style of the blog. This made it difficult for late entrants to identify problems to which their talents could be applied. There was also a natural fear that they might have missed an earlier discussion and that any contribution they made would be redundant. In open-source software development, this difficulty is addressed in part by using issue-tracking software to organize development around 'issues'—typically, bug reports or feature requests—giving late entrants a natural starting point, limiting the background material that must be mastered, and breaking the discussion down into modules. Similar ideas may be useful in future Polymath Projects.

Multidimensional Noughts and Crosses

To understand the density Hales-Jewett theorem (DHJ), imagine a multidimensional noughts-and-crosses (or tic-tac-toe) board, with k squares on a side (instead of the usual three), and in n dimensions rather than two. Any square in this board has n coordinates between 1 and k, so for instance if $k = 3$ and $n = 5$, then a typical point might be $(1,3,2,1,2)$. A line on such a board has coordinates that either stay the same from one point to the next, or go upwards or downwards. For instance, the three points $(1,2,3,1,3)$, $(2,2,3,2,2)$ and $(3,2,3,3,1)$, form a line. DHJ states that, for a very large number of dimensions, filling in even a tiny fraction of the board always forces a line to be filled in somewhere—there is no possible way of avoiding such a line. More than this, there is no way to avoid a 'combinatorial line', in which the coordinates that vary have to vary in the same direction (rather than some going up and some going down), as in the line $(1,2,3,1,1)$, $(2,2,3,2,2)$ and $(3,2,3,3,3)$. The initial aim of the polymath project was to tackle the first truly difficult case of DHJ, which is when $k = 3$.

Towards Open Science

The Polymath process could potentially be applied to even the biggest open problems, such as the million-dollar prize problems of the Clay Mathematics Institute in Cambridge, Massachusetts. Although the collaborative model might deter some people who hoped to keep all the credit for themselves, others could see it as their best chance of being involved in the solution of a famous problem.

Outside mathematics, open-source approaches have only slowly been adopted by scientists. One area in which they are being used is synthetic biology. DNA for the design of living organisms is specified digitally and uploaded to an online repository such as the Massachusetts Institute of Technology Registry of Standard Biological Parts. Other groups may use those designs in their laboratories and, if they wish, contribute improved designs back to the registry. The registry contains more than 3,200 parts, deposited by more than 100 groups. Discoveries have led to many scientific papers, and a 2008 study showed that most parts are not primitive but rather build on simpler parts (J. Peccoud *et al. PLoS ONE* **3**, e2671; 2008). Open-source biology and open-source mathematics thus both show how science can be done using a gradual aggregation of insights from people with diverse expertise.

Similar open-source techniques could be applied in fields such as theoretical physics and computer science, where the raw materials are informational and can be freely shared online. The application of open-source techniques to experimental work is more constrained, because control of experimental equipment is often difficult to share. But open sharing of experimental data does at least allow open data analysis. The widespread adoption of such open-source techniques will require significant cultural changes in science, as well as the development of new online tools. We believe that this will lead to the widespread use of mass collaboration in many fields of science, and that mass collaboration will extend the limits of human problem-solving ability.

Bridging the Two Cultures: Paul Valéry

PHILIP J. DAVIS

The conjunction of mathematics and poetry is alive and well. Thus, e.g., *Strange Attractors*, a stimulating anthology of poems that allude to mathematics (often very superficially) has just appeared.[1] Quite by accident I learned that Paul Valéry (1871–1945), the famous French poet, essayist, philosopher and aphorist, had left in his extensive notebooks many observations, impressions, ruminations and thoughts about mathematics. As opposed to the material in *Strange Attractors*, Valéry created no poems specifically about mathematics. What is remarkable in his case is the depth of his infatuation with the deepest aspects of mathematics, and his struggle to come to grips with them through his own poetic imagination and intuition.[2]

What, I wondered, was the significance of Valéry's innumerable jottings about mathematics? He is not mentioned in any history or philosophy of mathematics of which I am aware; his name attaches to no mathematical ideas, constructions, theorems, processes. Was he then, merely expressing, as a private citizen so to speak, his feelings about a subject for which he had a deep attraction; or was he putting forward a claim as having given birth to a unique vision as to what mathematics was all about? The question is relevant because Valéry's work has been the subject of thousands of articles and theses, and though he is known now in France to the general public largely as a university (Montpellier–Valéry), his writings are the object of continued studies by specialists. Thus, the University of Newcastle upon Tyne houses a Centre for Valéry Studies headed up by Prof. Brian Stimson.

Valéry's notebooks (*Cahiers*) have all been published. I opened one at random (Vol. V) and found strewn among the author's jottings on innumerable topics, mathematical equations, notations, computations, figures, together with animadversions on the same. Are these, I wondered, simply mathematical graffiti? Was he trying to play the mathematician and floundering around with difficult problems (e.g., the four color problem) or was he merely a commentator looking down with love and admiration from the

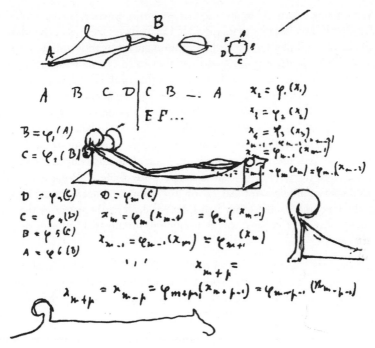

FIGURE 1. Jottings extracted from *Cahier V*, p. 200.

Olympian heights on a topic that seems to have engaged his thoughts continually? To appreciate and make sense of this material is not easy; one must go into the author's head and language, get under his skin. A knowledge of his total *oeuvre* is really required.

To help me arrive at an answer, I turn to a lengthy article by Judith Robinson, a scholar of French literature and a specialist on Valéry which treats the mathematical jottings in his *Cahiers* in some depth. Robinson boned up on mathematical ideas and was able to connect the dots of the dispersed jottings, figures, computations, into a continuous narrative. In what follows, when I use the name Valéry, I really mean the Valéry as seen principally through the imaginative eyes of Robinson.

Valéry was much concerned with mathematics and physics and in his mind the two are almost inextricable. He read extensively; he knew the mathematician Emile Borel and the physicists Jean Perrin and Paul Langevin. He corresponded with Jacques Hadamard. He admired Riemann and Cantor and his admiration of Poincaré was unbounded. He kept Poincaré's books beside his bed, rereading them constantly. Mathematics, he wrote, became his opium. Valéry was "greatly influenced in his thinking" by the developments relativity, in atomic and quantum physics, in particular by Heisenberg's Uncertainty Principle. He has many references to the ideas of mathematics and physics which he comprehended in an intuitive, impressionistic way and often employed metaphorically. His understanding was, according to Robinson, not superficial, not that of the mere amateur.

As befits a poet and a littérateur Valéry was much concerned with language. He questioned "ready-made" language, considered it impure, vague, imprecise, ambiguous, a stumbling block in perceiving and apprehending reality. He believed that our (natural) languages were responsible for the raising of stupid, meaningless, undefined, or often false questions such as "who created the world ?," "what is the purpose of life ?," "what is the relationship between mind and matter" or "the distinction between the knower and the known." On the other hand, he admired mathematics as the only precise, coherent, rigorous, and unambiguous intellectual language devised by humans.

> Perhaps the greatest virtue of mathematics as Valéry sees it is that it enables us not only to define our terms but also to relate them to one another in a logical and coherent way. Now this is exactly what ordinary language has great difficulty in doing.

Valéry's concern with language reminded me, by way of contrast, of the Neapolitan philosopher of language and anti-Cartesian Giambattista Vico (1668–1744.) My source for Vico is the book of Sir Isaiah Berlin.[3] Vico's ideas, when applied to mathematics, presage today's social constructivism.

> [Vico] concedes that mathematical knowledge is indeed wholly valid and its propositions are certain. But the reasons for this are only too clear: 'We demonstrate geometry because we make it'.

His catch phrase, often quoted, is "Verum esse ipse factum" (Truth itself is constructed).

I do not know whether Valéry read Vico; nonetheless, a remark of Valéry that he begins

> to inquire by what sign we recognize that a given object is or is not made by man (*Man and the Sea Shell*)

leads me to think that a study comparing and contrasting the ideas of Vico and those of Valéry would be of significance. I myself am sympathetic towards Vico's ideas, for while admitting the precise, coherent, rigorous, and unambiguous nature of mathematical language, such language lies totally embedded in natural languages (e.g., English) with all their impurities. Strip mathematical texts of all descriptions, explications, instructions given in natural language, and a page of naked mathematical symbols would be incomprehensible.

Berlin also remarks that

Political courage was no more characteristic of Vico than of Leibniz or a good many other scholars and philosophers of the age . . .[4]

These men were above the fray, and to this list we may safely add Valéry. Even as Leibniz dreamed of a formal language—a *characteristica universalis*—within which calculations of mathematical, scientific, legal and metaphysical concepts could be expressed and related problems solved, Valéry dreamed of a language, an "algebra de l'esprit," an *arithmetica universalis* that would be a useful instrument in the study of brain function, a language in which "the physical and the psychological could be compared"; a language into which "propositions of every kind could be translated." Turning to the influence of other contemporaries, Valéry admired the logic of Russell, and one can see the relationship between Valéry's thought and that of Wittgenstein. What interested him primarily was not the object, but the relationship between objects, and in the flexibility and generality of such relationships that he discovered in various geometries and in the ideas of Riemann. These were ideas that he hoped could be expressed in notational terms. Despite these perceptions, I do not believe Valéry could be termed a logicist. I would call him a formalist.

Valéry did not go public with these thoughts; he committed them to his notebooks, perhaps in the hope that they would be interpreted by later scholars. The aspects of mathematics that he admired so much are today simply part of the working knowledge and attitudes of today's mathematicians. He did not contribute anything either to mathematics or physics or to their philosophy. But what is absolutely remarkable in this story is that a person so devoted to language and literature should have immersed himself deeply and knowingly and arrived at individual insights into questions of mathematics and physics. Valéry was called by some a new Leonardo; by others a philosopher for old ladies. Whatever.

The importance of Valéry for me is that his notebooks are proof positive that the "Two Cultures" of C.P. Snow, the literary and the scientific, often thought to be unbridgeable, can indeed be bridged if only in one direction. The American philosopher William James wrote "The union of the mathematician with the poet, fervor with measure, passion with correctness, this surely is the ideal." While the thoughts of Valéry are as close as have been come to the realization of this ideal in measure and passion, the bridge he constructed was personal and cannot easily be transmitted to the general readership.

Acknowledgements

I wish to thank Brian Stimpson for supplying me with valuable information and Charlotte Maday and Vera Steiner for their encouragement.

Notes

1. *Strange Attractors,* AK Peters, 2008.
2. A 2006 Stanford University thesis by Rima Joseph, *Paul Valéry parmi la mathematique*, considers the influence that Valéry's knowledge of mathematics may have had on his poetry: "The elucidation of such [mathematical] models used in metaphors by Valéry reveal mathematics as an aesthetic object ultimately forming his poetics."
3. *Vico and Herder*, Viking.
4. Loc. cit. p. 5.

References

Sir Isaiah Berlin, *Vico and Herder*, Viking, 1976.

J. M. Cocking, *Duchesne-Guillemin on Valéry,* Modern Language Review, 62, 1967, pp. 55–60.

Jaques Duchesne-Guillemin, *Etudes pour un Paul Valéry,* La Baconnière, Neuchatel, 1964.

Rima Joseph, *Paul Valéry parmi la mathematique*, Doctoral thesis, Stanford University, 2006.

Walter Putnam, *Paul Valéry Revisited*, Twayne Publishers, 1995.

Judith Robinson, *Language Physics and Mathematics in Valéry's Cahiers.* Modern Language Review, Oct 1960, pp. 519–536.

Judith Robinson: *L'Analyse de l'esprit dans les Cahiers de Valéry*, Paris, Corti, 1963.

Judith Robinson (ed.), *Fonctions de l'esprit: 13 savants redécouvrent Valéry*, Paris, Hermann, 1983. (Contributions by Jean Dieudonné, René Thom, André Lichnerowicz, Pierre Auger, Bernard d'Espagnat, Jacques Bouveress and Ily Prigogine.)

Michel Sirvent, *Chiffrement, déchiffrement: de Paul Valéry à Jean Ricardou*, French Review 66:2 December 1992, pp. 255–266.

Paul Valéry, *History and Politics*, Vol 10, Bollingen, Pantheon, 1962.

Paul Valéry, *Cahiers 1894–1914, V*, Gallimard, 1994. Translations of various volumes forthcoming.

A Hidden Praise of Mathematics

ALICIA DICKENSTEIN

We have all given reasons to show how important pure mathematics is. In particular, I have often heard (and repeated) statements along the lines of "our abstract theories will eventually be useful to build important physical (chemical, biological, . . .) theories." But wouldn't it be great if someone like Albert Einstein himself said that?

Well, the news is he did—in fact on the very first page of his main article on the General Theory of Relativity, published in Annalen der Physik, 1916 [1]. But this page got lost in translation. . . .

The year 2005 brought a world celebration for the centennial anniversary of the *Annus Mirabilis* 1905, in which Einstein, at only 26, published four of his most famous papers. Buenos Aires, where I live and teach, was not an exception, and a series of lectures were organized [2]. I was invited to present a public 60-minute lecture, in which my aim was to sketch the evolution of the abstract ideas in geometry that Einstein had at his disposal to describe his brilliant theory of relativity. I am not an expert in the subject, so I spent several months collecting information.

One of my first "moves" was to get the Dover book on physics "The Principle of Relativity" [3]. This inexpensive and great collection of original papers on the special and general theory of relativity by A. Einstein, H.A. Lorentz, H. Minkowski, and H. Weyl, with notes by A. Sommerfeld, is the most common source for Einstein's article [1]. This review paper presents and explains for the first time the whole theory of relativity in a consistent, systematic and comprehensive way.

Some time later I discovered the fantastic online archives [4] containing the original manuscripts that Einstein donated to the Hebrew University of Jerusalem. There, I found the jewel which is featured as figure 1: a facsimile of Einstein's handwritten first page of his breakthrough paper, which, as is said in [5], "is arguably the most valuable Einstein manuscript in existence."

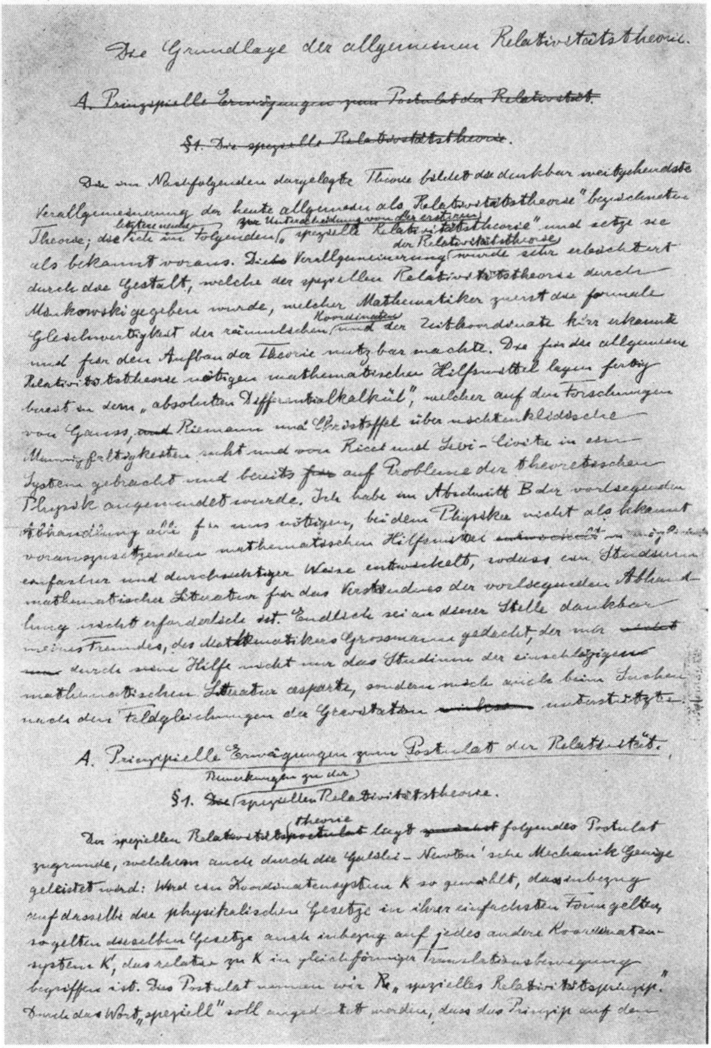

Die Grundlage der allgemeinen Relativitätstheorie.

~~A. Prinzipielle Erwägungen zum Postulat der Relativität.~~

~~§1. Die spezielle Relativitätstheorie.~~

Die im Nachfolgenden dargelegte Theorie bildet die denkbar weitgehendste Verallgemeinerung der heute allgemein als "Relativitätstheorie" bezeichneten Theorie; die letztere nenne ich im folgenden zur Unterscheidung von der ersteren "spezielle Relativitätstheorie" und setze sie als bekannt voraus. Die Verallgemeinerung der Relativitätstheorie wurde sehr erleichtert durch die Gestalt, welche der speziellen Relativitätstheorie durch Minkowski gegeben wurde, welcher Mathematiker zuerst die formale Gleichwertigkeit der räumlichen Koordinaten und der Zeitkoordinate klar erkannte und für den Aufbau der Theorie nutzbar machte. Die für die allgemeine Relativitätstheorie nötigen mathematischen Hilfsmittel lagen fertig bereit in dem „absoluten Differentialkalkül", welcher auf den Forschungen von Gauss, Riemann und Christoffel über nichteuklidische Mannigfaltigkeiten ruht und von Ricci und Levi-Civita in ein System gebracht und bereits auf Probleme der theoretischen Physik angewendet wurde. Ich habe im Abschnitt B der vorliegenden Abhandlung alle für uns nötigen, bei dem Physiker nicht als bekannt vorauszusetzenden mathematischen Hilfsmittel in möglichst einfacher und durchsichtiger Weise entwickelt, sodass ein Studium mathematischer Literatur für das Verständnis der vorliegenden Abhandlung nicht erforderlich ist. Endlich sei an dieser Stelle dankbar meines Freundes, des Mathematikers Grossmann gedacht, der mir durch seine Hilfe nicht nur das Studium der einschlägigen mathematischen Literatur ersparte, sondern mich auch beim Suchen nach den Feldgleichungen der Gravitation unterstützte.

A. Prinzipielle Erwägungen zum Postulat der Relativität.

§1. Bemerkungen zu der speziellen Relativitätstheorie.

Der speziellen Relativitätstheorie liegt folgendes Postulat zugrunde, welchem auch durch die Galilei-Newton'sche Mechanik Genüge geleistet wird: Wird ein Koordinatensystem K so gewählt, dass inbezug auf dasselbe die physikalischen Gesetze in ihrer einfachsten Form gelten, so gelten dieselben Gesetze auch inbezug auf jedes andere Koordinatensystem K', das relativ zu K in gleichförmiger Translationsbewegung begriffen ist. Dies Postulat nennen wir „spezielles Relativitätsprinzip." Durch das Wort „speziell" soll angedeutet werden, dass das Prinzip auf den

FIGURE 1. The original manuscript written by Albert Einstein is part of the Schwadron Autograph Collection, no. 31, Jewish National and University Library, Jerusalem. ©The Hebrew University of Jerusalem. Reprinted with permission.

I do not speak German, but I could of course easily recognize on this handwritten page the names of several mathematicians plus five occurrences of words starting with "Mathemat . . . ." I was amazed since I did not remember anything similar in the popular Dover book. Then, I realized that the Albert Einstein archives also offer an online English translation (extracted from [6]), which includes this first page and acknowledges the fact that it was missing in the circulating English version.

These are Albert Einstein's words on the first page of his most important paper on the theory of relativity:

> The theory which is presented in the following pages conceivably constitutes the farthest-reaching generalization of a theory which, today, is generally called the "theory of relativity"; I will call the latter one—in order to distinguish it from the first named—the "special theory of relativity," which I assume to be known. The generalization of the theory of relativity has been facilitated considerably by Minkowski, a mathematician who was the first one to recognize the formal equivalence of space coordinates and the time coordinate, and utilized this in the construction of the theory. The mathematical tools that are necessary for general relativity were readily available in the "absolute differential calculus," which is based upon the research on non-Euclidean manifolds by Gauss, Riemann, and Christoffel, and which has been systematized by Ricci and Levi-Civita and has already been applied to problems of theoretical physics. In section B of the present paper I developed all the necessary mathematical tools— which cannot be assumed to be known to every physicist—and I tried to do it in as simple and transparent a manner as possible, so that a special study of the mathematical literature is not required for the understanding of the present paper. Finally, I want to acknowledge gratefully my friend, the mathematician Grossmann, whose help not only saved me the effort of studying the pertinent mathematical literature, but who also helped me in my search for the field equations of gravitation.

So, indeed, he was not only paying homage to the work of the differential geometers who had built the geometry theories he used as the basic material for his general physical theory, but he also acknowledged H. Minkowski's idea of a four dimensional "world," with space and time coordinates. In fact, Einstein is even more clear in his recognition of the work of Gauss, Riemann, Levi-Civita and Christoffel in [7], where one could, for instance, read, "Thus it is that mathematicians long ago solved the formal problems to which we are led by the general postulate of relativity."

After Einstein's theory of special relativity was published in 1905, several mathematicians—including D. Hilbert and F. Klein—became very interested in his developments. While spacetime can be viewed as a consequence of Einstein's theory, it was first explicitly proposed mathematically by H. Minkowski [8]. Minkowski delivered a famous lecture at the University of Köln on 21 September 1908, in which he showed how the theory of special relativity follows naturally from just a simple fundamental hypothesis about the metric in spacetime. It is interesting to note that Einstein's first reaction had been to qualify this contribution as a "superfluous learnedness", and to critically assert that "since the mathematicians have attacked the relativity theory, I myself no longer understand it any more" [9]. This opinion changed soon, in part due to his mathematical friend M. Grossmann.

Einstein's acknowledgment of the importance of abstract mathematics is also shown in his reaction to Emmy Noether's paper [10]. In this paper, Noether was able to find a deep connection between differentiable symmetries of the action of a physical system and conservation laws, under the form known as Noether's theorem (see the article [11] by Nina Byers for a full account). It was her work in the theory of invariants which led to formulations for several concepts in the general theory of relativity. When Einstein received Noether's paper he wrote to Hilbert, "Yesterday I received from Miss Noether a very interesting paper on invariant forms. I am impressed that one can comprehend these matters from so general a viewpoint. It would not have done the old guard at Göttingen any harm had they picked up a thing or two from her . . ." [11]. (Despite Hilbert's and Klein's many efforts, Noether was for many years prevented from having an official position at Göttingen because she was a woman.)

In 1935, Einstein wrote an obituary [12] for Noether, who had emigrated to the U.S. in 1933 following her dismissal from her academic position because she was Jewish. Einstein once again praises Emmy Noether and shows the great appreciation that he came to develop for mathematics. In this obituary he writes, "Within the past few days a distinguished mathematician, Professor Emmy Noether, formerly connected with the University of Göttingen and for the past two years at Bryn Mawr College, died in her fifty-third year. In the judgment of the most competent living mathematicians, Fräulein Noether was the most significant creative mathematical genius thus far produced since the higher education of women began. In the realm of algebra, in which the most gifted mathematicians have been busy for centuries, she discovered methods which have proved of enormous importance in the development of the present-day younger generation of mathematicians. Pure mathematics is, in its way, the poetry of logical ideas. One seeks the most general ideas of operation which will bring together in simple, logical and unified form the largest possible circle of formal relationships. In this effort toward logical beauty spiritual formulas are discovered necessary for the deeper penetration into the laws of nature."

Was It a Conspiracy?

But, why was the first page by Einstein, containing his ebullient praise of mathematics, missing in the English version? Was it a conspiracy? I don't think so.

The translation was made in fact by the mathematical physicist G.B. Jeffery together with W. Perrett, a lecturer in German. I contacted the expert Tilman Sauer, who published an historical account of Einstein's 1916 paper, where he lists all the existing versions [13].

Sauer offered me the following clarifications: "The original publication of the paper on general relativity appeared in the journal "Annalen der Physik" and was also reprinted separately by the publisher J.A. Barth. An early English translation by S.N. Bose, University of Calcutta Press, 1920, does include the first page. The existence of this translation was apparently unbeknownst to Einstein. . . ." The first page is also missing in the very popular German collection of papers on relativity edited by Otto Blumenthal and published by Teubner.

A third edition of the Blumenthal collection was planned in 1919, which would include some of Einstein's later papers about general relativity. However, in a letter from 26 February 1919 [4, Nr. 41-991], Barth vetoed republication of the 1916 Annalen paper, since it still had several hundred copies of the separate printing of the paper in stock. On 12 December 1919, Barth retracted his veto since his stock of copies had almost been sold [4, Nr. 41-993] and on 21 December 1919, Einstein sent his only remaining copy of the Annalen version to Teubner for inclusion in the new edition of the Blumenthal collection: "With the same post I am sending you the only copy of my Annalen-paper that I still have. I ask you to use this copy for the printing, since on page 60 and 63 a few minor errors are corrected". . . . Now, apparently, Teubner sent the offprint that the typesetters had used (the page breaks are different!) back to Einstein, since the latter wrote on 20 April 1920 to Max Born [14]: "My last copy of the requested article is going out to you in the same post. It had been botched up like that at Teubner's press." According to the letters exchanged by Jeffery and Einstein ([15], [4, Nr. 13 424], [4, Nr. 13 426]), this was the German version used for the English translation.

And Sauer concludes, "My conjecture thus is that it was the typesetters from Teubner publishing house who missed to include the first page: maybe

Einstein's copy was frail and the first page was (almost?) loose and fell off? It's sort of curious that we seem to have no evidence that Einstein was even aware that Teubner had reprinted his 1916 review paper without the first page."

It's funny to think that only one deteriorated copy of this seminal paper was left, and that this may have been the reason to keep hidden this praise of mathematics and mathematicians.

Acknowledgments

I am grateful to David Eisenbud for listening to my story and suggesting to write it for the Bulletin. I also thank my friends Michael Singer, Daniel Arias and Adrián Paenza, for their thorough reading of the first version of this note.

References

[1] A. Einstein: *Die Grundlage der allgemeinen Relativitätstheorie*, Annalen der Physik, 49, 769–822, 1916.

[2] http://www.universoeinstein.com.ar/einstein.htm (in Spanish).

[3] H.A. Lorentz, A. Einstein, H. Minkowski, and H. Weyl, *The Principle of Relativity*, Dover Publications, New York, 1952. Republication of the 1923 translation first published by Methuen and Company, Ltd., translated from the text as published in the German collection "Des Relativitatsprinzip", Teubner, 4th ed., 1922.

[4] Albert Einstein Online Archives: http://www.alberteinstein.info/.

[5] http://www.alberteinstein.info/gallery/gtext3.html.

[6] Doc. 30 in: *The collected papers of Albert Einstein, Vol. 6, The Berlin Years: Writings, 1914–1917A,* J. Kox et al. (Eds.), A. Engel (Translator), E. Schucking (Consultant), Princeton University Press, 1996. MR1402241 (97f:01032).

[7] A. Einstein: Chapter 24 in: *Relativity: The Special and General Theory.* New York: Henry Holt, 1920; Bartleby.com, 2000. www.bartleby.com/173/.

[8] H. Minkowski, *Raum und Zeit,* 80. Versammlung Deutscher Naturforscher (Köln, 1908). English translation in [1].

[9] A. Pais: *Subtle Is the Lord: The Science and the Life of Albert Einstein,* Oxford University Press, 1983. MR690419 (84j:01072).

[10] E. Noether: *Invariante Variationsprobleme*, Nachr. d. König. Gesellsch. d. Wiss. zu Göttingen, Math-phys. Klasse, 235–257 (1918). English translation: *Invariant Variation Problems* by M. A. Tavel. Available at: http://arxiv.org/PS_cache/physics/pdf/0503/0503066vl.pdf MR0406752 (53:10538).

[11] N. Byers: *E. Noether's Discovery of the Deep Connection Between Symmetries and Conservation Laws,* in: Israel Mathematical Conference Proceedings, Vol. 12, 1999. Available at: http://www.physics.ucla.edu/cwp/articles/noether.asg/noether.html. MR1665436 (99m:58001).

[12] A. Einstein: *E. Noether's Obituary,* New York Times, May 5, 1935.

[13] T. Sauer: *Albert Einstein, Review paper on General Relativity Theory (1916)*, Chapter 63 in: Ivor Grattan-Guiness's "Landmark Writings in Western Mathematics". Available at: http://arxiv.org/abs/physics/0405066.

[14] Doc. 382 in: *The Collected Papers of Albert Einstein, Vol. 9: The Berlin Years: Correspondence, January 1919–April 1920*, Diana Kormos Buchwald et al. (Eds.), Princeton University Press, 2004. MR2265422 (2008c:83002).

[15] Doc. 230 in: *The Collected Papers of Albert Einstein, Volume 10, The Berlin Years: Correspondence, May–December 1920, and Supplementary Correspondence, 1909–1920*, Diana Kormos Buchwald et al. (Eds.), Princeton University Press, 2006.

Mathematics and
Its Applications

Mathematics and the Internet:
A Source of Enormous Confusion
and Great Potential

WALTER WILLINGER, DAVID L. ALDERSON,
AND JOHN C. DOYLE

For many mathematicians and physicists, the Internet has become a popular real-world domain for the application and/or development of new theories related to the organization and behavior of large-scale, complex, and dynamic systems. In some cases, the Internet has served both as inspiration and justification for the popularization of new models and mathematics within the scientific enterprise. For example, scale-free network models of the preferential attachment type [8] have been claimed to describe the Internet's connectivity structure, resulting in surprisingly general and strong claims about the network's resilience to random failures of its components and its vulnerability to targeted attacks against its infrastructure [2]. These models have, as their trademark, power-law type node degree distributions that drastically distinguish them from the classical Erdős-Rényi type random graph models [13]. These "scale-free" network models have attracted significant attention within the scientific community and have been partly responsible for launching and fueling the new field of *network science* [42, 4].

To date, the main role that mathematics has played in network science has been to put the physicists' largely empirical findings on solid grounds by providing rigorous proofs of some of their more highly publicized claims [14, 15, 16, 23, 11, 25]. The alleged scale-free nature of the Internet's topology has also led to mathematically rigorous results about the spread of viruses over scale-free graphs of the preferential attachment type, again with strong and unsettling implications such as a zero epidemic threshold [11, 25]. The relevance of the latter is that in stark contrast to more homogeneous graphs, on scale-free networks of the preferential attachment type, even viruses with small propagation rates have a chance to cause an epidemic, which is about as bad as it can get from an Internet security perspective. More recently, the realization that large-scale, real-world networks such as the Internet evolve over time has motivated the mathematically challenging problem of developing a theory of graph sequences and graph limits [17, 19, 20]. The underlying idea is that properly defined graph limits can be expected to represent viable models for some of the enormous dynamic graph structures that arise in real-world applications and seem too unwieldy to be described via more direct or explicit approaches.

The generality of these new network models and their impressive predictive ability notwithstanding, surprisingly little attention has been paid in the mathematics and physics communities to parallel developments in the Internet research arena, where the various non-rigorous and rigorous results derived from applying the scale-free modeling paradigm to the Internet have been scrutinized using available measurements or readily available domain knowledge. A driving force behind these Internet-centric validation efforts has been the realization that—because of its engineered architecture, a thorough understanding of its component technologies, and the availability of extensive (but not necessarily very accurate) measurement capabilities—the Internet provides a unique setting in which most claims about its properties, structure, and functionality can be unambiguously resolved, though perhaps not without substantial efforts. In turn, models or theories that may appeal to a more mathematically inclined researcher because of their simplicity or generality, but result in incorrect, misleading, or wrong claims about the Internet, can and will be identified and labeled accordingly, but it may take considerable time (and efforts) to expose their specious nature.

In this article, we take a closer look at what measurement-based Internet research in general, and Internet-specific validation efforts in particular, have to say about the popular scale-free modeling paradigm and the flurry of mathematical developments it has inspired. In particular, we illustrate why and how in the case of the Internet, scale-free network models of the preferential attachment type have become a classic lesson in how errors of various forms occur and can add up to produce results and claims that create excitement among non-networking researchers, but quickly collapse under scrutiny with real data or when examined by domain experts. These opposite reactions have naturally been a source of great confusion, but the main conclusion is neither controversial nor should it come as a big surprise: the scale-free modeling paradigm is largely inconsistent with the engineered nature of the Internet and the design constraints imposed by existing technology, prevailing economic conditions, and practical considerations concerning network operations, control, and management.

To this end, we document the main sources of errors regarding the application of the scale-free modeling approach to the Internet and then present an alternative approach that represents a drastic departure from traditional network modeling. In effect, we motivate here the development of a novel modeling approach for Internet-like systems that (1) respects the highly designed nature of the network; (2) reflects the engineering intuition that exists about a great many of its parts; (3) is fully consistent with a wide range of measurements; and (4) outlines a mathematical agenda that is more challenging, more relevant, and ultimately more rewarding than the type of mathematics motivated by an alluring but largely misguided approach to Internet modeling based on scale-free graphs of the preferential attachment type. In this sense, this article demonstrates the great potential that the Internet has for the development of new, creative, and relevant mathematical theories, but it is also a reminder of a telling comment attributed to S. Ulam [12] (slightly paraphrased, though), who said *"Ask not what mathematics can do for [the Internet]; ask what [the Internet] can do for mathematics."*

The Scale-free Internet Myth

The story recounted below of the scale-free nature of the Internet seems convincing, sound, and almost too good to be true. Unfortunately, it turned out to be a complete myth, but has remained a constant source of enormous confusion within the scientific community.

Somewhat ironically, the story starts with a highly cited paper in the Internet research arena by Faloutsos et al. [27]. Relying on available measurements and taking them at face value, the paper was the first to claim that the (inferred) node degree distributions of the Internet's router-level topology as well as AS-level topology are power-law distributions with estimated α-parameters between 1 and 2. To clarify, by *router-level topology*, we mean the Internet's physical connectivity structure, where nodes are physical devices such as routers or switches, and links are the connections between them. These devices are further organized into networks known as *Autonomous Systems (ASes)*, where each AS is under the administrative control of a single organization such as an Internet Service Provider (ISP), a company, or an educational institution. The relationships among ASes, when organized as a graph, produce what is known as the Internet's *AS-level topology*. Note that a link between two nodes in the AS-level topology represents a type of business relationship (either peering or customer-provider). Also, in contrast to the router-level topology that is inherently physical, the AS topology is a logical construct that reflects the Internet's administrative boundaries and existing economic relationships.

These reported power-law findings for the Internet were quickly picked up by Barabási et al., who were already studying the World Wide Web (WWW) and then added the Internet to their growing list of real-world network structures with an apparently striking common characteristic; that is, their vertex connectivities (described mathematically in terms of node *degrees*) *"follow a scale-free power-law distribution"* [8, 3]. This property is in stark contrast to the Poissonian nature of the node degrees resulting from the traditional Erdős-Rényi random graphs [13] that have been the primary focus of mathematical graph theory for the last 50 years. Naturally, it has fueled the development of new graph models that seek to capture and reproduce this ubiquitously reported power-law relationship, thereby arguing in favor of these models as more relevant for representing real-world network structures than the classical random graph models. In fact, much of the initial excitement in the nascent field of network science can be attributed to an early and appealingly simple class of network models that was proposed by Barabási and Albert [8] and turned out to have surprisingly strong predictive capabilities.

In short, Barabási and Albert [8] described a network growth model in which newly added vertices connect preferentially to nodes in the existing graph that are already well connected. This *preferential attachment* mechanism had been studied over the previous 75 years by Yule [54], Luria and Delbrück [38], and Simon [49], but it was its rediscovery and application to networks by Barabási and Albert that recently popularized it. Although many variants of the basic Barabási-Albert construction have been proposed and studied, we will focus in the following on the original version described in [8], mainly because of its simplicity and because it already captures the most important properties of this new class of networks, commonly referred to as *scale-free networks*. The term scale-free derives from the simple observation that power-law node degree distributions are free of scale—most nodes have small degree, a few nodes have very high degree, with the result that the average node degree is essentially non-informative. A detailed discussion of the deeper meanings often associated with scale-free networks is available in [34]. To avoid confusion and to emphasize the fact that preferential attachment is just one of many other mechanisms that is capable of generating scale-free graphs (i.e., graphs with power-law node degree distributions), we will refer here to the network models proposed in [8] as *scale-free networks of the preferential attachment (PA) type* and show an illustrative toy example with associated node degree distribution in Figure 1.

The excitement generated by this new class of models is mainly due to the fact that, despite being generic and largely oblivious to system-specific details, they share some key properties that give them remarkable predictive

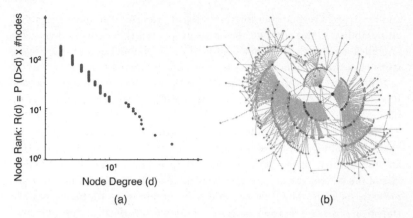

(a)   (b)

FIGURE 1. Scale-free networks of the preferential attachment type. (b) A toy example of a scale-free network of the preferential attachment type generated to match a power-law type degree distribution (a).

power. These properties were originally reported in [2], put on mathematically solid footing by Bollobás and Riordan in [14, 15, 16], and explain the key aspects of the structure and behavior of these networks. For one, a hallmark of their structure is the presence of "hubs"; that is, centrally located nodes with high connectivity. Moreover, the presence of these hubs makes these networks highly vulnerable to attacks that target the hub nodes. At the same time, these networks are extremely resilient to attacks that knock out nodes at random, since a randomly chosen node is likely to be one of the low-degree nodes that constitute the bulk of the nodes in the power-law node degree distribution, and the removal of such a node has typically minimal impact on the network's overall connectivity or performance.

This property—simultaneous resilience to random attacks but high vulnerability to targeted worst-case attacks (i.e., attacks against the hub nodes)—featured prominently in the original application of scale-free networks of the PA type to the Internet [2]. The underlying argument follows a very traditional and widely-used modeling approach. First, as reported in [27], the Internet has node degrees that follow a power-law distribution or are scale-free. Second, scale-free networks of the PA type are claimed to be valid models of the Internet because they are capable of reproducing the observed scale-free node degree distributions. Lastly, when abstracted to a scale-free model of the PA type, the Internet automatically inherits all the emergent features of the latter, most notably the presence of hub nodes that are critical to overall network connectivity and performance and are largely responsible for the network's failure tolerance and attack vulnerability. In this context, the latter property has become known as the "Achilles' heel of the Internet" and has been highly publicized as a success story of network science—the discovery of a fundamental weakness of the Internet that went apparently unnoticed by the engineers and researchers who have designed, deployed, and studied this large-scale, critical complex system.

The general appeal of such surprisingly strong statements is understandable, especially given the simplicity of scale-free networks of the PA type and the fact that, as predictive models, they do not depend on the particulars of the system at hand, i.e., underlying technology, economics, or engineering. As such, they have become the embodiment of a highly popular statistical physics-based approach to complex networks that aims primarily at discovering properties that are universal across a range of very diverse networks. The potential danger of this approach is that the considered abstractions represent simplistic toy models that are too generic to reflect features that are most important to the experts dealing with these individual systems (e.g., critical functionality).

Deconstructing the Scale-free Myth

Given that the scale-free story of the Internet is grounded in real measurement data and based on a widely-accepted modeling approach, why is it so far from the truth? To explain and trace the various sources of errors, we ask the basic question; i.e., *"Do the available measurements, their analysis, and their modeling efforts support the claims that are made in* [2]?" To arrive at a clear and simple answer to this question, we address below the issues of data hygiene and data usage, data analysis, and mathematical modeling (including model selection and validation).

KNOW YOUR DATA

A very general but largely ignored fact about Internet-related measurements is that what we can measure in an Internet-like environment is typically not the same as what we really want to measure (or what we think we actually measure). This is mainly because as a decentralized and distributed system, the Internet lacks a central authority and does not support third-party measurements. As a result, measurement efforts across multiple ASes become nontrivial and often rely on engineering hacks that typically do not yield the originally desired data but some substitute data. Moreover, using the latter at face value (i.e., as if they were the data we originally wanted)

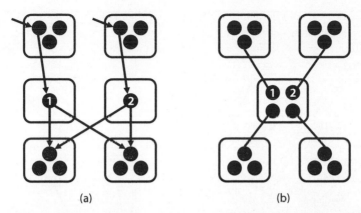

(a)　　　　　　　　　　　　　　　(b)

FIGURE 2. The IP alias resolution problem. Paraphrasing Fig. 4 of [50], `traceroute` does not list routers (boxes) along paths but IP addresses of input interfaces (circles), and alias resolution refers to the correct mapping of interfaces to routers to reveal the actual topology. In the case where interfaces 1 and 2 are aliases, (b) depicts the actual topology while (a) yields an "inflated" topology with more routers and links than the real one.

and deriving from them results that we can trust generally involves a leap of faith, especially in the absence of convincing arguments or evidence that would support an "as-is" use of the data.

Internet-specific connectivity measurements provide a telling example. To illustrate, consider the data set that was used in [27] to derive the reported power-law claim for the (inferred) node degrees of the Internet's router-level topology.[1] That data set was originally collected by Pansiot and Grad [44] for the explicitly stated purpose *"to get some experimental data on the shape of multicast trees one can actually obtain in [the real] Internet ..."* [44]. The tool of choice was `traceroute`, and the idea was to run `traceroute` between a number of different host computers dispersed across the Internet and glue together the resulting Internet routes to glean the shape of actual multicast trees. In this case, the engineering hack consisted of relying on `traceroute`, a tool that was never intended to be used for the stated purpose, and a substantial leap of faith was required to use Pansiot and Grad's data set beyond its original purpose and rely on it to infer the Internet's router-level topology [27].

For one, contrary to popular belief, running `traceroute` between two host computers does *not* generate the list of compliant (i.e., Internet Protocol (IP)-speaking) routers encountered en route from the source to the destination. Instead, since IP routers have multiple interfaces, each with its own IP address, what `traceroute` really generates is the list of (input interface)

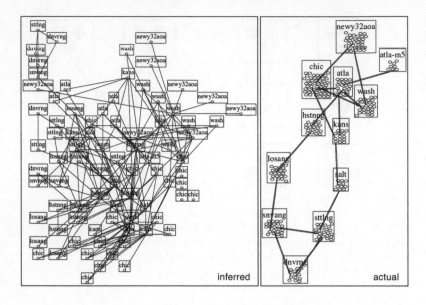

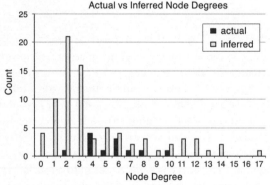

FIGURE 3. The IP alias resolution problem in practice. This is reproduced from [48] and shows a comparison between the Abilene/Internet2 topology inferred by Rocketfuel (left) and the actual topology (top right). Rectangles represent routers with interior ovals denoting interfaces. The histograms of the corresponding node degrees are shown in the bottom right plot. © 2008 ACM, Inc. Included here by permission.

IP addresses, and a very common property of traceroute-derived routes is that one and the same router can appear on different routes with different IP addresses. Unfortunately, faithfully mapping interface IP addresses to routers is a difficult open problem known as the *IP alias resolution problem* [51, 28], and despite continued research efforts (e.g., [48, 9]), it has remained a source of significant errors. While the generic problem is illustrated in Figure 2, its impact on inferring the (known) router-level topology of an actual network (i.e., Abilene/Internet2) is highlighted in Figure 3—the inability to solve the alias resolution problem renders in this case the inferred topology irrelevant and produces statistics (e.g., node degree distribution) that have little in common with their actual counterparts.

Another commonly ignored problem is that traceroute, being strictly limited to IP or layer-3, is incapable of tracing through opaque layer-2 clouds that feature circuit technologies such as *Asynchronous Transfer Mode (ATM)* or *Multiprotocol Label Switching (MPLS)*. These technologies have the explicit and intended purpose of hiding the network's physical infrastructure from IP, so from the perspective of traceroute, a network that runs these technologies will appear to provide direct connectivity between routers that are separated by local, regional, national, or even global physical network infrastructures. The result is that when traceroute encounters one of these opaque layer-2 clouds, it falsely "discovers" a high-degree node that is really a logical entity—a network potentially spanning many hosts or great distances—rather than a physical node of the Internet's router-level topology. Thus, reports of high-degree hubs in the core of the router-level Internet, which defy common engineering sense, can often be easily identified as simple artifacts of an imperfect measurement tool. While Figure 4(a) illustrates the generic nature of this problem, Figure 4(b) illuminates its impact in the case of an actual network (i.e., AS3356 in 2002), where the inferred topology with its highly connected nodes says nothing about the actual physical infrastructure of this network but is a direct consequence of traceroute's inability to infer the topology of an MPLS-enabled network.

We also note that from a network engineering perspective, there are technological and economic reasons for why high-degree nodes in the core of the router-level Internet are nonsensical. Since a router is fundamentally limited in terms of the number of packets it can process in any time interval, there is an inherent tradeoff in router configuration: it can support either a few high-throughput connections or many low-throughput connections. Thus, for any given router technology, a high-connectivity router in the core will either have poor performance due to its slow connections or be prohibitively expensive relative to other options. Conversely, if one deploys high-degree devices at the router-level, they are necessarily located at the edge of the network where the technology exists to multiplex a large number of relatively low-bandwidth links. Unfortunately, neither the original traceroute-based study of Pansiot and Grad nor any of the larger-scale versions that were subsequently performed by various network research

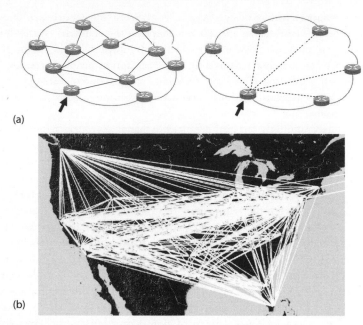

(a)

(b)

FIGURE 4. How traceroute detects fictitious high-degree nodes in the network core. (a.) The actual connectivity of an opaque Layer-2 cloud, i.e., a router-level network running a technology such as ATM or MPLS (left) and the connectivity inferred by traceroute probes entering the network at the marked router (right). (b) The Rocketfuel-inferred backbone topology of AS3356 (Level3), a Tier-1 Internet service provider and leader in the deployment of MPLS (reproduced from [50]) © 2002 ACM, Inc. Included here by permission.

groups have the ability to detect those actual high-degree nodes. The simple reason is that these traditional traceroute studies lack access to a sufficient number of participating host computers in any local end-system to reveal their high connectivity. Thus, the irony of traceroute is that the high-degree nodes it detects in the network core are necessarily fictitious and represent entire opaque layer-2 clouds, and if there are actual high-degree nodes in the network, existing technology relegates them to the edge of the network where no generic traceroute-based measurement experiment will ever detect them.

Lastly, the nature of large-scale traceroute experiments also makes them susceptible to a type of measurement bias in which some points of the network are oversampled, while others are undersampled. Ironically, although this failure of traceroute experiments has received the most attention in the theoretical computer science and applied mathematics communities [32, 1] (most likely, because this failure is the most amenable to mathematical treatment), it is the least significant from a topology modeling perspective.

In view of these key limitations of traceroute, it should be obvious that starting with the Pansiot and Grad data set, traceroute-based measurements cannot be taken at face value and are of no or little use for inferring the Internet's router-level topology. In addition, the arguments provided above show why domain knowledge in the form of such traceroute-specific "details" like IP aliases or layer-2 technology matters when dealing with issues related to data hygiene and why ignoring those details prevents us from deriving results from such data that we can trust. Ironically, Pansiot and Grad [44] detailed many of the above-mentioned limitations and shortcomings of their measurements. Unfortunately, [27] failed to revive these issues or recognize their relevance. Even worse, the majority of subsequent papers in this area typically cite only [27] and no longer [44].

KNOW YOUR STATISTIC

The inherent inability of traceroute to reveal unambiguously the actual node degree of any router (i.e., the number of different interfaces) due to the IP alias resolution problem, combined with the fundamental difficulties of the tool to correctly infer even the mere absence or presence of high-degree nodes (let alone their actual degrees) makes it impossible to describe accurately statistical entities such as node degree distributions. Thus, it should come as no surprise that taking traceroute-derived data sets "as is" and then making them the basis for any fitting of a particular parameterized distribution (e.g., power-law distribution with index α as in [27]) is statistical "overkill", irrespective of how sophisticated a fitting or corresponding parameter estimation technique has been used. Given the data's limitations, even rough rules-of-thumb such as a Pareto-type 80/20 rule (i.e., 80% of the effects come from 20% of the causes) cannot be justified with any reasonable degree of statistical confidence.

It is in this sense that the claims made in [27] and subsequent papers that have relied on this data set are the results of a data analysis that is not commensurate with the quality of the available data. It is also a reminder that there are important differences between analyzing high-quality and low-quality data sets, and that approaching the latter the same way as the former is not only bad statistics but also bad science, and doing so bolsters the popular notion that "there are lies, damned lies, and statistics." Unfortunately, the work required to arrive at this conclusion is hardly glamorous or newsworthy, especially when compared to the overall excitement generated by an apparent straight-line behavior in the easily obtainable log-log plots of degree vs. frequency. Even if the available measurements were amenable to such an analysis, these commonly-used and widely-accepted log-log plots are not only highly non-informative, but have a tendency to obscure power-law relationships when they are genuine and fabricate them when they are absent (see for example [34]). In the case of the data set at hand, the latter observation is compounded by the unreliable nature of the `traceroute`-derived node degree values and shows why the power-law claims for the vertex connectivities of the Internet's router-level topology reported in [27] cannot be supported by the available measurements.

WHEN MODELING IS MORE THAN DATA-FITTING

We have shown that the data set used in [27] turns out to be thoroughly inadequate for deriving and modeling power-law properties for the distribution of node degrees encountered in the Internet's router-level topology. As a result, the sole argument put forward in [2] for the validity of the scale-free model of the PA type for the Internet is no longer applicable, and this in turn reveals the specious nature of both the proposed model and the sensational features the Internet supposedly inherits from the model.

Even if the node degree distribution were a solid and reliable statistic, who is to say that matching it (or any other commonly considered statistics of the data) argues for the validity of a proposed model? In the case of scale-free models of the PA type, most "validation" follows from the ability of a model to replicate an observed degree distribution or sequence. However, it is well known in the mathematics literature that there can be many graph realizations for any particular degree sequence [47, 29, 35, 10] and there are often significant structural differences between graphs having the same degree sequence [6]. Thus, two models that match the data equally well with respect to some statistics can still be radically different in terms of other properties, their structures, or their functionality. A clear sign of the rather precarious current state of network-related modeling is that the same underlying data set can give rise to very different, but apparently equally "good" models, which in turn can give rise to completely opposite scientific claims and theories concerning one and the same observed phenomenon. Clearly, modeling and especially model validation has to mean more than being able to match the data if we want to be confident that the results that we drive from our models are valid.

At this point, it is appropriate to recall a quote attributed to G. E. P. Box, who observed that "*All models are wrong, but some models are useful.*" Without being more specific about which models are deemed useful and why, this comment is of little practical value. A more constructive piece of advice that is more directly aligned with what we envision modeling should mean in the presence of imprecise data is from B. B. Mandelbrot [39], who observed *"If exactitude is elusive, it is better to be approximately right than certifiably wrong."*

For complex network systems whose measured features suffer from the types of fundamental ambiguities, omissions, and/or systematic errors outlined above, we argue that network modeling must move beyond efforts that merely match particular statistics of the data. Such efforts are little more than exercises in data-fitting and are particularly ill-advised whenever the features of interest cannot be inferred with any reasonable statistical confidence from the currently available measurements. For systems such as the router-level Internet, we believe this to be a more scientifically grounded and constructive modeling approach. For one, given the known deficiencies in the available data sets, matching a particular statistic of the data may be precisely the wrong approach, unless that statistic has been found to be largely robust with respect to these deficiencies. Moreover, it eliminates the arbitrariness associated with determining which statistics of the data to focus on. Indeed, it treats all statistics equally. A model that is "approximately right" can be expected to implicitly match most statistics of the data (at least approximately).

If we wish to increase our confidence in a proposed model, we ought also to ask what new types of measurements are either already available (but have not been used in the present context) or could be collected and used for validation. Here, by "new" we do not mean "same type of measurements as before, just more." What we mean are completely new types of data, with very different semantic content, that have played no role whatsoever in the entire modeling process up to this point. A key benefit of such an approach is that the resulting measurements are used primarily to "close-the-loop", as advocated in [53], and provide a statistically clean separation between the data used for model selection and the data used for model validation—a feature that is alien to most of today's network-related models. However, a key question remains: *What replaces data-fitting as the key ingredient and driver of the model selection and validation process so that the resulting models are approximately right and not certifiably wrong?* The simple answer is: *rely on domain knowledge and exploit the details that matter when dealing with a highly engineered system such as the Internet.* Note that this answer is in stark contrast to the statistical physics-based approach that suggests the development of a system such as the Internet is governed by robust self-organizing phenomena that go beyond the particulars of the individual systems (of interest) [8].

A FIRST-PRINCIPLES APPROACH TO INTERNET MODELING

If domain knowledge is the key ingredient to build "approximately right" models of the Internet, what exactly is the process that helps us achieve our goal? To illustrate, we consider again the router-level topology of the Internet, or more specifically, the physical infrastructure of a regional, national, or international Internet Service Provider (ISP).

The first key observation is that the way an ISP designs its physical infrastructure is certainly not by a series of (biased) coin tosses that determine whether or not two nodes (i.e., routers) are connected by a physical link, as is the case for the scale-free network models of the PA type. Instead, ISPs design their networks for a purpose; that is, their decisions are driven by objectives and reflect tradeoffs between what is feasible and what is desirable. The mathematical modeling language that naturally reflects such a decision-making process is *constrained optimization*. Second, while in general it may be difficult if not impossible to define or capture the precise meaning of an ISP's purpose for designing its network, an objective that expresses a desire to provide connectivity and an ability to carry an expected traffic demand efficiently and effectively, subject to prevailing economic and technological constraints, is unlikely to be far from the "true" purpose. In view of this, we are typically not concerned with a network design that is "optimal" in a strictly mathematical sense and is also likely to be NP-hard, but in a solution that is *"heuristically optimal"* in the sense that it results in "good" performance. That is, we seek a solution that captures by and large what the ISP can afford to build, operate, and manage (i.e., economic considerations), given the hard constraints that technology imposes on the network's physical entities (i.e., routers and links). Such models have been discussed in the context of highly organized/optimized tolerances/tradeoffs (HOT) [18, 26]. Lastly, note that in this approach, randomness enters in a very specific manner, namely in terms of the uncertainty that exists about the "environment" (i.e., the traffic demand that the network is expected to carry), and the heuristically optimal network designs are expected to exhibit *strong robustness properties* with respect to changes in this environment.

Figure 5 shows a toy example of an ISP router-level topology that results from adopting the mathematical modeling language of constrained optimization and choosing a candidate network as a solution of an heuristically optimal network design problem. Despite being a toy example, it is rich enough to illustrate the key features of our engineering-driven approach to network modeling and to contrast it with the popular scale-free network models of the PA type. Its toy nature is mainly due to a number of simplifying assumptions we make that facilitate the problem formulation. For one, by simply equating throughput with revenues, we select as our objective

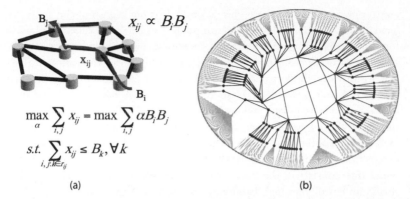

(a)   (b)

FIGURE 5. Generating networks using constrained optimization. (a) Engineers view network structure as the solution to a design problem that measures performance in terms of the ability to satisfy traffic demand while adhering to node and arc capacity constraints. (b) A network resulting from heuristically optimized trade-offs (HOT). This network has very different structural and behavioral properties, even when it has the same number of nodes, links, and degree distribution as the scale free network depicted in Fig. 1.

function the maximum throughput that the network can achieve for a given traffic demand and use it as a metric for quantifying the performance of our solutions. Second, considering an arbitrary distribution of end-user traffic demand x_i, we assume a *gravity model* for the unknown traffic demand; that is, assuming shortest-path routing, the demands are given by the traffic matrix X, where for the traffic X_{ij} between routers i and j we have $X_{ij} = \alpha x_i x_j$, for some constant α. Lastly, we consider only one type of router and its associated technologically feasible region; that is, (router degree, router capacity)-pairs that are achievable with the considered router type (e.g., CISCO 12416 GSR), and implicitly avoid long-haul connections due to their high cost.

The resulting constrained optimization problem c an be written in the form

$$\max_\rho \sum_{i,j} X_{i,j}$$

$$s.t \quad RX \leq B$$

where X is the vector obtained by stacking all the demands $X_{ij} = \rho x_i x_j$; R is the routing matrix obtained by using standard shortest path routing and defined by $R_{kl} = 1$ or 0, depending on whether or not demand l passes through router k; and B is the vector consisting of the router degree-bandwidths constraints imposed by the technologically feasible region of the router at hand. While all the simplifying assumptions can easily be relaxed to allow for more realistic objective functions, more heterogeneity in the constraints, or more accurate descriptions of the uncertainty in the environment, Figure 5 illustrates the key characteristics inherent in a heuristically optimal solution of such a problem. First, the cost-effective handling of end user demands avoids long-haul connections (due to their high cost) and is achieved through traffic aggregation starting at the edge of the network via the use of high-degree routers that support the multiplexing of many low-bandwidth connections. Second, this aggregated traffic is then sent toward the "backbone" that consists of the fastest or highest-capacity routers (i.e., having small number of very high-bandwidth connections) and that forms the network's mesh-like core. The result is a network that has a more or less pronounced backbone, which is fed by tree-like access networks, with additional connections at various places to provide a degree of redundancy and robustness to failures.

What about power-law node degree distributions? They are clearly a non-issue in this engineering-based first-principles approach, just as they should be, based on our understanding illustrated earlier that present measurement techniques are incapable of supporting them. Recognizing their irrelevance is clearly the beginning of the end of the scale-free network models of the PA type as far as the Internet is concerned. What about replacing power-laws by the somewhat more plausible assumption of high variability in node degrees? While the answer of the scale-free modeling approach consists of tweaks to the PA mechanism to enforce an exponential cut-off of the power-law node degree distribution at the upper tail, the engineering-based approach demystifies high-variability in node degrees altogether by identifying its root cause in the form of high variability in end-user bandwidth demands (see [33] for details). In view of such a simple physical explanation of the origins of node degree variability in the Internet's router-level topology, Strogatz' question, paraphrasing Shakespeare's Macbeth, ". . . power-law scaling, full of sound and fury, signifying nothing?" [52] has a resounding affirmative answer.

Great Potential for Mathematics

Given the specious nature of scale-free networks of the PA type for modeling Internet-related connectivity structures, their rigorous mathematical treatment and resulting highly-publicized properties have lost much of their luster, at least as far as Internet matters are concerned. Considering again our example of the router-level Internet, neither the claim of a hub-like core, nor the asserted error tolerance (i.e., robustness to random component failure) and attack vulnerability (i.e., Achilles' heel property), nor the often-cited zero epidemic threshold property hold. In fact, as illustrated with our HOT-based network models, intrinsic and unavoidable tradeoffs between network performance, available technology, and economic constraints necessarily result in network structures that are in all important ways exactly the opposite of what the scale-free models of the PA type assert. In this sense, the HOT toy examples represent a class of network models for the Internet that are not only consistent with various types of measurements and in agreement with engineering intuition, but whose rigorous mathematical treatment promises to be more interesting and certainly more relevant and hence more rewarding than that of the scale-free models of the PA type.[2]

THE INTERNET'S ROBUST YET FRAGILE NATURE

Because high-degree nodes in the router-level Internet can exist only at the edge of the network, their removal impacts only local connectivity and has little or no global effect. So much for the widely-cited discovery of the Internet's Achilles' heel! More importantly, the Internet is known to be extremely robust to component failures, but this is *by design*[3] and involves as a critical ingredient the Internet Protocol (IP) that "sees failures and routes traffic around them." Note that neither the presence of protocols nor their purpose play any role in the scale-free approach to assessing the robustness properties of the Internet. At the same time, the Internet is also known to be very fragile, but again in a sense that is completely different from and has nothing in common with either the sensational Achilles' heel claim or the zero epidemic threshold property, both of which are irrelevant as far as the actual Internet is concerned. The network's true fragility is due to an original trust model[4] that has been proven wrong almost from the get-go and has remained broken ever since. While worms, viruses, or spam are all too obvious and constant reminders of this broken trust model, its more serious and potentially lethal legacy is that it facilitates the malicious exploitation or hijacking of the very mechanisms (e.g., protocols) that ensure the network's impressive robustness properties. This "robust yet fragile" tradeoff is a fundamental aspect of an Internet architecture whose basic design dates back some 40 years and has enabled an astonishing evolution from a small research network to a global communication infrastructure supporting mission-critical applications.

One of the outstanding mathematical challenges in Internet research is the development of a theoretical foundation for studying and analyzing this robustness-fragility tradeoff that is one of the single most important characteristics of complexity in highly engineered systems. To date, this tradeoff has been largely managed with keen engineering insights and little or no theoretical backing, but as the Internet scales even further and becomes ever more heterogeneous, the need for a relevant mathematical theory replacing engineering intuition becomes more urgent. The difficulties in developing such a theory are formidable as the "typical" behavior of a system such as the Internet is often quite simple, inviting naïve views and models like the scale-free network models of the PA type that ignore any particulars of the underlying system, inevitably cause confusion, result in misleading claims, and provide simple explanations that may look reasonable at first sight but turn out to be simply wrong. Only extreme circumstances or rare accidents not easily replicable in laboratory experiments or simulations reveal the enormous internal complexity in systems such as the Internet, and any relevant mathematical theory has to respect the underlying architectural design and account for the various protocols whose explicit purpose is in part to hide all the complexity from the user of the system [5].

NETWORK DYNAMICS AND SYSTEM FUNCTION

Real networks evolve over time in response to changes in their environment (e.g., traffic, technology, economics, government regulation), and currently proposed network models such as the scale-free models of the PA type cannot account for such interactions. They either ignore the notion of network *function* (i.e., the delivery of traffic) altogether or treat networks as strictly open-loop systems in which modeling exists largely as an exercise in data-fitting. In stark contrast to the scale-free models of the PA type, the proposed HOT-based network models make the dependence of network structure on network traffic explicit. This is done by requiring as input a traffic demand model in the form of a traffic matrix (e.g., gravity model). A particular network topology is "good" only if it can deliver traffic in a manner that satisfies demand. When viewed over time, changes in the environment (e.g., traffic demands), constraints (e.g., available technologies), or objectives (e.g., economic conditions) are bound to impact the structure of the network, resulting in an intricate feedback loop between network traffic and network structure.

   The task at hand is to develop a mathematical framework that enables and supports modeling network evolution in ways that account for this feedback loop between the structure of the networks and the traffic that they carry or that gets routed over them. This new modeling paradigm for networks is akin to recent efforts to model the network-wide behavior of TCP or the TCP/IP protocol stack as a whole: the modeling language is (constrained) optimization; a critical ingredient is the notion of separation of time scales; heuristic solution methods (with known robustness properties to changes in the environment) are preferred over mathematically optimal solution techniques (which are likely to be NP-hard); and the overall goal is to transform network modeling from an exercise in data-fitting into an exercise in reverse-engineering. In this sense, relevant recent theoretical works includes *network utility maximization* (e.g., see [30, 36, 37]), *layering as optimization decomposition* (e.g., see [21, 22]), and *the price of anarchy* (e.g., see [45]).

In view of this objective, developing a mathematical framework for studying sequences and limits of graphs that arise in a strictly open-loop manner (e.g., see [17, 19, 20]), while of independent mathematical interest, is of little relevance for studying and understanding real-world networks such as the Internet, unless it is supported by strong and convincing validation efforts. This difference in opinions is fully expected: while mathematicians and physicists tend to view the enormous dynamic graph structures that arise in real-world applications as too complex to be described by direct approaches and therefore invoke randomness to model and analyze them, Internet researchers generally believe they have enough domain knowledge to understand the observed structures in great detail and tend to rely on randomness for the sole purpose of describing genuine uncertainty about the environment. While both approaches have proven to be useful, it is the responsibility of the mathematician/physicist to convince the Internet researcher of the relevance or usefulness of their modeling effort. The scale-free models of the PA type are an example where this responsibility has been badly lacking.

MULTISCALE NETWORK REPRESENTATIONS

Multiscale representations of networks is an area where Ulam's paraphrased quote "*Ask not what mathematics can do for [the Internet]; ask what [the Internet] can do for mathematics*" is highly appropriate. On the one hand, there exists a vast literature on mathematical multi-resolution analysis (MRA) techniques and methodologies for studying complex objects such as high-dimensional/semantic-rich data and large-scale structures. However, much less is known when it comes to dealing with highly irregular domains such as real-world graph structures or with functions or distributions defined on those domains. In fact, from an Internet perspective, what is needed is an MRA specifically designed to accommodate the vertical (i.e., layers) and horizontal (i.e., administrative or geographic domains) decompositions of Internet-like systems and capture in a systematic manner the "multi-scale" nature of the temporal, spatial, and functional aspects of network traffic over corresponding network structures. In short, the mathematical challenge consists of developing an MRA technology appropriate for dealing with meaningful multi-scale representations of very large, dynamic, and diverse Internet-specific graph structures; for exploring traffic processes associated with those structures; and for studying aggregated spatio-temporal network data representations and visual representations of them.

The appeal of an Internet-specific MRA is that the Internet's architecture supports a number of meaningful and relevant multi-scale network representations with associated traffic processes. For example, starting with our example of the router-level Internet (and associated hypothetical traffic matrix), aggregating routers and the traffic they handle into Points-of-Presences, or PoPs, yields the PoP-level Internet and PoP-level traffic matrix. Aggregating PoPs and the traffic they handle into Autonomous Systems (ASes) or domains produces the AS-level Internet and corresponding AS-level traffic matrix. Aggregating even further, we can group Ases that belong to the same Internet Service Provider (ISP) or company/institution and obtain the ISP-level Internet. While the router- and PoP-level Internet are inherently physical representations of the Internet, the AS- and ISP-level structures are examples of logical or virtual constructs where nodes and links say little or nothing about physical connectivity. At the same time, the latter are explicit examples that support a meaningful view of the Internet as a "network of networks" (see below). With finer-resolution structures and traffic matrices also of interest and of possible use (e.g., BGP prefix-level, IP address-level), the expectations for an Internet-specific MRA technique are that it is capable of recovering these multiple representations by respecting the architectural, administrative, and technological aspects that give rise to this natural hierarchical decomposition and representation of the Internet. While traditional wavelet-based MRA techniques have proven to be too rigid and inflexible to meet these expectations, more recent developments concerning the use of *diffusion wavelets* (e.g., see [40, 41]) show great promise and are presently explored in the context of Internet-specific structures.

NETWORKS OF NETWORKS

Changing perspectives, we can either view the Internet as a "network of networks" (e.g., AS-level Internet) or consider it as one of many networks that typically partake in the activities of enterprises: transportation of energy, materials, and components; power grid; supply chains, and control of transportation assets; communication and data networks. The networks' activities are correlated because they are invoked to support a common task, and the networks are interdependent because the characteristics of one determine the inputs or constraints for another. They are becoming even more correlated and interdependent as they shift more and more of their controls to be information-intensive and data-network-based. While this "networks of networks" concept ensures enormous efficiency and flexibility, both technical and economical, it also has a dark side—by requiring increasingly complex design processes, it creates vastly increased opportunities for potentially catastrophic failures, to the point where national and international critical infrastructure systems are at risk of large-scale disruptions due to intentional attacks, unintentional (but potentially devastating) side effects, the possibility of (not necessarily deliberate) large cascading events, or their growing dependence on the Internet as a "central nervous system".

This trend in network evolution poses serious questions about the reliability and performability of these critical infrastructure systems in the absence of an adequate theory [46]. Thus the long-term goal of any mathematical treatment of networked systems should be to develop the foundation of a nascent theory in support of such a "networks of networks" concept. To this end, the Internet shows great promise to serve as a case study to illustrate how early verbal observations and arguments with deep engineering insight have led via an interplay with mathematics and measurements to increasingly formal statements and powerful theoretical developments that can be viewed as a precursor of a full-fledged "network of networks" theory.

Conclusion

Over the last decade, there has been a compelling story articulated by the proponents of network science. Advances in information technology have facilitated the collection of petabyte scale data sets on everything from the Internet to biology to economic and social systems. These data sets are so large that attempts even to visualize them are nontrivial and often yield nonsensical results. Thus the "Petabyte Age" requires new modeling approaches and mathematical techniques to identify hidden structures, with the implication that these structures are fundamental to understanding the systems from which the vast amounts of measurements are derived. In extreme cases, this perspective suggests that the ubiquity of petabyte scale data on *everything* will fundamentally change the role of experimentation in science and of science as a whole [7].

In this article we have presented a retrospective view of key issues that have clouded the popular understanding and mathematical treatment of the

Internet as a complex system for which vast amounts of data are readily available. Foremost among these issues are the dangers of taking available data "at face value" without a deeper understanding of the idiosyncracies and ambiguities resulting from domain-specific collection and measurement techniques. When coupled with the naïve but commonly-accepted view of validation that simply argues for replicating certain statistical features of the observed data, such an "as is" use of the available data reduces complex network modeling to mere "data fitting", with the expected and non-informative outcome that given sufficient parameterization, it is always possible to match a model to any data set without necessarily capturing any underlying hidden structure or key functionality of the system at hand.

For systems whose measured features are subject to fundamental ambiguities, omissions, and/or systematic errors, we have proposed an alternative approach to network modeling that emphasizes data hygiene (i.e., practices associated with determining the quality of the available data and assessing their proper use) and uses constrained optimization as modeling language to account for the inherent objectives, constraints, and domain-specific environmental conditions underlying the growth and evolution of real-world complex networks. We have shown that in the context of the router-level Internet, this approach yields models that not only respect the forces shaping the real Internet but also are robust to the deficiencies inherent in available data.

In this article, the Internet has served as a clear case study, but the issues discussed apply more generally and are even more pertinent in contexts of biology and social systems, where measurement is inherently more difficult and more error prone. In this sense, the Internet example serves as an important reminder that despite the increasing ubiquity of vast amounts of available data, the "Garbage In, Gospel Out" extension of the phrase "Garbage In, Garbage Out" remains as relevant as ever; no amount of number crunching or mathematical sophistication can extract knowledge we can trust from low-quality data sets, whether they are of petabyte scale or not. Although the Internet story may seem all too obvious in retrospect, managing to avoid the same mistakes in the context of next-generation network science remains an open challenge. The consequences of repeating such errors in the context of, say, biology are potentially much more grave and would reflect poorly on mathematics as a discipline.

Notes

1. While the arguments and reasons differ for the data sets used in [27] to derive the power-law claim for the Internet's AS-level topology, the bottom line is the same—the

available measurements are not of sufficient quality for the purpose for which they are used in [27] (see for example [31]).

2. The Fall 2008 Annual Program of the Institute for Pure and Applied Mathematics (IPAM) on "Internet Multi-Resolution Analysis: Foundations, Applications, and Practice" focused on many of the challenges mentioned in this section; for more details, check out http://www.ipam.ucla.edu/programs/mra2008/.

3. Being robust to component failures was the number one requirement in the original design of the Internet [24].

4. The original Internet architects assumed that all hosts can be trusted [24].

References

[1] D. ACHLIOPTAS, A. CLAUSET, D. KEMPE, and C. MOORE, On the bias of traceroute sampling, *Journal of the ACM* (to appear), 2008.

[2] R. ALBERT, H. JEONG, and A.-L. BARABÁSI, Error and attack tolerance of complex networks, *Nature* **406** (2000).

[3] R. ALBERT and A.-L. BARABÁSI, Statistical mechanics of complex networks, *Rev. Mod. Phys.* **74** (2002).

[4] D. L. ALDERSON, Catching the "Network Science" Bug: Insight and Opportunities for the Operations Researchers, *Operations Research* **56**(5) (2009), 1047–1065.

[5] D. L. ALDERSON and J. C. DOYLE, Contrasting views of complexity and their implications for network-centric infrastructures, *IEEE Trans. on SMC-A* (submitted), 2008.

[6] D. ALDERSON and L. LI, Diversity of graphs with highly variable connectivity, *Phys. Rev. E* **75**, 046102, 2007.

[7] C. ANDERSON, The end of theory, *Wired Magazine* **16**, July 2008.

[8] A.-L. BARABÁSI and R. ALBERT, Emergence of scaling in random networks, *Science* **286** (1999).

[9] A. BENDER, R. SHERWOOD, and N. SPRING, Fixing ally's growing pains with velocity modeling, *Proc. ACM IMC*, 2008.

[10] E. BENDER and E. R. CANFIELD, The asymptotic number of labeled graphs with given degree sequences, *J. of Comb. Theory A* **24** (1978), 296–307.

[11] N. BERGER, C. BORGS, J. T. CHAYES, and A. SABERI, On the spread of viruses on the Internet, *Proc. SODA'05*, 2005.

[12] W. A. BEYER, P. H. SELLERS, and M. S. WATERMAN, Stanislaw M. Ulam's contributions to theoretical theory, *Letters in Mathematical Physics* **10** (1985), 231–242.

[13] B. BOLLOBÁS, *Random Graphs*, Academic Press, London, 1985.

[14] B. BOLLOBÁS and O. RIORDAN, Mathematical results on scale-free random graphs, *Handbook of Graphs and Networks* (S. Bornholdt and H. G. Schuster, eds.), Wiley-VCH, Weinheim, 2002.

[15] ———, Robustness and vulnerability of scale-free graphs, *Internet Mathematics* **1**(1) (2003), 1–35.

[16] ———, The diameter of a scale-free random graph, *Combinatorica* **24**(1) (2004), 5–34.

[17] C. BORGS, J. T. CHAYES, L. LOVÁSZ, V. T. SÓS, B. SZEGEDY, and K. VESZTERGOMBI, Graph limits and parameter testing, *Proc. STOC'06*, 2006.

[18] J. M. CARLSON and J. C. DOYLE, Complexity and robustness, *Proc. Nat. Acad. of Sci. USA* **99** (2002), 2538–2545.

[19] J. T. CHAYES, C. BORGS, L. LOVÁSZ, V. T. SÓS, and K. VESZTERGOMBI, Convergent sequences of dense graphs I: Subgraph frequencies, metric properties and testing,

preprint, https://research. microsoft.com/en-us/um/people/jchayes/Papers/Conv-Metric.pdf, 2006.

[20] ————, Convergent sequences of dence graphs II: Multiway cuts and statistical physics, preprint, https://research.microsoft.com/en-us/um/people/jchayes/Papers/ConRight.pdf, 2007.

[21] M. CHIANG, S. H. LOW, A. R. CALDERBANK, and J. C. DOYLE, Layering as optimization decomposition, *Proc. of the IEEE* **95** (2007).

[22] L. CHEN, S. H. LOW, M. CHIANG, and J. C. DOYLE, Cross-layer congestion control, routing and scheduling design in ad-hoc wireless networks, *Proc. IEEE INFOCOM'06*, 2006.

[23] F. CHUNG and L. LU, The average distance in a random graph with given expected degrees, *Internet Math.* **1** (2003), 91–113.

[24] D. D. CLARK, The design philosophy of the Darpa Internet protocol, *ACM Computer Communication Review* **18**(4) (1988), 106–114.

[25] R. DURRETT, *Random Graph Dynamics,* Cambridge University Press, New York, 2007.

[26] A. FABRIKANT, E. KOUTSOUPIAS, and C. PAPADIMITRIOU, Heuristically optimized trade-offs: A new paradigm for power-laws in the internet, *Proc. ICALP,* 2002, 110–122.

[27] M. FALOUTSOS, P. FALOUTSOS, and C. FALOUTSOS, On power-law relationships of the Internet topology, *ACM Comp. Comm. Review* **29**(4) (1999).

[28] M. H. GUNES and K. SARAC, Resolving IP aliases in building traceroute-based Internet maps, *IEEE/ACM Trans. Networking* (to appear) 2008.

[29] S. HAKIMI, On the realizability of a set of integers as degrees of the vertices of a linear graph, *SAM. J. Appl. Math.* **10** (1962), 496–506.

[30] F. P. KELLY, A. MAULLOO, and D. TAN, Rate control in communication networks: Shadow prices, proportional fairness stability, *Journal of the Operational Research Society* **49** (1998), 237–252.

[31] B. KRISHNAMURTHY and W. WILLINGER, What are our standards for validation of measurement-based networking research? *Proc. HotMetrics'08,* 2008.

[32] A. LAKHINA, J. W. BYERS, M. CROVELLA, and P. XIE, Sampling biases in IP topology measurements, *IEEE INFOCOM 2003.*

[33] L. LI, D. ALDERSON, W. WILLINGER, and J. C. DOYLE, A first principles approach to understanding the Internet's router-level topology, *Proc. ACM SIGCOMM'04* **34**(4) (2004), 3–14.

[34] L. LI, D. ALDERSON, J. C. DOYLE, and W. WILLINGER, Towards a theory of scale-free graphs: Definitions, properties, and implications. *Internet Mathematics* **2**(4) (2005), 431–523.

[35] S. Y. R. LI, Graphic sequences with unique realization, *J. Combin. Theory B* **19** (1975), 42–68.

[36] S. H. LOW, A duality model of TCP and queue management algorithms, *IEEE/ACM Trans. on Networking* **11**(4) (2003), 525–536.

[37] J. WANG, L. LI, S. H. LOW, and J. C. DOYLE, Cross-layer optimization in TCP/IP networks, *IEEE/ACM Trans. on Networking* **13** (2005), 582–595.

[38] S. E. LURIA and M. DELBRÜCK, Mutations of bacteria from virus sensitivity to virus resistance, *Genetics* **28** (1943), 491–511.

[39] B. B. MANDELBROT, *Fractals and Scaling in Finance,* Springer-Verlag, New York, 1997.

[40] M. MAGGIONI, A. D. SZLAM, R. R. COIFMAN, and J. C. BRENNER, Diffusion-driven multiscale analysis on manifolds and graphs: Top-down and bottom-up constructions, *Proc. SPIE Wavelet XI,* 5914, 2005.

[41] M. MAGGIONI, J. C. BRENNER, R. R. COIFMAN, and A. D. SZLAM, Biorthogonal diffusion wavelets for multiscale representations on manifolds and graphs, *Proc. SPIE Wavelet XI,* 5914, 2005.

[42] National Research Council Report, *Network Science,* National Academies Press, Washington, 2006.

[43] R. OLIVEIRA, D. PEI, W. WILLINGER, B. ZHANG, and L. ZHANG, In search of the elusive ground truth: The Internet's AS-level connectivity structure, *Proc. ACM SIGMETRICS,* 2008.

[44] J.-J. PANSIOT and D. GRAD, On routes and multicast trees in the Internet, *ACM Computer Communication Review* 28(1) (1998).

[45] C. H. PAPADIMITRIOU, Algorithms, games, and the Internet, *Proc. STOC'01,* 2001.

[46] President's Commission on Critical Infrastructure Protection. Tech. Report, The White House, 1997.

[47] H. J. RYSER, Combinatorial properties of matrices of zeroes and ones, *Canad. J. Math.* 9 (1957), 371–377.

[48] R. SHERWOOD, A. BENDER, and N. SPRING, Dis-Carte: A disjunctive Internet cartographer, *Proc. ACM SIGCOMM,* 2008.

[49] H. A. SIMON, On a class of skew distribution functions, *Biometrika* 42 (1955), 425–440.

[50] N. SPRING, R. MAHAJAN, and D. WETHERALL, Measuring ISP topologies with Rocketfuel, *Proc. ACM SIGCOMM,* 2002.

[51] N. SPRING, M. DONTCHEVA, M. RODRIG, and D. WETHERALL, How to Resolve IP Aliases, *UW CSE Tech. Report* 04-05-04, 2004.

[52] S. STROGATZ, Romanesque networks, *Nature* 433 (2005).

[53] W. WILLINGER, R. GOVINDAN, S. JAMIN, V. PAXSON, and S. SHENKER, Scaling phenomena in the Internet: Critically examining criticality, *Proc. Nat. Acad. Sci.* 99 (2002), 2573–2580.

[54] G. YULE, A mathematical theory of evolution based on the conclusions of Dr. J.C. Willis, F.R.S. *Philosophical Transactions of the Royal Society of London (Series B)* 213 (1925), 21–87.

The Higher Arithmetic:
How to Count to a Zillion without
Falling Off the End of the Number Line

BRIAN HAYES

In 2008 the National Debt Clock in New York City ran out of digits. The billboard-size electronic counter, mounted on a wall near Times Square, overflowed when the public debt reached $10 trillion, or 10^{13} dollars. The crisis was resolved by squeezing another digit into the space occupied by the dollar sign. Now a new clock is on order, with room for growth; it won't fill up until the debt reaches a quadrillion (10^{15}) dollars.

The incident of the Debt Clock brings to mind a comment made by Richard Feynman in the 1980s—back when mere billions still had the power to impress:

> There are 10^{11} stars in the galaxy. That used to be a *huge* number. But it's only a hundred billion. It's less than the national deficit! We used to call them astronomical numbers. Now we should call them economical numbers.

The important point here is not that high finance is catching up with the sciences; it's that the numbers we encounter everywhere in daily life are growing steadily larger. Computer technology is another area of rapid numeric inflation. Data storage capacity has gone from kilobytes to megabytes to gigabytes, and the latest disk drives hold a terabyte (10^{12} bytes). In the world of supercomputers, the current state of the art is called petascale computing (10^{15} operations per second), and there is talk of a coming transition to exascale (10^{18}). After that, we can await the arrival of zettascale (10^{21}) and yottascale (10^{24}) machines—and then we run out of prefixes!

Even these numbers are puny compared with the prodigious creations of pure mathematics. In the 18th century the largest known prime number had 10 digits; the present record-holder runs to almost 13 million digits. The value of π has been calculated to a trillion digits—a feat at once magnificent and mind-numbing. Elsewhere in mathematics there are numbers so big that even trying to describe their size requires numbers that are too big to describe. Of course none of these numbers are likely to turn up in everyday chores such as balancing a checkbook. On the other hand, logging into a bank's web site involves doing arithmetic with numbers in the vicinity of 2^{128}, or 10^{38}. (The calculations take place behind the scenes, in the cryptographic protocols meant to ensure privacy and security.)

Which brings me to the main theme of this column: Those streams of digits that make us so dizzy also present challenges for the design of computer hardware and software. Like the National Debt Clock, computers often set rigid limits on the size of numbers. When routine calculations begin to bump up against those limits, it's time for a rethinking of numeric formats and algorithms. Such a transition may be upon us soon, with the approval in 2008 of a revised standard for one common type of computer arithmetic, called floating point. Before the new standard becomes too deeply entrenched, perhaps it's worth pausing to examine a few alternative schemes for computing with astronomical and economical and mathematical numbers.

Numerical Eden

In their native habitat—which is *not* the digital computer—numbers are boundless and free-ranging. Along the real number line are infinitely many integers, or whole numbers. Between any two integers are infinitely many rational numbers, such as $3/2$ and $5/4$. Between any two rationals are infinitely many irrationals—numbers like $\sqrt{2}$ or π.

The reals are a Garden of Eden for doing arithmetic. Just follow a few simple rules—such as not dividing by zero—and these numbers will never lead you astray. They form a safe, closed universe. If you start with any set of real numbers, you can add and subtract and multiply all day long—and divide, too, except by zero—and at the end you'll still have real numbers. There's no risk of slipping through the cracks or going out of bounds.

Unfortunately, digital computers exist only outside the gates of Eden. Out here, arithmetic is a treacherous process. Even simple counting can get you in trouble. With computational numbers, adding 1 over and over eventually brings you to a largest number—something unknown in mathematics. If you try to press on beyond this limit, there's no telling what will happen. The next number after the largest number might be the smallest number; or it might be something labeled ∞; or the machine might sound an alarm, or die in a puff of smoke.

This is a lawless territory. On the real number line, you can always rely on principles like the associative law: $(a + b) + c = a + (b + c)$. In some versions of computer arithmetic, that law breaks down. (Try it with $a = 10^{30}$, $b = -10^{30}$, $c = 1$.) And when calculations include irrational numbers—well, irrationals just don't exist in the digital world. They have to be approximated by rationals—the very thing they are defined not to be. As a result, mathematical identities such as $\left(\sqrt{2}\right)^2 = 2$ are not to be trusted.

Bignums

The kind of computer arithmetic that comes closest to the mathematical ideal is calculation with integers and rationals of arbitrary size, limited only by the machine's memory capacity. In this "bignum" arithmetic, an integer is stored as a long sequence of bits, filling up as much space as needed. A rational number is a pair of such integers, interpreted as a numerator and a denominator.

A few primitive computers from the vacuum-tube era had built-in hardware for doing arithmetic on integers of arbitrary size, but our sophisticated modern machines have lost that capability, and so the process has to be orchestrated by software. Adding two integers proceeds piece by piece, starting with the least-significant bits and working right to left, much as a paper-and-pencil algorithm sums pairs of digits one at a time, propagating any carries to the next column. The usual practice is to break up the sequence of bits into blocks the size of a machine register—typically 32 or 64 bits. Algorithms for multiplication and division follow similar principles; operations on rationals require the further step of reducing a fraction to lowest terms.

Looking beyond integers and rationals, there have even been efforts to include irrational numbers in exact computations. Of course there's no hope of expressing the complete value of π or $\sqrt{2}$ in a finite machine, but a program can calculate the values incrementally, supplying digits as they are needed—a strategy known as lazy computing. For example, the assertion $\pi < 3.1414$ could be tested—and shown to be false—by generating the first five decimal digits of π. Another approach is to treat irrational numbers as unevaluated units, which are carried through the computation from start to finish as symbols; thus the circumference of a circle of unit radius would be given simply as 2π.

The great virtue of bignum arithmetic is exactness. If the machine ever gives an answer, it will be the right answer (barring bugs and hardware failures). But there's a price to pay: You may get no answer at all. The program could run out of memory, or it could take so long that it exhausts human patience or the human lifespan.

For some computations, exactness is crucial, and bignum arithmetic is the only suitable choice. If you want to search for million-digit primes, you have to look at every last digit. Similarly, the security module in a web browser must work with the exact value of a cryptographic key.

For many other kinds of computations, however, exactness is neither needed nor helpful. Using exact rational arithmetic to calculate the interest on a mortgage loan yields an unwieldy fraction accurate to hundreds of decimal places, but knowing the answer to the nearest penny would suffice. In many cases the inputs to a computation come from physical measurements accurate to no more than a few significant digits; lavishing exact calculations on these measurements cannot make them any more accurate.

What's the Point?

Most computer arithmetic is done not with bignums or exact rationals but with numbers confined to a fixed allotment of space, such as 32 or 64 bits. The hardware operates on all the bits at once, so arithmetic can be very fast. But an implacable law governs all such fixed-size formats: If a number is represented by 32 bits, then it can take on at most 2^{32} possible values. You may be able to choose *which* 2^{32} values are included, but there's no way to increase the size of the set.

For 32-bit numbers, one obvious mapping assigns the 2^{32} bit patterns to the integers from 0 through 4,294,967,295 (which is $2^{32} - 1$). The same range of integers could be shifted along the number line, or the values could be scaled to cover a smaller numeric range in finer increments (perhaps 0.00 up to 42,949,672.95) or spread out over a wider range more sparsely. Arithmetic done in this style is known as "fixed point," since the position of the decimal point is the same in all numbers of a given class.

Fixed-point arithmetic was once the mainstay of numerical computing, and it still has a place in certain applications, such as high-speed signal processing. But the dominant format now is floating point, where the decimal point (or binary point) can be moved around to represent a wide range of magnitudes. The floating-point format is based on the same idea as scientific notation. Just as we can write a large number succinctly as 6.02×10^{23}, floating-point arithmetic stores a number in two parts: the significand (6.02 in this example) and the exponent (23).

Designing a floating-point format entails a compromise between range and precision. Every bit allocated to the significand doubles its precision; but the bit has to be taken from the exponent, and it therefore reduces the range by half. For 32-bit numbers the prevailing standard dictates a 24-bit significand and an 8-bit exponent; a few stray bits are lost to storing signs and marking special cases, leaving an effective range of 2^{-126} up to 2^{127}. In decimal notation the largest representable number is about 3×10^{38}. Standard 64-bit numbers allocate 53 bits to the significand and 11 to the exponent, allowing a range up to about 10^{308}.

The idea of floating-point arithmetic goes back to the beginning of the computer age, but it was widely adopted only in the 1980s. The key event was the drafting of a standard, approved by the Institute of Electrical and Electronic Engineers (IEEE) in 1985. This effort was led by William Kahan of the University of California, Berkeley, who remains a strong advocate of the technology.

Early critics of the floating-point approach worried about efficiency and complexity. In fixed-point arithmetic, many operations can be reduced to a single machine instruction, but floating-point calculations are more involved. First you have to extract the significands and exponents, then operate on these pieces separately, then do some rounding and adjusting, and finally reassemble the parts.

The answer to these concerns was to implement floating-point algorithms in hardware. Even before the IEEE standard was approved, Intel designed a floating-point coprocessor for early personal computers. Later generations incorporated a floating-point unit on the main processor chip. From the programmer's point of view, floating-point arithmetic became part of the infrastructure.

Safety in Numbers

It's tempting to pretend that floating-point arithmetic is simply real-number arithmetic in silicon. This attitude is encouraged by programming languages that use the label *real* for floating-point variables. But of course floating-point numbers are *not* real numbers; at best they provide a finite model of the infinite real number line.

Unlike the real numbers, the floating-point universe is not a closed system. When you multiply two floating-point numbers, there's a good chance that the product—the *real* product, as calculated in real arithmetic—will not be a floating-point number. This leads to three kinds of problems.

The first problem is rounding error. A number that falls between two floating-point values has to be rounded by shifting it to one or the other of the nearest representable numbers. The resulting loss of accuracy is usually small and inconsequential, but circumstances can conspire to produce numerical disasters. A notably risky operation is subtracting one large quantity

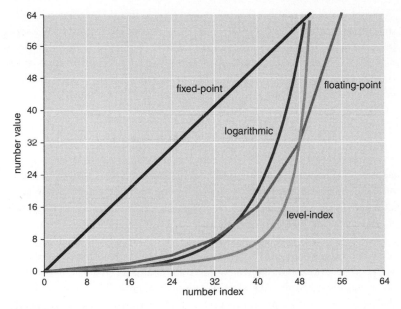

FIGURE 1.

from another, which can wipe out all the significant digits in the small difference. Textbooks on numerical analysis are heavy with advice on how to guard against such events; mostly it comes down to "Don't do that."

The second problem is overflow, when a number goes off the scale. The IEEE standard allows two responses to this situation. The computer can halt the computation and report an error, or it can substitute a special marker, "∞," for the oversize number. The latter option is designed to mimic the properties of mathematical infinities; for example, $\infty + 1 = \infty$. Because of this behavior, floating-point infinity is a black hole: Once you get into it, there is no way out, and all information about where you came from is annihilated.

The third hazard is underflow, where a number too small to represent collapses to zero. In real arithmetic, a sequence like $1/2, 1/4, 1/8, \ldots$ can go on indefinitely, but in a finite floating-point system there must be a smallest nonzero number. On the surface, underflow looks much less serious than overflow. After all, if a number is so small that the computer can't distinguish it from zero, what's the harm of making it exactly zero? But this reasoning is misleading. In the exponential space of floating-point numbers, the distance from, say, 2^{-127} to zero is exactly the same as the distance from 2^{127} to infinity. As a practical matter, underflow is a frequent cause of failure in numerical computations.

Problems of rounding, overflow and underflow cannot be entirely avoided in any finite number system. They can be ameliorated, however, by adopting a format with higher precision and a wider range—by throwing more bits at the problem. This is one approach taken in a recent revision of the IEEE standard, approved in June 2008. It includes a new 128-bit floating-point format, supporting numbers as large as $2^{16,383}$ (or about $10^{4,932}$).

Tapering Off, or Rolling Off a Log

By now, IEEE floating-point methods are so firmly established that they often seem like the *only* way to do arithmetic with a computer. But many alternatives have been discussed over the years. Here I shall describe two of them briefly and take a somewhat closer look at a third idea.

The first family of proposals might be viewed more as an enhancement of floating point than as a replacement. The idea is to make the trade-off between precision and range an adjustable parameter. If a calculation does not require very large or very small numbers, then it can give more bits to the significand. Other programs might want to sacrifice precision in order to gain wider scope for the exponent. To make such flexibility possible, it's necessary to set aside a few bits to keep track of how the other bits are allocated. (Of course those bookkeeping bits are thereby made unavailable for *either* the exponent or the significand.)

A scheme of this kind, called tapered floating point, was proposed as early as 1971 by Robert Morris, who was then at Bell Laboratories. A decade later, more elaborate plans were published by Shouichi Matsui and Masao Iri of the University of Tokyo and by Hozumi Hamada of Hitachi, Ltd. More recently, Alan Feldstein of Arizona State University and Peter R. Turner of Clarkson University have described a tapered scheme that works exactly like a conventional floating-point system except when overflow or underflow threaten.

The second alternative would replace numbers by their logarithms. For example, in a decimal version of the plan the number 751 would be stored as 2.87564, since $10^{2.87564} = 751$. This plan is not as radical a departure as it might seem, because floating-point is already a semi-logarithmic notation: The exponent of a floating-point number is the integer part of a logarithm. Thus the two formats record essentially the same information.

If the systems are so similar, what's gained by the logarithmic alternative? The motive is the same as that for developing logarithms in the first place: They facilitate multiplication and division, reducing those operations to addition and subtraction. For positive numbers a and b, $\log(ab) = \log(a) + \log(b)$. In general, multiplying takes more work than adding, so this substitution is a net gain. But there's another side to the coin: Although logarithms make multiplying easy, they make adding hard. Computing $a + b$ when you have only $\log(a)$ and $\log(b)$ is not straightforward. For this reason logarithmic arithmetic is attractive mainly in specialized areas such as image processing where multiplications tend to outnumber additions.

On the Level

The third scheme I want to mention here addresses the problem of overflow. If you are trying to maximize the range of a number system, an idea that pops up quite naturally is to replace mere exponents with towers of exponents. If 2^N can't produce a number large enough for your needs, then try

$$2^{2^N} \text{ or } 2^{2^{2^N}} \text{ or } 2^{2^{2^{2^N}}} .$$

(Whatever the mathematical merits of such expressions, they are a typographical nightmare, and so from here on I shall replace them with a more convenient notation, invented by Donald E. Knuth of Stanford University: $2\uparrow 2\uparrow 2\uparrow 2\uparrow N$ is equivalent to the last of the three towers shown above. It is to be evaluated from right to left, just as the tower is evaluated from top to bottom.)

Number systems based on iterated exponentiation have been proposed several times; for example, they are mentioned by Matsui and Iri and by Hamada. But one particular version of the idea, called the level-index system, has been worked out with such care and thoughtful analysis that it deserves closer attention. Level-index arithmetic is a lost gem of computer science. It may never make it into the CPU of your laptop, but it shouldn't be forgotten.

The scheme was devised by Charles W. Clenshaw and Frank W. J. Olver, who first worked together (along with Alan Turing) in the 1940s at the National Physical Laboratory in Britain. They proposed the level-index idea in the 1980s, writing a series of papers on the subject with several other colleagues, notably Turner and Daniel W. Lozier, now of the National Institute of Standards and Technology (NIST). Clenshaw died in 2004; Olver is now at the University of Maryland and NIST, and is co-editing with Lozier a new version of the *Handbook of Mathematical Functions* by Abramowitz and Stegun.

Iterated exponentials can be built on any numeric base; most proposals have focused on base 2 or base 10. Clenshaw and Olver argue that the best base is e, the irrational number usually described as the base of the natural logarithms or as the limiting value of the compound-interest formula

$(1 + 1/n)^n$; numerically e is about 2.71828. Building numbers on an irrational base is an idea that takes some getting used to. For one thing, it means that almost all numbers that have an exact representation are irrational; the only exceptions are 0 and 1. But there's no theoretical difficulty in constructing such numbers, and there's a good reason for choosing base e.

In the level-index system a number is represented by an expression of the form $e \uparrow e \uparrow \ldots \uparrow e \uparrow m$, where the m at the end of the chain is a fractional quantity analogous to the mantissa of a logarithm. The number of up-arrows—or in other words the height of the exponential tower—depends on the magnitude of the number being represented.

To convert a positive number to level-index form, we first take the logarithm of the number, then the logarithm of the logarithm, and so on, continuing until the result lies in the interval between 0 and 1. Counting the successive logarithm operations gives us the *level* part of the representation; the remaining fraction becomes the *index*, the value of m in the expression above. The process is defined by the function $f(x)$:

$$\text{if } 0 \leq x < 1 \text{ then } f(x) = x$$
$$\text{else } f(x) = 1 + f(\ln(x)).$$

Here's how the procedure applies to the national-debt amount at the beginning of this piece:

$$\ln(10{,}659{,}204{,}157{,}341) = 29.9974449$$
$$\ln(29.9974449) = 3.40111221$$
$$\ln(3.40111221) = 1.22410249$$
$$\ln(1.22410249) = 0.20220791$$

We've taken logarithms four times, so the level is 4, and the fractional amount remaining becomes the index. Thus the level-index form of the national debt is 4.20220791 (which seems a lot less worrisome than $10,659,204,157,341).

The level-index system accommodates *very* large numbers. Level 0 runs from 0 to 1, then level 1 includes all numbers up to e. Level 2 extends as far as $e \uparrow e$, or about 15.2. Beyond this point, the growth rate gets steep. Level 3 goes up to $e \uparrow e \uparrow e$, which is about 3,814,273. Continuing the ascent through level 4, we soon pass the largest 64-bit floating-point number, which has a level-index value of about 4.63. The upper boundary of level 4 is a number with 1.6 million decimal digits. Climbing higher still puts us in the realm of numbers where even a description of the size is hopelessly impractical. Just seven levels are enough to represent all distinguishable level-index numbers. Thus only three bits need to be devoted to the level; the rest can be used for the index.

What about the other end of the number scale—the very small numbers? The level-index system is adequate for many purposes in this region, but a variation called symmetric level-index provides additional precision close to zero. In this scheme a number x between 0 and 1 is denoted by the level-index representation of $1/x$.

Apart from its wide range, the level-index system has some other distinctive properties. One is smoothness. For floating-point numbers, a graph of the magnitudes of successive numbers is a jointed sequence of straight lines, with an abrupt change of slope at each power of 2. The corresponding graph for the level-index system is a smooth curve. For iterated exponentials this is true only in base e, which is the reason for choosing that base.

Olver also points out that level-index arithmetic is a closed system, like arithmetic with real numbers. How can that be? Since level-index numbers are finite, there must be a largest member of the set, and so repeated additions or multiplications should eventually exceed that bound. Although this reasoning is unassailable, it turns out that the system does not in fact overflow. Here's what happens instead. Start with a number x, then add or multiply to generate a new larger x, which is rounded to the nearest level-index number. As x grows very large, the available level-index values become sparse. At some point, the spacing between successive level-index values is greater than the change in x caused by addition or multiplication. Thereafter, successive iterations of x round to the same level-index value.

This is not a perfect model of unbounded arithmetic. In particular, the process is not reversible: A long series of $x + 1$ operations followed by an equal number of $x - 1$s will not bring you back to where you started, as it would on the real number line. Still, the boundary at the end of the number line seems about as natural as it can be in a finite system.

Shaping a Number System

Is there any genuine need for an arithmetic that can reach beyond the limits of IEEE floating point? I have to admit that I seldom write a program whose output is a number greater than 10^{38}. But that's not the end of the story.

A program with inputs and outputs of only modest size may nonetheless generate awkwardly large intermediate values. Suppose you want to know the probability of observing exactly 1,000 heads in 2,000 tosses of a fair coin. The standard formula calls for evaluating the factorial of 2,000, which is $1 \times 2 \times 3 \times \ldots \times 2,000$ and is sure to overflow. You also need to calculate $(1/2)^{2,000}$, which could underflow. Although the computation *can* be successfully completed with floating-point numbers—the answer is about 0.018—it requires careful attention to cancellations and reorderings of the operations. A number system with a wider range would allow a simpler and more robust approach.

In 1993 Lozier described a more substantial example of a program sensitive to numerical range. A simulation in fluid dynamics failed because of severe floating-point underflow; redoing the computation with the symmetric version of level-index arithmetic produced correct output.

Persuading the world to adopt a new kind of arithmetic is a quixotic undertaking, like trying to reform the calendar or replace the QWERTY keyboard. But even setting aside all the obstacles of history and habit, I'm not sure how best to evaluate the alternatives in this case. The main conceptual question is this: Since we don't have enough numbers to cover the entire number line, what is the best distribution of the numbers we *do* have? Fixed-point systems sprinkle them uniformly. Floating-point numbers are densely packed near the origin and grow farther apart out in the numerical hinterland. In the level-index system, the core density is even greater, and it drops off even more steeply, allowing the numbers to reach the remotest outposts.

Which of these distributions should we prefer? Perhaps the answer will depend on what numbers we need to represent—and thus on how quickly the national debt continues to grow.

Bibliography

• Clenshaw, C. W., and F. W. J. Olver. 1984. Beyond floating point. *Journal of the Association for Computing Machinery* 31:319–328.
• Clenshaw, C. W., F. W. J. Olver, and P. R. Turner. 1989. Level-index arithmetic: An introductory survey. In *Numerical Analysis and Parallel Processing: Lectures Given at the Lancaster Numerical Analysis Summer School, 1987*, pp. 95–168. Berlin: Springer-Verlag.
• Hamada, Hozumi. 1987. A new real number representation and its operation. In *Proceedings of the Eighth Symposium on Computer Arithmetic*, pp. 153–157. Washington, D.C.: IEEE Computer Society Press.
• Lozier, D. W., and F. W. J. Olver. 1990. Closure and precision in level-index arithmetic. *SIAM Journal on Numerical Analysis* 27:1295–1304.
• Lozier, Daniel W. 1993. An underflow-induced graphics failure solved by SLI arithmetic. In *Proceedings of the 11th Symposium on Computer Arithmetic*, pp. 10–17. Los Alamitos, Calif.: IEEE Computer Society Press.
• Matsui, Shourichi, and Masao Iri. 1981. An overflow/underflow-free floating-point representation of numbers. *Journal of Information Processing* 4:123–133.
• Turner, Peter R. 1991. Implementation and analysis of extended SLI operations. In *Proceedings of the 10th Symposium on Computer Arithmetic*, pp. 118–126. Los Alamitos, Calif.: IEEE Computer Society Press.

Knowing When to Stop:
How to Gamble If You Must—The
Mathematics of Optimal Stopping

Theodore P. Hill

Every decision is risky business. Selecting the best time to stop and act is crucial. When Microsoft prepares to introduce Word 2020, it must decide when to quit debugging and launch the product. When a hurricane veers toward Florida, the governor must call when it's time to stop watching and start evacuating. Bad timing can be ruinous. Napoleon learned that the hard way after invading Russia. We face smaller-consequence stopping decisions all the time, when hunting for a better parking space, responding to a job offer or scheduling retirement.

The basic framework of all these problems is the same: A decision maker observes a process evolving in time that involves some randomness. Based only on what is known, he or she must make a decision on how to maximize reward or minimize cost. In some cases, little is known about what's coming. In other cases, information is abundant. In either scenario, no one predicts the future with full certainty. Fortunately, the powers of probability sometimes improve the odds of making a good choice.

While much of mathematics has roots that reach back millennia to Euclid and even earlier thinkers, the history of probability is far shorter. And its lineage is, well, a lot less refined. Girolamo Cardano's famed 1564 manuscript *De Ludo Aleae*, one of the earliest writings on probability and not published until a century after he wrote it, primarily analyzed dice games. Although Galileo and other 17th-century scientists contributed to this enterprise, many credit the mathematical foundations of probability to an exchange of letters in 1654 between two famous French mathematicians, Blaise Pascal and Pierre de Fermat. They too were concerned with odds and dice throws—for example, whether it is wise to bet even money that a pair of sixes will occur in 24 rolls of two fair dice. Some insisted it was, but the true probability of a double six in 24 rolls is about 49.1 percent.

That correspondence inspired significant advances by Abraham de Moivre, Christiaan Huygens, Siméon Poisson, Jacob Bernoulli, Pierre-Simon Laplace and Karl Friedrich Gauss into the 19th century. Still, for a long time there was no formal definition of probability precise enough for use in mathematics or robust enough to handle increasing evidence of random phenomena in science. Not until 1933, nearly four centuries after Cardano, did the Russian mathematician Andrey Kolmogorov put probability theory on a formal axiomatic basis, using the emerging field we now call measure theory.

The history of optimal-stopping problems, a subfield of probability theory, also begins with gambling. One of the earliest discoveries is credited to the eminent English mathematician Arthur Cayley of the University of Cambridge. In 1875, he found an optimal stopping strategy for purchasing lottery tickets. The wider practical applications became apparent gradually. During World War II, Abraham Wald and other mathematicians developed the field of statistical sequential analysis to aid military and industrial decision makers faced with strategic gambles involving massive amounts of men and material. Shortly after the war, Richard Bellman, an applied mathematician, invented dynamic programming to obtain optimal strategies for many other stopping problems. In the 1970s, the theory of optimal stopping emerged as a major tool in finance when Fischer Black and Myron Scholes discovered a pioneering formula for valuing stock options. That transformed the world's financial markets and won Scholes and colleague Robert Merton the 1997 Nobel Prize in Economics. (Black had died by then.)

The Black-Scholes formula is still the key to modern option pricing, and the optimal-stopping tools underlying it remain a vigorous area of research in academia and industry. But even elementary tools in the theory of optimal stopping offer powerful, practical and sometimes surprising solutions.

The Marriage Problem

Suppose you decide to marry, and to select your life partner you will interview at most 100 candidate spouses. The interviews are arranged in random order, and you have no information about candidates you haven't yet spoken to. After each interview you must either marry that person or forever lose the chance to do so. If you have not married after interviewing candidate 99, you must marry candidate 100. Your objective, of course, is to marry the absolute best candidate of the lot. But how?

This problem has a long and rich history in the mathematics literature, where it is known variously as the marriage, secretary, dowry or best-choice problem. Certainly you can select the very best spouse with probability at least 1/100, simply by marrying the first person. But can you do better? In fact, there is a simple rule that will guarantee you will marry the absolute best more than one-third of the time. And the rule can be transferred to other scenarios.

As an enlisted man in the U.S. Air Force during the Vietnam era, John Elton, now a Georgia Institute of Technology mathematician, transformed the marriage problem into a barracks moneymaking scheme. Elton asked his fellow airmen to write down 100 different numbers, positive or negative, as large or small as they wished, on 100 slips of paper, turn them face down on a table and mix them up. He would bet them he could turn the slips of paper over one at a time and stop with the highest number. He convinced them it was "obvious" that the chance of him winning was very small, so he asked ten dollars if he won and paid them one dollar if he lost. There was no shortage of bettors. Even though my friend lost nearly two-thirds of the games, he won more than one-third of them. And with the 10-1 odds, he raked in a bundle. How?

First, note that there is a very simple strategy for winning more than one-fourth of the time, which would already put him ahead. Call an observed number a "record" if it is the highest number seen so far. Suppose you turn over half the numbers—or interview the first 50 marriage candidates—and never stop, no matter how high the number. After that you stop with the first record you see. If the second-highest number in the 100 cards happens to be in the first 50 you look at, and the highest in the second half—which happens 1 in 4 times   then you win. That strategy is good, but there is an even better one. Observe only 37 cards (or potential partners) without stopping and then stop with the next record. John Gilbert and Frederick Mosteller of Harvard University proved that this strategy is best and guarantees stopping with the best number about 37 percent of the time. In fact, observing $N/e \cong 0.370$ of the candidates, where N is the total number of candidates and e is the base of the natural logarithms, $e = 2.71828 \ldots$, guarantees winning with probability more than $1/e > 0.36$, no matter how many cards or candidates there are. (Note that the "observe half the numbers" strategy clearly wins with probability at least ¼, also independent of the number of cards.)

Sometimes the goal is to stop with one of the best k of N candidates. That is, you win if you stop with any one of the highest k numbers. In the Olympics or in horse racing, for example, the objective often is the $k = 3$ case—to win a medal or to show—rather than the all-or-nothing $k = 1$ goal of a gold medal or a win, which is much riskier. The optimal strategy

optimal strategy (win with probability more than 1/3)

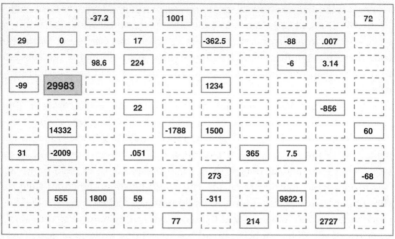

(1) turn over 37 cards without stopping and identify the highest number

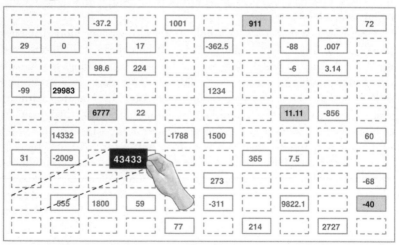

(2) continue to turn over cards and stop when a "record" is found

FIGURE 1. Stopping strategies don't guarantee victory, but they can create dependable odds. When looking for the highest number out of a group of 100, examine 37 options and then select the best seen after that. You'll succeed 37 percent of the time. Image by Tom Dunne and Stephanie Freese.

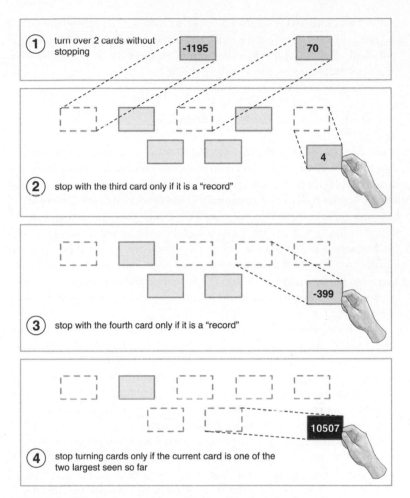

FIGURE 2. If the goal is to stop with one of the best k values, there are ways to improve the chances there too. In a scenario when one of the best two choices are desired out of seven options, you can win two-thirds of the time with this approach. Image by Stephanie Freese.

for stopping with one of the best k is similar to stopping with the best. First, you should observe a fixed number of candidates without ever stopping, thereby obtaining a baseline to work with. Then for another certain fixed length of time, stop if you see a record. Since it is the best seen so far, it is somewhat likely to be one of the best k. If no record appears during that stretch, then continue to the next stage where you stop with one of the highest two numbers for a fixed period of time, and so on. For $k = 2$, this method guarantees a better than 57 percent chance of stopping with one of the two best even if there are a million cards. For small N, the probability is quite high. Figure 2 illustrates the optimal strategy for $N = 7$ and $k = 2$, which guarantees a win two-thirds of the time.

$N = 2$ *Surprise*

Now, suppose you must decide when to stop and choose between only two slips of paper or two cards. You turn one over, observe a number there and then must judge whether it is larger than the hidden number on the second. The surprising claim, originating with David Blackwell of the University of California, Berkeley, is that you can win at this game more than half the time. Obviously you can win exactly half the time by always stopping with the first number, or always stopping with the second, without even peeking. But to win more than half the time, you must find a way to use information from the first number to decide whether or not to stop. (Readers take comfort: When mathematicians first heard this claim, many of us found it implausible.)

Here is one stopping rule that guarantees winning more than half the time. First, generate a random number R according to a standard Gaussian (bell-shaped) curve by using a computer or other device. Then turn over one of the slips of paper and observe its number. If R is larger than the observed number, continue and turn over the second card. If R is smaller, quit with the number observed on the first card. How can such a simple-minded strategy guarantee a win more than half the time?

If R is smaller than each of the two written numbers, then you win exactly half the time ($p/2$ of the unknown probability p in Figure 3); if it is larger than both, you again win half that time ($q/2$ of q, also in Figure 3). But if R falls *between* the two written numbers, which it must do with strictly positive probability (since the two numbers are different and the Gaussian distribution assigns positive probability to every interval) then you win all the time. This gives you the edge you need, since $p/2 + q/2 + 1 - p - q$ is greater than ½, because $1 - p - q$ is greater than zero. For example, if the two hidden numbers are 1 and π, this Gaussian method yields a value for p about .8413 and q about .0008, so the probability that it will select the larger number is more than 57 percent.

Of course if the number writer knows this Gaussian strategy, he can make your winnings as close to ½ as he wants by writing numbers that are very close.

If the number writer is not completely free to pick any number, but instead is required to choose an integer in the range $\{1, 2, \ldots, 100\}$, say,

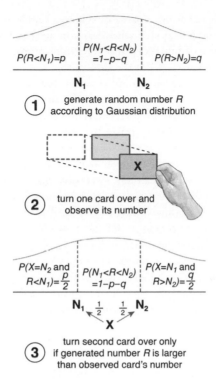

FIGURE 3. Even when choosing between just two options you can succeed more than half the time—that is, if you have access to a means to generate Gaussian random numbers. Image by Stephanie Freese.

then he cannot make your probability of winning arbitrarily close to ½. In this case it also seems obvious that the number-writer would never write a 1, since if you turn over a 1, you will always win by not stopping. But if he never writes a 1, he then would never write a 2 either since he never wrote a 1, and so on *ad absurdum*. Interested readers are invited to discover for themselves the optimal strategy in this case, and the amount more than ½ one can guarantee to win on the average.

Backward Induction

At the opposite end from having no information about future values is having full information—that is, complete information about the probabilities and the exact values of all potential future observations. In the spirit of Cardano, Fermat and Pascal's discoveries about probability with dice centuries ago, let's consider a game of rolling a standard, fair, six-sided die at most five times. You may stop whenever you want and receive as a reward the number of Krugerrands corresponding to the number of dots shown on the die at the time you stop. (At the time of this writing, one Krugerrand is worth $853.87.) Unlike the no-information marriage problems, here everything is known. The values at each roll will be 1, 2, 3, 4, 5, or 6, and the probability of each number on each roll is one-sixth. The objective is to find the stopping rule that will maximize the number of Krugerrands you can expect to win on average.

If you always stop with the first roll, for example, the winnable amount is simply the expected value of a random variable that takes the values 1, 2, 3, 4, 5, and 6 with probability 1/6 each. That is, one-sixth of the time you will win 1, one-sixth of the time you will win 2, and so on, which yields the expected value $1(1/6) + 2(1/6) + 3(1/6) + 4(1/6) + 5(1/6) + 6(1/6) = 7/2$. Thus if you always quit on the first roll, you expect to win 3.5 Krugerrands on the average. But clearly it is not optimal to stop on the first roll if it is a 1, and it is always optimal to stop with a 6, so already you know part of the optimal stopping rule. Should you stop with a 5 on the first roll? One powerful general technique for solving this type of problem is the method of *backward induction*.

Clearly it is optimal to stop on the first roll if the value seen on the first roll is greater than the amount expected if you do not stop—that is, if you continue to roll after rejecting the first roll. That would put you in a new game where you are only allowed four rolls, the expected value of which is also unknown at the outset. The optimal strategy in a four-roll problem, in turn, is to stop at the first roll if that value is greater than the amount you expect to win if you continue in a three-roll problem, and so on. Working down, you arrive at one strategy that you do know. In a one-roll problem there is only one strategy, namely to stop, and the expected reward is the expected value of one roll of a fair die, which we saw is 3.5. That information now yields the optimal strategy in a two-roll problem— stop on the first roll if the value is more than you expect to win if you continue, that is, more than 3.5. So now we know the optimal strategy for a two-roll problem—stop at the first roll if it is a 4, 5, or 6, and otherwise continue—and that allows us to calculate the expected reward of the strategy.

In a two-roll problem, you win 4, 5, or 6 on the very first roll, with probability 1/6 each, and stop. Otherwise (the half the time that the first roll was a 1, 2 or 3) you continue, in which case you expect to win 3.5 on the average. Thus the expected reward for the two-roll problem is $4(1/6) + 5(1/6) + 6(1/6) + (1/2)(3.5) = 4.25$. This now gives you the optimal strategy for a three-roll problem—namely, stop if the first roll is a 5 or 6 (that is, more than 4.25), otherwise continue and stop only if the second roll is a 4, 5, or 6, and otherwise proceed with the final third roll. Knowing this expected reward for three rolls in turn yields the optimal strategy for a four-roll problem, and so forth. Working backwards, this yields the optimal strategy in the original five-roll problem: Stop on the first roll only if it is a 5 or 6, stop on the second roll if it is a 5 or 6, on the third roll if it is a 5 or 6, the fourth roll if it is a 4, 5, or 6, and otherwise continue to the last roll. This strategy guarantees that you will win about 5.12 Krugerrands on average, and no other strategy is better. (So, in a six-roll game you should stop with the initial roll only if it is a 6.)

The method of backward induction is very versatile, and works equally well if the process values are not independent (as they were assumed to be in repeated rolls of the die), or if the objective is to minimize some expected value such as cost. Suppose that a company must purchase its weekly energy supplies on the previous Monday, Tuesday or Wednesday and the likelihood of future prices can be estimated based on past statistics. For example, if on Monday the decision maker has the opportunity to purchase energy for the following week at a cost of 100, she may know from past experience that there is a 50–50 chance that Tuesday's price will be 110, and otherwise it will be 90. Furthermore she knows that if it is 110 on Tuesday, then it will be 115 on Wednesday with probability $1/3$, and otherwise will be 100; and that if it is 90 on Tuesday, it is equally likely to be 100 or 85 on Wednesday. Using backward induction, the optimal rule on Tuesday is seen to be not to buy if the price is 110, since 110 is larger than the expected price $(1/3)(115) + (2/3)(100) = 105$ if she delays buying until Wednesday. Similarly, if the price Tuesday is 90, it is optimal to buy. Working backwards to Monday, since 100 is larger than the expected cost if she continues—namely, $(1/2)(105) + (1/2)(90) = 97.5$—it is optimal not to buy on Monday. Putting these together yields her optimal stopping strategy.

In the full-information case, with the objective of stopping with one of the largest k values, the best possible probabilities of winning were unknown for general finite sequences of independent random variables. Using both backward and forward induction, and a class of distributions called "Bernoulli pyramids," where each new variable is either the best or the worst seen thus far (for example, the first random variable is either $+1$ or -1 with certain probabilities, the second variable is $+2$ or -2, and so forth), Douglas Kennedy of the University of Cambridge and I discovered those optimal probabilities. We proved that for every finite sequence of independent random variables, there is always a stop rule that stops with the highest

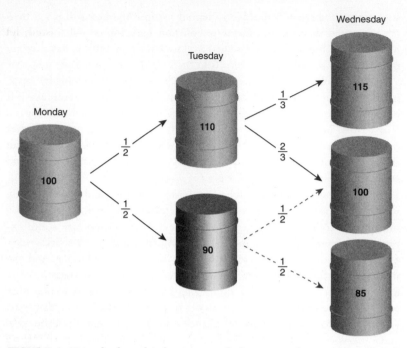

FIGURE 4. Using backward induction to calculate optimal stop rules isn't only helpful at the gaming table. It can help predict when you may be seeing your best price for a needed commodity in a fluctuating marketplace. In this example it is optimal not to buy Monday, to buy Tuesday only if the price is 90, and otherwise buy Wednesday (see text). Image by Stephanie Freese.

value with probability at least $1/e$, and a stop rule that stops with one of the two highest values with probability at least

$$e^{-\sqrt{2}}\left(1+\sqrt{2}\right) \cong 0.59,$$

which is the best you can do. (The probabilities for stopping with one of the k highest values have similar formulas.)

    Computers were not useful for solving that problem. In fact, all of the problems described in this article were solved using traditional mathematicians' tools—working example after example with paper and pencil; settling the case for two, three and then four unknowns; looking for patterns; waiting for the necessary *Aha!* insights; and then searching for formal proofs for each step. Computers are very helpful for after-the-fact applications of many results, such as backward induction. But in theoretical probability, computers often do not significantly aid the discovery process. To better understand this, reconsider the two card problem described above. How would one program a computer to imagine such a strategy existed, let alone to search the universe of ideas to find it?

Partial Information

The case of partial information is the most difficult. Usually one does not know how many applicants there will be for a job, nor the exact probabilities of future stock values. In these situations, one method of solution is to use tools from the theory of zero-sum, two-person games, where the stopping problem can be thought of as a decision maker playing against an opponent (often called Nature or God) who may assign the values and probabilities in any way she chooses.

Ulrich Krengel of the University of Göttingen and I used this technique to discover the optimal strategy in the so-called marriage problem where only a bound on the number of applicants is known. As a concrete example, consider the problem where the objective is to select the highest number from a hat containing at least one, and at most five, numbered cards (if you do not stop and there are no cards left, you lose.) We proved that the optimal strategy in this case is to stop with the first card with probability $26/75$ (which you may do using a random number generator or other method). If you do not stop with the first card, then you should continue to the second card, if there is one. If the second card's number is higher than the first card's number, stop with probability $26/49$. Otherwise, continue, stopping with the first record thereafter (or when you run out of cards or are forced to choose the number on the fifth card). This guarantees a probability of $26/75$ of stopping with the highest number, no matter how many cards Nature deposited in the hat.

There is no better strategy. We found the exact formulas and strategies for all possible bounds on the maximum number of cards and the winning probabilities are surprisingly high. For example, even if you only know that there are somewhere between 1 and 100 cards in the hat, it is still possible to win about 20 percent of the time. Exactly the same method can be employed to obtain optimal stopping rules in many real-life problems such as a situation where an employer wants to hire the very best salesperson available, and knows the maximum number of candidates available for the position, but does not know how many have already accepted other jobs.

In another type of stopping problem involving partial information, the observer knows the length of the sequence exactly (say, for example, the number of cards), but has only partial information about the random values on the cards. Instead of having no information at all, or knowing all the possible values and probabilities, he might know only the average value and standard deviation of each variable. In the case where variables are independent, Frans Boshuizen of the Free University of Amsterdam and I were able to use game theory and mass-shifting "balayage" (sweeping probability mass away from its center in both directions) arguments to determine the optimal stop rules, but those techniques fail for most other partial-information stopping problems.

Unsolved Problems

Although many stopping problems have been solved, there are still tantalizingly simple unsolved problems, even ones involving full information. My favorite is this: Toss a fair coin repeatedly and stop whenever you want, receiving as a reward the average number of heads accrued at the time you stop. If your first toss is a head, and you stop, your reward is 1 Krugerrand. Since you can never have more than 100 percent heads, it is clearly optimal to stop in that case. If the first toss is a tail, on the other hand, it is clearly best not to stop, since your reward would be zero. Suppose the first toss is a tail and the second a head. You may quit then and receive half a Krugerrand, or continue to toss. A moment's reflection shows that it is never optimal to stop with half a Krugerrand or less, since the law of large numbers says that the average number of heads will converge over time to 50 percent, and in doing so will oscillate above and below 50 percent repeatedly. Stopping with 50 percent or less is simply not greedy enough.

With a little more difficulty it can be shown that stopping with the third toss if you saw tail-head-head is optimal, and that stopping the very first time you observe more heads than tails is optimal for a while. But stopping the first time you have one more head than tails is not optimal forever. After a certain critical time you should only stop when you have two more heads than tails, and after a second critical time, stop only when you are three heads ahead, and so forth. The proof of this fact, which relies on the law of the iterated logarithm, is not easy, and the complete list of critical times is still not known. Backward induction will not work for this problem since there is no a priori end to the sequence and, hence, no future time to calculate backwards from. Even though Wolfgang Stadje of the University of Osnabrück has advanced this problem very recently, and despite the gains of a century of development of mathematical probability, the exact optimal rule for all sequences of heads and tails is unknown.

Still, the general field of optimal stopping, especially with its applications to financial markets, continues to develop at a rapid pace. In fact, some experts feel that the pace has been too quick and that computer models of option and derivative pricing (basic optimal-stopping problems) are the roots of the current economic crisis. But it is not the theory that is at fault. Like Steven Shreve of Carnegie Mellon University, I blame the decision makers' blind trust in computer model predictions. In fact, new ideas and discoveries in optimal stopping, including better estimates of the risk that mathematical models are wrong, are exactly what we need, not only for guidance on when to stop the bailouts, for example, but also for help with multiple other monumental problems, including when to stop using fossil fuels or stockpiling nuclear weapons.

Note

1. There is a math error in the last example, which appeared in the original, and which I left as in the original. The error was discovered and corrected by Professors Medina and Zeilberger (Rutgers), and appears online in the Amer Sci article. http://www.americanscientist.org/issues/feature/2009/2/knowing-when-to-stop/1

Bibliography

* Boshuizen, F., and T. Hill. 1992. Moment-based minimax stopping functions for sequences of random variables. *Stochastic Processes and Applications* 43:303–316.
* Bruss, F. T. 2000. Sum the odds to one and stop. *Annals of Probability* 28:1384–1391.
* Chow, Y., H. Robbins and D. Siegmund. 1991. *Great Expectations: The Theory of Optimal Stopping.* Mineola, N.Y.: Dover Publications.
* Cover, T. 1987. Pick the largest number, in *Open Problems in Communication and Computation,* eds. Cover, T., and B. Gopinath. New York: Springer-Verlag, p. 152.
* Dubins, L., and L. Savage. 1976. *Inequalities for Stochastic Processes: How to Gamble If You Must.* New York: Dover Publications.
* Ferguson, T. 1989. Who solved the secretary problem? *Statistical Science* 4:282–296.
* Freeman, P. R. 1983. The secretary problem and its extensions: A review. *International Statistical Review* 51:189–208.
* Hill, T., and U. Krengel. 1992. Minimax-optimal stop rules and distributions in secretary problems. *Annals of Probability* 19:342–353.
* Hill, T., and D. Kennedy. 1992. Sharp inequalities for optimal stopping with rewards based on ranks. *Annals of Applied Probability* 2:503–517.
* Jacka, S., et al. 2007. Optimal stopping with applications. *Stochastics* 79:1–4.
* Peskir, G., and Albert Shiryaev. 2006. *Optimal Stopping and Free Boundary Problems.* Boston: Birkhäuser.
* Samet, D., I. Samet and D. Schmeidler. 2004. One observation behind two-envelope problems. *American Mathematical Monthly* 111:347–351.
* Shreve, S. 2008. Don't blame the quants. Forbes.com. http://www.forbes.com/2008/10/07/securities-quants-models-oped-cx_ss_1008shreve.html.
* Stadje, W. 2008. The maximum average gain in a sequence of Bernoulli trials. *American Mathematical Monthly* 115:902–910.

Homology:
An Idea Whose Time Has Come

Barry A. Cipra

An old rap against biologists is that they chose a science as far from mathematics as they could get—only to wake up in an age of computer models and bioinformatics. Following a similar path of least resistance, stereotypical applied mathematicians steered clear of a subject called topology; to the extent that this is so, they could be in for a similar slow but rude awakening. In an invited presentation on sensor networks and topology at this year's SIAM Conference on Applications of Dynamical Systems, Robert Ghrist of the University of Pennsylvania showed how the softer side of geometry is infiltrating the analysis of complex systems.

In particular, Ghrist predicts that coming generations of applied mathematicians and other computational scientists and engineers will embrace a term now rarely uttered outside esoteric academic specialties: homology.

It's happened before, Ghrist points out. In the 1960s, differentiable manifolds were primarily the purview of pure theorists. By the 1980s, applied mathematicians were writing books on manifolds, stable and unstable. Nowadays, it's often engineers who wax eloquent on the nature of hetero- and homoclinic points. It's hard to escape an idea whose time has come.

Broadly speaking, homology is a mathematical mechanism for collating individually meaningless, local data into useful, global information. A famous example concerns triangulated surfaces: If each vertex of a triangulation (of a compact surface without boundary, to be precise) reports its "degree"— the number of edges emanating from it—then the number of vertices V and the total degree D can be combined into the interesting number $V - D/6$, which turns out to depend only on the surface, and not the specific triangulation. This number, known as the Euler characteristic, relates to the number of "holes" in the surface. (The formula for the Euler characteristic is more generally given as $V - E + F$, where E and F are the numbers of edges and

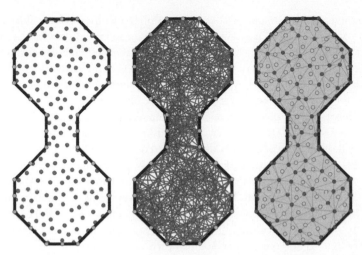

FIGURE 1. Sleep Mode. When sensors are scattered about a region (left), a homology computation can determine whether they provide complete coverage simply from knowledge of which sensors are within each other's sensing radius (center). The homology computation also suggests sensors that can be turned off without loss of coverage (right). From V. de Silva and R. Ghrist, "Coordinate-free Coverage in Sensor Networks with Controlled Boundaries via Homology," *International Journal of Robotics Research*, Vol. 25, No. 12, 2006, pages 1205–22.

faces when the surface is cut into bite-sized chunks, not necessarily triangles. When all chunks are triangles, $E = D/2$ and $F = D/3$.)

Collecting local data is what sensors do—even a telescope registers only the photons that enter its aperture. Sensors have become ubiquitous: You can't go into a public restroom in most places without the faucets, hand driers, and toilets taking note of your presence. Increasingly, sensors are networked, so that the left hand might know what the right hand is doing. And that's where things become potentially topological.

The standard engineering paradigm of sensor technology envisions more or less complete and exact knowledge of the locations of sensors and the data streams they are built to provide, with all this information centrally processed and controlled. Indeed, for certain sensor systems—say the traffic-detector loops put in place to coordinate rush-hour gridlock—such centralized planning makes sense. But this assumes essentially unlimited communications and computing power. What happens when sensors are constrained by cost or uncertainty?

Overlapping scraps of data can be stitched together to form a crazy quilt of insight. Take, for example, a network of randomly placed sensors, each able to identify any fellow sensors within some reasonable radius of its (unknown) position and to count the number of other nearby objects of interest, generically known as "targets." Ghrist and colleagues have shown how tools of topology, including souped up versions of the Euler characteristic, can answer a variety of such questions: Does the network cover the entire domain of interest? Are there obstacles, such as walls, that limit some sensors' view? How many "targets" are there altogether?

The technical answers depend, of course, on additional technical assumptions. In a 2006 paper that appeared in the *International Journal of Robotics Research*, Ghrist and Vin de Silva of Pomona College considered the coverage problem for polygonal domains in $\mathbb{R}^2$, with one set of "fencepost" sensors at the vertices of the polygon and another batch scattered about the interior. Their main result is that complete coverage of the domain can be verified by finding a certain nontrivial homology class. One useful corollary has to do with cost-cutting: Once you compute which sensors correspond to the crucial homology class, you can turn off or ignore the others. Alternatively, if you compute more than one representative of such a class, you've got yourself a fault-tolerant backup system. (See illustration.)

But how computable are such things? Topology has a reputation for being abstract. Indeed, the theory has a tendency to wander off into a thicket of sub- and superscripted notation, intricate terminology, and formulas that fill the better part of a blackboard. But linear programming, and linear algebra generally, have the same tendency. In fact, the whole point of algebraic topology is to provide an algebraic toolbox for objects whose plastic nature seemingly defies any effort to pin them down.

Homology groups in particular are eminently computable. As a postdoc with the computational topology group at Stanford University in the early 2000s, de Silva developed Plex, a suite of MATLAB routines for studying simplicial complexes, which, in rough algebraic-topology-speak, are networks. (The simplest kind of simplex is a clique within a graph, e.g., a tight grouping of sensors, each aware of the others' existence.) Plex, now in its third (Java) version, is available at the Stanford computational topology group's website, comptop.stanford.edu.

The extensive machinery of algebraic topology is giving computational scientists plenty of new tools to work with, Ghrist says. It has also kept them busy reading the instruction manuals. Applying homology is just step one. Wait until you get to the part on applied sheaf theory.

Mathematics Education

Adolescent Learning and Secondary Mathematics

ANNE WATSON

Introduction

To confirm the deepest thing in our students is the educator's special privilege. It demands that we see in the failures of adolescence and its confusions, the possibility of something untangled, clear, directed.
(Windle, 1988)

In this paper I develop my thinking about how learning secondary mathematics can relate to the adolescent project of negotiating adulthood. All too often it does not, yet the same kinds of adolescent autonomous thinking which so often lead to disaffection and rejection, not only of mathematics but of school and life more generally, can be embedded and enhanced positively within the teaching and learning of mathematics. I suggest parallels, and some kinds of tasks which enhance the adolescent project through mathematics, and mathematics through adolescence.

My personal rationale for taking this view lies in my work as head of a mathematics department during the late 80s and early 90s in a school which served a socio-economically disadvantaged area. In this school, academic achievement was generally low. Nevertheless, our mathematics results were usually the best of all subjects, sometimes ahead and sometimes just behind those achieved in creative arts, and nearly 100% of the cohort, including persistent absentees, achieved some kind of mathematics grade. I take little pride in this, however, because the teaching methods we used, while appealing to the students and to us at the time, did not enable students to do as well as those at nearby middle class schools. While students were learning *some* mathematics, we nurtured and depended on their powers of exploration, application, and their natural enquiry to achieve success. We were praised for this in some quarters, because we had developed ways of working on mathematics *in* school which were similar to the ways in which quantitative problems arise *out* of school. However, we did not enable the majority of students to contact the essentially abstract, structural, understandings which characterise the subject in its entirety. We did little to help them engage out of school with new ways of thinking. In Vygotsky's terms, we failed to support them in engaging with the scientific concepts which can only be deliberately taught (Kozulin, 1986, p. xxxiii). Since then my work has focused on how students with low achieving backgrounds can engage with mathematical thinking, without limiting them to merely being better at workaday thinking in mathematical contexts (Watson, 2006).

Adolescent Concerns

There is widespread agreement that adolescents are broadly concerned with the development of identity, belonging, being heard, being in charge, being supported, feeling powerful, understanding the world, and being able to argue in ways which make adults listen (Coleman & Hendry, 1990). Naming these as adolescent concerns assumes that Western psychological understandings of adolescence can be taken to be universal, which may not be the case due to the interaction of physiological and cultural factors, but I adopt this understanding for the purposes of this paper. Adolescents engage with these concerns through interaction with peers. This was regarded by Vygotsky as the 'leading activity' of adolescence (Elkonin, 1972), that is the activity which leads the course of development but is not necessarily the only influential activity. Nevertheless, as he saw it, peer interaction is the context in which adolescents work out their relationships with others, adults, the world and themselves. They do this by employing their new-to-them ability to engage in formal-logical thinking, so that they become capable of self-analysis, and analysis of other situations, as internalisations of social consciousness develops with peers (Karpov, 2003).

This new-to-them facility is attributed by Piaget to a biological development of the cortex, yet different societies appear to engender different kinds of world view, different forms of mature argument, and different kinds of abstraction. Vygotsky (1986) recognises the role of physiological maturity, and emphasises that the use of the maturing brain is influenced in how the notion of self is worked out by relations in the social world. Thus the biological ability to understand things in more complex, abstract, ways is not the most important influence on learning. Rather, it is how adolescents use adult behaviour and interaction as mediators of their main activity with peers that influences the development of identity and self-knowledge.

School classrooms are thus very important places for adolescents, because it is there that adult behaviour dominates, while the majority present are adolescents whose behaviour and interactions are constrained by adult goals and expectations. It is an easy step from this analysis to ask for more discussion, more work in groups, and more attention to students' ideas and their social world to produce more engaged learners.

'Cognitive bullying' is not too strong a term to describe the kind of teaching which ignores or negates the way a student thinks, imposes mental behaviour which feels unnatural and uncomfortable, undermines students' thoughtful efforts to make sense, causes stress, and is repeated over time, possibly with the backing of the system or institution. To be forced to revisit sites of earlier failure by, for example, doing fractions during grades 6, 7, 8 and 9, can, with this perspective, be seen as cognitive bullying which is at best marginally productive and at worst emotionally damaging. Students can become trapped into repeated failure with no way out except to adopt negative behaviour or to accept such treatment compliantly. However, compliance does not imply mental comfort or lack of anxiety. As well as repetitious failure, much mathematics teaching involves showing students how to manipulate and adapt abstract ideas—ways of thinking which are often in conflict with their intuitive notions. Even in the most therapeutic of classrooms this requires ways of seeing and paying attention which are contrary to what the student does naturally—against intuition. Thus for some students the mathematics classroom is a site where natural ways of thinking and knowing are constantly overridden by less obvious ideas. Without time to make personal adjustments, many students give up attempts to make personal sense of what they are offered, and instead rely on a disconnected collection of rules and methods. The term 'cognitive bullying' can therefore be taken to mean that students' own ways of thinking are constantly ignored or rejected, the mathematical experiences which generate fear and anxiety are constantly revisited for repetition, and they are expected to conform to methods and meanings which they do not understand.

This state of affairs reaches a climax in adolescence, when examinations become high-stakes, major curriculum topics become less amenable to concrete and diagrammatic representations, full understanding often depends on combining several concepts which, it is assumed, have been learnt earlier, and adolescents are developing a range of serious disruption habits. Those whose thinking never quite matches what the teacher expects, but who never have the space, support and time to explore why, can become disaffected at worst, and at best come to rely on algorithms. While all mathematics students and mathematicians rely on algorithmic knowledge sometimes, learners for whom that is the only option are dependent on the authority of the teacher, textbooks, websites and examiners for affirmation. Since a large part of the adolescent project is the development of autonomous identity, albeit in relation to other groups, something has to break this tension—and that can be a loss of self-esteem, rejection of the subject, or adoption of disruptive behaviour (Coleman & Hendry, 1990, pp. 70, 155).

Enquiry Methods

Stoyanova (2007) reports on the evaluation of an extensive mathematics curriculum project in which students used tasks which encouraged enquiry, investigation, problem posing, and other features commonly associated with so-called 'reform' and enquiry methods. Results of analyzing test answers from 1600 students showed very few clear connections between aspects of the programme and mathematical achievement during adolescence. Most notably: problem-posing, checking by alternative methods, asking 'what if . . . ?' questions, giving explanations, testing conjectures, checking answers for reasonableness, and splitting problems into subproblems were associated with higher achievement in year 10, while using general problem-solving strategies, making conjectures, and sharing strategies were not. Use of 'real life' contexts was negatively associated with achievement at this level. Other interesting aspects of her results include that teachers' beliefs about learners' ability to learn turned out to be very important, a finding repeated elsewhere (e.g. Watson & De Geest, 2005), and that explicit teaching of strategies was not found to be associated with achievement, either positively or negatively.

Realistic tasks can provide contexts for enquiry and often enquiry methods of teaching and learning are recommended for adolescent learning. Historically, mathematics has been inspired by observable phenomena, and mathematicians develop new ideas by exploring and enquiring into phenomena in mathematics and elsewhere. It is also possible to conjecture relationships from experience with examples, and thus get to know about general behaviour. But mathematics is not *only* an empirical subject at school level; indeed it is not *essentially* empirical. Its strength and power are in its abstractions, its reasoning, and its hypotheses about objects which only exist in the mathematical imagination. Enquiry alone cannot fully justify results and relationships, nor can decisions be validated by enquiry alone. Many secondary school concepts are beyond observable manifestations, and beyond everyday intuition. Indeed, those which cause most difficulty for learners and teachers are those which require rejection of intuitive sense and reconstruction of new ways of acting mathematically.

School mathematics as a human activity has at least two dimensions, that of horizontal and vertical mathematisations (Treffers, 1987).

In horizontal mathematization, the students come up with mathematical tools which can help to organize and solve a problem located in a real-life situation. Vertical mathematization is the process of reorganization within the mathematical system itself, like, for instance, finding shortcuts and discovering connections between concepts and strategies and then applying these discoveries (Van den Heuvel-Panhuizen, 2007).

This shift from horizontal to vertical mathematisation has to be structured through careful task design; it does not happen automatically. A Vygotskian view would be that this kind of shift necessarily involves disruption of previous notions, challenges intuitive constructs, and replaces them with new ways of thinking appropriated by learners as tools for new kinds of action in new situations.

Grootenboer and Zevenbergen (2007) work within a social paradigm which emphasizes agency and identity, yet found that they had to go beyond Burton's analysis of professional mathematical activity to be informative about what actions learners had to make when working mathematically. In their study, learners had to identify patterns, construct generalizations, use examples to test hypotheses, and identify limits to make progress with a task. By naming these actions, they show how important it is that the intellectual demands of mathematics itself should be taken into account when thinking about teaching and learning. Many of the features of the programme evaluated by Stoyanova (2007) could be described as offering agency, but it was seen that only some of them led to improved mathematics learning. The effective features all engaged adolescents in exercising power in relation to new mathematical experiences, new forms of mathematical activity, and being asked to use these, express these, and to display authority in doing so. For example, year 10 level adolescents responded well to being given authority in aspects of mathematical work: checking answers, giving explanations, asking new questions, testing hypotheses, and problem-posing. These actions appeal to the adolescent concerns of being in charge, feeling powerful, understanding the world, and being able to argue in ways which make adults listen. They offer more than belonging by doing what everyone else is doing, or being heard merely through sharing what has already been done.

Shifts of Mathematical Action

Analyses of student errors in mathematics (e.g. Booth, 1981; Hart, 1982; Ryan & Williams, 2007) suggest that many students get stuck using 'child methods', intuitive notions, invented algorithms which depend on left-to-right reading, or misapplication of verbal tricks. When these methods do not produce the right answers to school questions, often at the start of secondary school, these could well be a contributory factor to rejection of the mathematics curriculum.

The contradictions between intuitive, spontaneous, understandings and the scientific concepts of secondary mathematics can be the beginning of the end of mathematical engagement for adolescents. If they cannot understand the subject by *seeing* what it does and how it works, but instead have to believe some higher abstract authority that they do not understand, then the subject holds nothing for them. This analysis has contributed to the belief that mathematics need not be taught to everyone and that many adolescents only need to become functionally numerate (Bramall & White, 2000). But this view misses the point. The authority of mathematics does not reside in teachers or textbook writers but in the ways in which minds work with mathematics itself (Freudenthal, 1973, p. 147; Vergnaud, 1997). For this reason mathematics, like some of the creative arts, can be an arena in which the adolescent mind can have some control, can validate its own thinking, and can appeal to a constructed, personal, authority. But to do so in ways which are fully empowering has to take into account the new intellectual tools which simultaneously enable students to achieve in mathematics, and which develop further through mathematics (Stech, 2007). To understand this further I present key intellectual tools of the secondary curriculum as illustrations of what needs to be appropriated in order to engage with new kinds of mathematical understanding. Unsurprisingly, these turn out to be aspects which cause most difficulty and my argument is that it is likely that many teachers do not pay enough attention to these shifts as inherently difficult. Not only do these shifts represent epistemological obstacles, in Brousseau's terms, but they are also precisely those changes to new forms of action which constitute the scientific knowledge of mathematics—that which can only be learnt at school. It is inequitable to expect students to bring their everyday forms of reasoning to bear meaningfully on mathematical problems, when everyday forms do not enable these shifts to be made:

> Shift from looking for relationships, such as through pattern seeking, to seeing properties as defined by relationships;
> Shift from perceptual, kinaesthetic responses to mathematical objects to conceptual responses; and a related shift from intuitive to deductive reasoning;
> Shift from focusing on procedures to reflecting on the methods and results of procedures;

Shift from discrete to continuous ways of seeing, defining, reasoning
and reporting objects;
Shifts from additive to multiplicative and exponential understandings
of number;
Shift from assumptions of linearity to analysing other forms of
relationship;
Shift from enumeration to non-linear measure and appreciating
likelihood;
Shift from knowing specific aspects of mathematics towards relationships and derivations between concepts.

A shift towards seeing abstract patterns and structures within a complex
world is seen by many psychologists, not only those influenced by Piaget but
also the Vygotskian school, as typical of adolescent development (Coleman &
Hendry, 1990, p. 47; Halford, 1999). One of the ways in which these two
schools of thought differ is that Piaget appears to be saying that this shift
happens biogenetically, and new forms of learning follow. In mathematics,
however, it is common at all levels of competent study to move fluently
between concrete, diagrammatic and abstract approaches, and between examples and generalisations as appropriate, for the exploration being undertaken. Vygotsky's (1986) approach is that learners are capable of developing
abstract ideas but need interaction with expert others to achieve this for
themselves. He recognised that adolescence is a particularly appropriate
time for conversations and scaffolding towards abstraction to take place, and
that the biological and physical changes which occur at this age also relate to
making sense of self in society and self in relation to ideas (p. 107).

Shifts of Mathematical Action Compatible with Adolescence

Over three projects, IAMP, CMTP, and MkiTeR[1], I have observed many lessons with engaged adolescents in which a central feature appears to be the
introduction and use of new intellectual tools which reflect new-to-them
forms of mathematical action. For example, in one of the IAMP classrooms
a teacher would sometimes be very explicit about shifts: '*You have been doing
adding for years; we have to change to thinking about multiplication*'.

I shall now give some examples of tasks which generate and nurture
mathematical identity in adolescence, while staying focused on the secondary mathematics curriculum as the locus of new forms of thinking. The
point of showing these tasks is to demonstrate that, while exploratory and
realistic learning environments have much to offer adolescents, it is also
possible to structure short, tightly-focused, curriculum-led tasks in ways

that lead directly to higher levels of engagement and also employ the social and emotional modes of working that are widely desired.

Learner-Generated Examples

Students in a lesson were familiar with multiplying numbers and binomials by a grid method:

X	20	7
50	1000	350
9	180	63

X	z	+3
2z	$2z^2$	6z
−1	−z	−3

They had been introduced to numbers of the form $a \pm \sqrt{b}$. The teacher then asked them to choose pairs of values for a and b, and to use the grid method to multiply such numbers to try to get rational answers. Students worked together and began to explore. At the very least they practised multiplying irrationals of this form. Gradually, students chose to limit their explorations to focus on numbers like 2 and 3 and, by doing so, some realised that they did not need explicit numbers but something more structural which would 'get rid of' the roots through multiplication. Although during the lesson none found a way to do this, some carried on with their explorations over the next few days in their own time (Watson & Shipman, in press).

Tasks in which students gain technical practice while choosing their own examples, with the purpose of closing in on a particular property, relationship, or class of objects, can be adapted to most mathematical topics, and when the constraints in such tasks are incorporated into the goal, learners have to engage with new mathematical ideas.

Another and Another . . .

The following task-type also starts from learners' examples:

> Ask students to give you examples of something they know fairly well; then keep asking for more and more until they are pushing up against the limits of what they know.

> e.g. Give me a number between zero and a half; and another; and another . . .

> Now give me one which is between zero and the smallest number you have given me; and another; . . . and another . . .

Each student works on a personally generated patch, or in a place agreed by a pair or group. Teachers ensure there are available tools to aid the generation—in this case some kind of 'zooming-in' software, or mental imagery, would help.

This approach recognises learners' existing knowledge, and where they already draw distinctions; it then offers them opportunity to add more things to their personal example spaces, either because they have to make new examples in response to your prompts, or because they hear each other's ideas (Watson & Mason, 2006). Self esteem comes at first from the number of new examples generated, then from being able to describe them as a generality, and finally from being able to split them into distinct classes. This is not merely about 'sharing examples' but about adopting them and making claims about them; taking note of peers' contributions, mediated by an adult, and enhancing self-knowledge through this process.

Putting Exercise in Its Place

If getting procedural answers to exercises in textbooks is the focus of students' mathematical work (whether that was what the teacher intended or not) then shifts can be made to use this as merely the generation of raw material for future reflection. Many adolescents have their mathematical identities tied up with feeling good when they finish such work quickly, neatly and more or less correctly; others reject such work by delaying starting it, working slowly, losing their books and so on; still more can produce good-looking work which shows little understanding, their self-esteem tied up with form rather than function. Restructuring their expectations is, however, easy to do if new kinds of goal are explicated which expect reflective engagement, rather than finishing, so that new mathematical identities can develop which are more in tune with the self-focus of early adolescence while requiring new forms of action, reflection on results and processes.

Examples of different ways to use exercises are:

Do as many of these as you need to learn three new things; make up examples to show these three new things.

At the end of this exercise you have to show the person next to you, with an example, what you learnt.

Before you start, predict the hardest and easiest questions and say why; when you finish, see if your prediction was correct; make up harder ones and easier ones.

When you were doing question N, did you have to think more about: method, negative signs, correct arithmetical facts, or what? Can you make up examples which show that you understand the method without getting tied up with negative signs and arithmetic?

Rules versus Tools

Student-centred approaches often depend on choice of method, and this, of course, celebrates autonomy. However, mathematics is characterised by, among many other things, variation in the efficiency and relevance of methods. For example, 'putting a zero on the end when multiplying by ten' and 'change side-change sign' are fine so long as you know when to do these—and mathematicians do not abandon these ways of seeing. Rather than it being a *rule* it becomes a *tool* to be used when appropriate. Adolescents often cannot see why they should be forced to abandon methods and behaviour which have served them well in the past (repeated addition for multiplication; guessing and checking 'missing numbers'; and so on) to adopt algorithms or algebraic manipulations. On the other hand, they often choose autonomously to abandon past behaviour in the service of new goals. One way to work on this is to give a range of inputs and show students that they can decide which of their methods works best in which situation, and why. This leads to identifying methods which work in the greatest range of cases, and the hardest cases. These 'supermethods' need to be rehearsed so that they are ready to use when necessary, and have the status of tools, rather than rules; empowering rather than oppressive. One teacher I have observed calls these 'bits of technology' to emphasise that appreciating their usefulness may be delayed.

Abstract Mathematics and Adolescence

In this paper, I have advanced the idea that ways can be devised to teach all adolescents the scientific conceptualizations, and methods of enquiry, which characterise hard mathematics. I suggest that these cohere with and enhance many features of adolescent development. Moreover this can be achieved without resorting to cognitive bullying which is counterproductive and alienating, because the epistemological changes of activity embedded in mathematics are similar to the ways in which adolescents learn to negotiate with themselves, authority, and the world. Agency and identity do not have to be denied, but neither does abstract mathematics have to get lost in the cause of relevance and personal investigation.

In all of the above task-types, students create input which affects the direction of the lesson and enhances the direction of their own learning.

Classrooms in which these kinds of task are the norm provide recognition and value for the adolescent, a sense of place within a community, and a way to get to new places which can be glimpsed, but can only be experienced with help. To use the 'zone' metaphor—these tasks suggest that mathematical development, relevance, experience and conceptual understanding are all proximal zones, and that moves to more complex places can be scaffolded in communities by the way teachers set mathematical tasks.

Notes

1. Improving Mathematics Attainment (Esmee Fairbairn Grants 01-1415, 02-1424), Changes in Mathematics Teaching (Esmee Fairbaim Grant 05-1638) and Mathematics Knowledge in Teaching e-Research (John Fell Fund). The views expressed in this paper are mine and not those of the grantors.

Bibliography

Booth, L. (1981). Child-methods in Secondary Mathematics. *Educational Studies in Mathematics*, *12*(1), 29–41.

Bramall, S., & White, J. (Eds.). (2000). *Why Learn Maths?* London: Institute of Education.

Coleman, J., & Hendry, L. (1990). *The Nature of Adolescence* (3rd ed.). London: Routledge.

Elkonin, D. (1972). Toward the Problem of Stages in the Mental Development of the Child. *Soviet Psychology*, *10*, 225–251.

Freudenthal, H. (1973). *Mathematics as an Educational Task*. Dordrecht: Kluwer.

Grootenboer, P., & Zevenbergen, R. (2007). Identity and Mathematics: Towards a Theory of Agency in Coming to Learn Mathematics. In J. Watson & K. Beswick (Eds.), *Proceedings of the 30th Annual Conference of the Mathematics Education Research Group of Australasia* (pp. 335–344). Hobart, Tasmania.

Halford, G. (1999). The Development of Intelligence Includes the Capacity to Process Relations of Greater Complexity. In M. Anderson (Ed.), *The Development of Intelligence* (pp. 193–213). Hove, UK: Psychology Press.

Hart, K. (Ed.). (1982). *Children's Understanding of Mathematics 11–16*. London: John Murray.

Karpov, Y. (2003). Development Through Lifespan: A Neo-Vygotskian Approach. In A. Kozulin, B. Gindis, V. Ageyev, & S. Miller (Eds.), *Vygotsky's Educational Theory in Cultural Context* (pp. 138–155). Cambridge: Cambridge University Press.

Kozulin, A. (Ed.). (1986). Vygotsky in Context. In *Thought and language* (pp. xi–1xi). Cambridge, MA: MIT Press.

Ryan, J., & Williams, J. (2007). *Children's Mathematics 4–15: Learning from Errors and Misconceptions*. Maidenhead: Open University Press.

Stech, S. (2007). School Mathematics as a Developmental Activity. In A. Watson & P. Winbourne (Eds.), *New Directions for Situated Cognition in Mathematics Education* (pp. 13–30). New York: Springer.

Stoyanova, E. (2007). *Exploring the Relationships Between Student Achievement in Working Mathematically and the Scope and Nature of the Classroom Practices*. Paper presented at AAMT Conference (pp. 232–240). Hobart, Tasmania.

Treffers, A. (1987). *Three Dimensions: A Model of Goal and Theory Description in Mathematics Education: The Wishobas Project.* Dordrecht: Kluwer.

Van den Heuvel-Panhuizen, M. (1998). *Realistic Mathematics Education: Work in Progress.* Retrieved August 8, 2008, from http://www.fi.uu.nl/en/rme/.

Vergnaud, G. (1997). The Nature of Mathematical Concepts. In T. Nunes & P. Bryant (Eds.), *Learning and Teaching Mathematics: an International Perspective.* London: Psychology Press.

Vygotsky, L. (1978). *Mind and Society: The Development of Higher Psychological Processes.* Cambridge, MA: Harvard University Press.

Vygotsky, L. (1986). In A. Kozulin (Ed.), *Thought and Language.* Cambridge, MA: MIT Press.

Watson, A., & Shipman, S. (2008). Using Learner Generated Examples to Introduce New Concepts. *Educational Studies in Mathematics,* 69(2), 97–109.

Watson, A., & De Geest, E. (2005). Principled Teaching for Deep Progress: Improving Mathematical Learning Beyond Methods and Materials. *Educational Studies in Mathematics, 58,* 209–234.

Watson, A., & Mason, J. (2006). *Mathematics as a Constructive Activity.* Mahwah, NJ: Erlbaum.

Watson, A. (2006). *Raising Achievement in Secondary Mathematics.* Maidenhead: Open University Press.

Windle, B. (1988). In E. Perkins (Ed.), *Affirmation, Communication and Cooperation: Papers from the QSRE Conference on Education, July 1988.* London: The Religious Society of Friends.

Accommodations of Learning Disabilities in Mathematics Courses

KATHLEEN AMBRUSO ACKER,

MARY W. GRAY, AND BEHZAD JALALI

The requirement of the No Child Left Behind Act (NCLB) [27] that measures of academic progress be disaggregated by groups has renewed focus on the issue of accommodating students with disabilities. Although NCLB does not apply to postsecondary education, over the past fifteen years there has been a substantial increase in the attention directed to learning disabilities in this arena. In particular, questions have been raised by institutions of higher education as well as by testing bodies such as the College Board as to whether some recommended accommodations accomplish the purpose for which they were intended and whether they are fair to other students. However, aside from discussions in law reviews, little attention has been focused on whether the accommodations are legally required in a higher-educational setting. We address the legal framework, focusing on what constitutes a disability from a legal point of view and the nature and appropriateness of accommodations, noting where mathematics courses have been affected. Lastly, we will review suggestions for best practices in accommodating learning disabilities in the mathematics classroom in the light of legal requirements.

Legal Framework

Two federal laws affect college-level instruction: the Rehabilitation Act, Section 504 [39], which prohibits discrimination on the basis of disability by entities receiving federal funds, and the Americans with Disabilities Act (ADA) [2], which broadens the prohibition of discrimination by requiring that all services and places of public accommodation, including colleges and universities, be accessible to those with disabilities. As a result of several Supreme Court decisions narrowing the scope of protection, the ADA was amended, with changes effective from January 1, 2009. The Rehabilitation Act was also amended to conform. Some state laws may impose additional requirements, but we deal only with the federal context. The Office of Civil Rights of the U.S. Department of Education (OCR) issues letters of interpretation in response to both complaints and inquiries concerning disability laws.[1] These are intended to provide guidance; although they do not have the force of law, they are given substantial deference by courts should litigation develop.

Both the ADA and the Rehabilitation Act are broadly applicable to many aspects of higher education, but this paper concentrates on colleges and universities and to a lesser extent on testing bodies such as the College Board, the National Board of Medical Examiners, and various state boards of bar examiners.

The ADA (as amended) states that:

The term "disability" means, with respect to an individual—

(A) a physical or mental impairment that substantially limits one or more major life activities of such individual;

(B) a record of such impairment; or

(C) being regarded as having such an impairment.[2]

Focusing for the most part on (A), this paper primarily examines what constitutes a legal disability in the learning disability context and what accommodations may be required.

To begin, the statutory definition of "substantially limited" is not particularly helpful in that it equates "substantially limited" to the term "materially restricted" without further explanation or clarification. However, the amendments of 2009 do define disability more broadly than courts had done in the recent past. The amended ADA states that "major life activities include, but are not limited to, caring for oneself, performing manual tasks, seeing, hearing, eating, sleeping, walking, standing, lifting, bending, speaking, breathing, learning, reading, concentrating, thinking, communicating and working" [2].

The standards of measurement of impairment are also altered by the amendments, which specifically state that whether one is impaired should be judged without respect to any amelioration. Thus, for example, one whose diabetes is controlled by medication would still be considered disabled for the purposes of protection under the ADA.[3] Whether the impairment must be in comparison to the average person's ability or to that of a person of similar skills and training has been an issue. For example, should a mathematics graduate student's reading ability be compared to that of an average member of the general public or to that of other mathematics graduate students? The courts have generally adopted an average person standard.[4]

"Learning disabilities" are generally defined to be specific difficulties in learning when the student is generally of average or above-average intelligence [47]; that is, the student's performance on some aspect of learning is substantially below what would be expected at a given age and IQ. Exactly what this means has been the subject of much controversy, not to mention litigation, but for the purpose of discussion we accept this characterization. Note also that Attention Deficit Disorder (ADD) and Attention Deficit Hyperactivity Disorder (ADHD) are conditions that some label learning disabilities and others denominate as a separate category of disability that may be eligible for accommodation. For the purposes of this paper both conditions are considered learning disabilities. Learning disabilities affect education at all levels and in all subjects; in particular, dyscalculia, dyslexia, ADD, and ADHD create difficulties for students to understand and apply mathematics appropriately.

Under the ADA and the Rehabilitation Act, any "otherwise qualified" disabled person is entitled to "reasonable" accommodations in order to provide access to education. As might be expected, what "otherwise qualified" and "reasonable" mean has been the subject of much litigation. However, more basic is whether a "learning disability" is a disability for purposes of either act. If not, post-secondary accommodations are not legally mandated, although, of course, a college or university may choose to offer them. It is also important to understand that in order for an adverse action (such as dismissal from a program) to violate the ADA or Rehabilitation Act, there must be a causal connection between it and the disability or perceived disability [4]. For example, many court decisions have noted that a student's failure or dismissal resulting from an inability to meet academic standards even with reasonable accommodations does not constitute discrimination on the basis of a disability.

There are some who contend that "learning disability is merely a subjective social construct that is inherently tied to underlying politics," and indeed there is some disagreement about how learning disabilities are characterized [21], [47]. There has been some movement away from the traditional "discrepancy" measurement of learning disability to one that assesses the struggling student's response to high-quality general education instruction or focuses on an absolute low level of achievement. Such a standard would shift from individual identification to a rulelike process that would not have a "bright" child who performs at a mediocre level classified as learning disabled. More important, however, for the present discussion is the tendency of the courts to declare that under some circumstances "learning disabilities" may not be disabilities for the purpose of the ADA and the

Rehabilitation Act and the effect the amendments to the ADA might have on this trend.

Higher Education vs. K–12 Requirements

In the college setting many difficulties result from failures to distinguish what might have been required at the K–12 level and what is required in the postsecondary context under a different legal framework [26]. Students in K–12 education are protected under the Individuals with Disabilities Education Act (IDEA) [19], whose provisions are designed to guarantee successful outcomes for the disabled, whereas the Rehabilitation Act and the ADA are focused on guaranteeing access. Thus, in K–12 education students with disabilities are entitled to an Individualized Education Program (IEP) developed jointly by their teachers and special education professionals and in consultation with their parents. The goal of an IEP is to assure that the student has a chance to achieve academically in an education setting commensurate with his abilities. The proportion of children classified as learning disabled in the K–12 system has grown enormously in the past several decades, as evidenced by the fact that one in eleven college freshmen self-identify as having a disability, over 50 percent of which are described as "learning disabilities," an increase by as much as threefold over the past twenty years [41]. As a result, IEP beneficiaries have come to expect similar accommodations in college and indeed on licensing exams such as those required for students seeking to become physicians or lawyers. There has been a great deal of litigation about the sufficiency of school districts' individualized plans and also a great deal of controversy about learning disability classification being "affirmative action" for the middle class, as parents and students seek accommodations to secure advantages for their children [23], [42].

At the K–12 level, schools have the responsibility to identify students' disabilities as well as to work cooperatively to develop appropriate accommodations. At the postsecondary level it is up to students to make a focused request to the appropriate administrative office to receive accommodations for their disabilities and to provide appropriate documentation supporting such requests for accommodations [6].

Institutions of higher learning are also constrained by the Health Insurance Portability and Accountability Act of 1996 (HIPAA),[5] which specifically prohibits medical information to be included as part of an educational record, and the Family Education Rights and Privacy Act (FERPA), an act designed to protect the privacy of a student's educational records. "Educational records" as defined by FERPA are records "(1) directly related to a student; and (2) maintained by an educational agency or institution or by a party acting for the agency or institution."[6] Furthermore, the definition explicitly excludes inclusion of records made by physicians, psychiatrists, or psychologists. In other words, records from specialists who diagnose and treat learning disabilities cannot be included as part of the student's education records. Thus, even if the institution requires documentation from specialists to substantiate a student's request for accommodation, this information cannot be shared directly with faculty members. To facilitate open discourse between faculty and students learning to be self-advocating, administrations at many postsecondary schools provide students with appropriately documented learning disabilities a letter detailing what accommodations the student needs. Thus, schools do not contact faculty on behalf of the students, and it is the responsibility of the student to disclose to faculty, on an as-needed basis, the need for accommodations.

Major Life Activity

A fundamental question, one which has not always been addressed by the courts and hardly at all by colleges and universities, is to what extent a documented learning disability might limit an activity and whether that activity is a major life activity.

A key Supreme Court case, Sutton v. United Air Lines [45], involved twin sisters who applied to be pilots. Although they were severely nearsighted, corrected by glasses their vision was 20-20; nonetheless, they were denied employment based on their eyesight. The Supreme Court decided that with the accommodation of glasses they were not disabled and hence were not entitled to the protection of antidiscrimination legislation. This opened the door to a series of decisions greatly limiting the scope of disability protections, an outcome that the 2009 amendments were enacted to reverse. Nonetheless, ultimately the amendments would most likely not have assisted the sisters, because the airline could very likely justify the requirement for 20-20 uncorrected vision as business related by citing what might happen in an emergency situation to the glasses or lenses on which a pilot was relying.[7]

The Supreme Court further limited the scope of the ADA and the Rehabilitation Act in Toyota v. Williams [46] when it overturned a lower court decision that had found that a woman who was unable to perform specific tasks in one job was entitled to reassignment to another job that she was able to carry out successfully. The court said that the specific set of tasks was not a major life activity and hence her limitation in performing them was not covered by the ADA. This is the sort of result that the amendments to the ADA seek to reverse.

Until the Sutton and Toyota cases the courts had generally assumed that the plaintiff was disabled and then examined whether or not there had been discrimination on the basis of that disability, in particular whether reasonable accommodations had been made (see, e.g., [31], [52]). Subsequently the courts have engaged in detailed individual assessments of whether a "major life activity" was implicated and, if so, whether it was "substantially limited". This is likely to continue under the amended statute, albeit under relaxed standards [37].

Applying this back to an educational setting, we see that there are many cases that examine accommodations where the courts have declared that the fact that the examinees have done as well as they have by getting into college, law school, or medical school with accommodations demonstrates that they are not disabled. One example is the case of McGuiness v. University of New Mexico School of Medicine [25]. Essentially McGuiness had completed several degrees, including a bachelor of science in chemistry and biology as well as a doctorate in psychology, prior to entering medical school. Throughout his educational experience McGuiness worked through anxiety in taking chemistry and mathematics classes without accommodations. However, his medical school grades did not meet the standards set by the medical school. The courts determined that his inability to satisfy medical school requirements did not substantially limit the broader life activity of "working", as it excluded him only from certain limited types of jobs. Furthermore, the court held that his claim of anxiety did not meet the definition of disability under either the ADA standard or the Rehabilitation Act.

In a licensing exam case at the district court level, Price v. National Board of Medical Examiners [35], the "learning" involved was declared to be a "major life activity", but since the plaintiff's prior academic record, achieved without accommodation, was above average, the court held that Price was not disabled and hence not entitled to protection. Whether these cases would have been decided differently under the 2009 amendments is not entirely clear, but it would appear that they would not have been, since neither of the plaintiffs would have met the broadened definition of disability.

Although the amended ADA leaves open the ability of the courts to find that those who have achieved an advanced level without accommodation are not disabled and hence not entitled to the particular accommodation they might now seek, it augurs a different outcome for those who have achieved their current status with accommodations. In particular, they cannot now be deemed not disabled because the ameliorating effects of accommodations have reduced or eliminated any impairment they might otherwise experience. In other words, they are entitled to continued assistance in the form of appropriate accommodations, but it may be questionable whether the impairment they face without accommodations constitutes a limitation in a major life activity.

The underlying issue not directly addressed by the ADA amendments is, as noted above, whether when deciding that a person is impaired, should the comparison be with the average person or with a group of peers (e.g., other students in the same academic program) [4], [35], [44]. Anyone who has progressed to taking a bar exam or even being admitted to college, for example, might on the basis of that achievement be held to be not impaired in the major life activity of "learning" or of "working" since most jobs certainly do not require passing a bar exam and many do not require a postsecondary education. Thus a student could be able without assistance to perform at the level of an "average person" and hence not be considered legally disabled, but not be successful in a program for law students, medical students, or other professionals.

In the case of Singh v. George Washington University (GWU) [44], Carolyn Singh was dismissed from George Washington's medical school due to poor academic progress, primarily on timed multiple-choice exams. Singh's suit against GWU asserted that they did not accommodate her claimed learning disability. After a complicated pair of rulings in the lower court, the appellate court remanded the case for consideration of whether Singh was legally disabled, mandating the "average person" rather than other medical students as the appropriate standard for comparison. It also found that test taking is not in and of itself a major life activity but rather a component of "learning," so that the determination of impairment should be with regard to the totality of "learning."

Otherwise Qualified

The concept of "otherwise qualified" is intertwined in a complicated way with the notion of "reasonableness" of accommodations. Although others have attempted to identify gaps in existing law and to consider them in the context of real-world implementation by educators and test administrators [41], we confine our discussion to the existing situation and what may develop under the newly amended ADA and Rehabilitation Act.

A person with a disability who can perform the "essential functions" of a job or meet the requirements for services with "reasonable accommodations" can be considered "otherwise qualified." Therefore, if students are unable to make satisfactory progress with reasonable accommodations, it would appear that they are not otherwise qualified, the case of a blind person seeking employment as a bus driver being an extreme example. As noted above, there are many cases involving medical students who do well with or without accommodations until they reach the clinical stage of their training. Then, in spite of repeated accommodations having to do with stretched-out scheduling of clinical rotations, repeated attempts at exams, special supervision and other adjustments to their program, they are unable to achieve a standard acceptable to their institutions. In Falcone v. University of Minnesota [11, p. 160] the court said "the University is not required to tailor a program in which Falcone could graduate with a medical degree without establishing the ability to care for patients." Similarly, in Powell v. National Board of Medical Examiners and the University of Connecticut [34], the court determined that the plaintiff was not entitled to the protection of the ADA, as even with many accommodations she could not do the required clinical rotations.

In a relatively early lower court case, Pandazides v. Virginia Board of Education [29], the court said that an accommodation must not "fundamentally alter the measurement of the skills or knowledge the examination is intended to test." The outcome of Falchenberg v. New York State Department of Education [10] was similar. Falchenberg had received multiple accommodations concerning time requirements and a reader and a scribe on an exam required for employment as a New York State teacher; however, Falchenberg, a dyslexic, was required to spell and punctuate correctly on her own. Falchenberg felt that this stipulation did not amount to a reasonable accommodation for her disability. Since proper grammar and spelling were integral components of the exam,[8] the court found that Falchenberg was looking for an accommodation that would fundamentally alter the purpose of the exam and thus ruled in favor of the Department of Education. In other words, even with reasonable accommodations there was no way Falchenberg could be considered otherwise qualified.

Appropriate Accommodations

If a student is not "disabled" under the terms of the law, academic judgment as to what is an "appropriate accommodation" may not have legal significance. However, under the amended ADA, courts are more likely, but not certain, to classify a traditional "learning disability" as a legal disability. Moreover, universities may choose to offer accommodations to those with learning disabilities despite the question of legal obligation. As stated earlier, OCR's stance on treatment of disabled students is given deference should litigation develop.

The test for reasonableness of requested accommodations rests either on whether the accommodation was administratively or financially burdensome or on whether it required a fundamental alteration in an educational program. The issue of administrative or financial burden has arisen in part because, unlike at the K–12 level, where the school system is responsible for providing sufficient services to ensure an appropriate education, in higher education the student assumes responsibility. It is generally the case that post-secondary students must provide documentation by professionals at their own cost. This potentially creates a barrier for low-income but undiagnosed learning-disabled students who struggle to achieve academically. In addition, learning-disabled students often bear the cost burden of private tutors, although many schools provide free tutoring support for all students by way of open labs.

In determining whether a requested accommodation is reasonable, the totality of circumstances must be considered [53]. Administrative or financial burdens may also arise concerning such accommodations as adjustments in exam schedules, provisions for conducting an oral exam, and developing alternative exam forms. Generally, however, the question of the burden has not been as significant as has whether the proposed accommodation would alter the course or the program to the extent that it is unrecognizable. Thus the key to the determination of the appropriateness of an accommodation is usually whether the requested accommodation requires a fundamental alteration in an academic program or lowers standards, in which case the accommodation is not required.[9]

A case often cited as validating the accommodation of learning-disabled students, Guckenberger v. Boston University [17], is worth examining. The provost of the university became concerned about the substantial increase in the number of learning-disabled students receiving accommodations, some of which he felt were inappropriate. After some intemperate remarks from the provost characterizing learning-disabled students as slackers and instituting a requirement for revalidation of documentation of learning disabilities under new standards, a suit was filed by students seeking accommodations, asking for, among other things, a waiver of any courses in mathematics and foreign languages required for graduation.

Although few examples of requests for waiver of the mathematics requirement and no documentation of the waiver being granted were presented, the mathematics department had agreed to allow students to choose an alternative course to meet the requirement. Among the possible substitutes listed were Anthropology of Money, Economics of Less Developed Regions, and Introduction to Environmental Science. Instead of a foreign language students were permitted to take such courses as African Colonial History and Arts of Japan. The court held that the plaintiffs had failed to present scientific evidence that any learning disability was sufficiently severe to preclude sufficient proficiency with appropriate accommodations short of substitution of courses. In the case of the foreign language requirement the court opined that no course in English could substitute fully for the foreign language requirement but that the university had not established the essentiality of such a requirement in the degree program at issue. The actual result of the case was to remove the authority to decide on accommodations from the provost, to modify the documentation of disabilities requirements, and to establish a faculty committee to study the issue of whether eliminating the foreign language requirement would constitute an undesirable alteration in the academic program leading to a degree in the Boston University College of Liberal Arts.[10]

The Boston University case highlights the significance of careful consideration by a university as to what accommodations are reasonable. In response to a complaint at another institution, the Office of Civil Rights reviewed the situation of a woman with severe dyscalculia who enrolled in a mathematics course that had been required by her choice of major. Despite using all services available to her through the college, she failed the course. She then petitioned the college to take a course substitution (an option unavailable to her) or to waive the mathematics requirement. The petition was denied and the student was told to retake the course. An investigation by OCR found that the college did not consider the course substitution as a possibility because their policy on course substitutions was undeveloped and in general course substitutions were not granted. OCR determined that this lack of consideration for this sort of academic adjustment was a violation of ADA. They also found that the school did not present evidence why the mathematics class was an essential requirement of the course of study, nor was there evidence of a collegiate dialogue debating whether granting the course substitution would then be considered a fundamental alteration of the program of study. OCR also noted, "Absolute rules against any particular form of academic adjustment or accommodation are disfavored by the law."[11] The view of the OCR, if not necessarily of the courts, may indicate that more flexibility may be required of an institution at the undergraduate level than in graduate programs, particularly medicine, where the stakes of lowered standards may be greater.

For the learning disabled, the most common accommodation requested is increased time for exams, although provision of a note taker, access to a faculty member's lecture notes, oral instead of written examinations, audio or video recordings of lectures, adjustments in course loads, extension of deadlines, and an isolated place in which to take exams are also commonly prescribed. A more unusual accommodation at American University was the rescheduling of a special section of a mathematics class for learning-disabled students to 11:10 a.m. rather than the "early" hour of 9:55.[12]

In many universities, determination of what constitutes an "appropriate accommodation" is done exclusively by special education professionals in an office of disability services or similar unit.[13] While such experts have a role to play in, for example, dealing with the issue of documentation and matching accommodations to amelioration of a disability, it is not clear whether those outside the discipline in question should decide whether a requested accommodation might require a fundamental alteration in a course or program or create unfairness. For example, the University of California statement on Practices for the Documentation and Accommodation of Students with Learning Disabilities states in part:

> It is the responsibility of a Learning Disabilities Specialist, the Program Director, or other staff member designated by the Director to determine appropriate accommodations and services. This determination will be made after interviewing the student and reviewing the information furnished by the diagnosing professional(s).[14]

No mention is made of consultation with the instructors. Since privacy requirements may preclude making clear to instructors why students need accommodation, it may be difficult to formulate accommodations that ameliorate the disability without fundamentally altering the course.

Nonetheless, often accommodations are presented to the instructor of a course without consultation either as to the administrative burden or the alteration in the fundamental nature of the course or program. However, were either of these effects found to result from the proposed accommodations, the accommodations would unlikely be deemed reasonable by the courts. In general there has been great reluctance from courts to decide academic issues such as whether the nature of a course has been altered [8], [53]. Courts have been clear that an institution need not lower its standards as it defines them in order to accommodate disabilities [10], [52]. For example, the Betts court [4] noted that teachers do not have to grant accommodation requests that in their opinion substantially alter the fundamental aspects of the coursework.

Whether the academic freedom of an instructor to determine how to conduct a course absent a showing of a fundamental alteration of the course or lowering of standards trumps the requirement to make accommodations has not been tested in the courts. In a situation involving a mathematics course at the University of California, Berkeley, the Department of Education's Office of Civil Rights declared that an instructor's academic freedom claim did not supersede the ADA's requirement to make reasonable accommodations.[15] A subsequently filed suit was settled before going to trial.

Traditionally, when working with a student who has academic difficulties, regardless of the origin, most faculty members strive to do what is best for the students in the professional judgment of the faculty member. When presented a list of accommodations from the administration accompanied by mandates to follow, some faculty feel alienated from the process, as well as experiencing a diminished sense of academic freedom [20], [40] and a concern for fairness to other students.

Are Accommodations Fair?

A discussion of the fairness of an accommodation for a learning-disabled student begins by looking at how it may affect other students [21], [22], [42].[16] As noted, a common accommodation is granting students additional time to work on exams. Time might not be considered a skill that a test is intended to measure but rather as incidental to the form the test takes, so that extending time does not significantly alter the academic requirement. But can quick thinking be fundamental? Some cases have found that it may be essential in making a medical diagnosis.[17] In the medical school context courts have been very clear that certain accommodations may so impede a student's training as to endanger future patients. Although the seriousness or immediacy of harm may be less in other situations, the argument of the necessity of the quickness of judgment can be compelling. In fact, courts have generally deferred to academic judgments about whether certain accommodations are "reasonable" as they did in the Falchenberg case described above. However, if time is not an essential element of the test taking, should not all students be permitted extra time?

Some students require use of technology to complete an exam. For example, a student who may have difficulty with physical transcription may be granted the use of a computer to complete an essay for an exam. In the context of a mathematics course, one could ask whether allowing those diagnosed as having a learning disability to use a calculator when other students are not permitted to do so is fair. Would the use of a calculator alter the skill a test is intended to measure? Are the accommodated students as well qualified as those who have met the requirements without accommodations? And who decides these issues—a learning service office, the relevant department, the course instructor, or the school administration? The Office of Civil Rights conducted an investigation after a student complained of alleged discrimination when prevented from using a calculator during the mathematics placement exam despite being diagnosed with dyscalculia. This resulted in the vice chancellor of the system reminding college presidents that calculators were indeed allowed for all mathematics examinations, placement or otherwise, provided the student was appropriately classified as learning disabled.[18] However, whether this was a considered academic judgment or not is unclear. What might happen should a student without a documented learning disability challenge the differential treatment is also not clear.

As an "appropriate accommodation" some students request video and/ or audio taping of the course lecture. This presents two challenges for consideration. First, if a student records the class, who protects the privacy of the other students in the class? Consider then the student who does not feel comfortable participating in a class where recording occurs. In essence, is it fair to "accommodate" one student while unintentionally discriminating against another? Currently no cases of record to date have dealt with this issue. Parkland College policy regarding audio-taped lectures, like that of many institutions, gives the professor permission to tell the class that recording will be occurring but does not prescribe how to deal with objections to such a policy.[19] Second, faculty members have long maintained copyrighted ownership of course materials unless otherwise specified by contract with their associated academic institution. Tapes would be included under course materials protected by copyright law.[20] Again, as is common practice, Parkland College policy allows the instructor to ask the student to sign a taping agreement noting copyright and requiring permission of the instructor for derivative dissemination. California State University at San Bernadino provides each student requesting accommodation a handbook that specifically states that audio tapes must be disposed of at the end of the semester.[21] Wallace Community College informs students that they cannot share tapes with nonstudents, agencies, or media, but they do have the option of donating the tapes back to Disability Support Services.[22] However, little in-depth attention has been given to intellectual property rights in general when considering accommodation requests.

There have been arguments made that the use of standardized examinations is per se discriminatory.[23] As long as they are not the only criterion for success, their use generally has been upheld. Ninety percent of those receiving accommodations on standardized tests have been diagnosed with learning disabilities rather than other disabilities such as physical limitations [41]. Is the fairness problem resolved by "flagging" exams taken under accommodations or courses in which exam modifications or other accommodations were made? It used to be the case that SATs were so flagged, and LSATs and medical school exams still are [41]. Obviously, flagging identifies a person as disabled and could result in discrimination.[24] On the other hand, in the absence of flagging, it could be argued that inaccurate pictures of qualifications are presented,[25] which may prove to be unfair to the accommodated students if they are unable to carry out academic programs or jobs for which they have allegedly qualified, as well as being unfair to their competitors.

It could also be said that "rewarding" disabilities creates an incentive for people to define themselves as disabled, thus constituting a moral hazard [23]. Even if not a moral hazard, do the accommodations unfairly ameliorate a disability? Given that the diagnostic regimen required to establish a learning disability at the collegiate level can be expensive, does that mean that students from lower-income families are unfairly disadvantaged? Should there be the possibility for all students to be tested at the university's cost? Or should anyone who wants accommodations get them? There are many features of higher education that disadvantage low-income students; is this another to be lived with, or should there be expanded legal protection for students who might potentially benefit from a diagnosis they cannot afford to obtain? Beyond the question of whether accommodations might constitute alterations in the fundamental nature of a program, we can ask, do the accommodations really address the disability? For example, is more time for mathematics exams really needed to accommodate slowness in reading, given the limited amount of reading normally required in mathematics exams?[26] If a student has difficulty writing, more time may directly ameliorate the limitation. However, there is also anecdotal evidence that excessive time for an exam may in fact be detrimental if eventual fatigue causes students to alter work that was earlier correctly completed.

Consider the following: if a course would be fundamentally altered by an accommodation, then the accommodation is not appropriate. Students routinely transfer credits from one institution to another. Is it possible that upon review some of the credits would not be transferrable because the student received accommodations that, if provided by the second institution, would have fundamentally altered the second institution's course?

Learning Disabilities and Mathematics

Among examples of learning disabilities are some which are mathematics specific, such as dyscalculia, and others that impact the ability to learn mathematics, such as dyslexia. Dyscalculia is an umbrella term used to describe a collection of challenges students encounter when solving mathematics problems. For example, some students lack number sense, others cannot interpret graphs, and yet others cannot solve problems that rely on sequencing or algorithms for their solution; i.e., they can't solve an equation for x [49]. The degree to which students have one or more of these deficiencies varies. Dyslexic students may have issues reading word problems and number transposition, and since math can be considered a language, decoding of characters may become problematic [43].

Above we note that often in higher education instructors are told to accommodate without necessarily being given specific information as to why an accommodation might be necessary or being consulted as to whether it might fundamentally alter the course requirements. This is unfortunate, particularly for mathematics instructors. Teachers who have knowledge about learning styles and how their students learn may be able to adapt their instruction without compromising the standards of the course, so that students can construct mathematical knowledge for themselves in spite of disabilities [13], [32]. In fact, many "accommodations" are simply techniques that may enhance the learning of all students: supplemental notes, online access to classroom material, audio or video recordings available for replaying as required, access to tutors, ample in-person and/or virtual office hours.

Calculator use in the mathematics classroom, as we have already noted above, is not always left up to the instructor, but perhaps an instructor might dictate the type of calculator to be used and still appropriately accommodate learning-disabled students. Mathematics is cumulative by nature. One has to learn to count before adding or subtracting. One has to understand the order of operations before learning to solve equations. Basic calculators perform the four fundamental operations of addition, subtraction, multiplication, and division; but advances in technology have created calculators that will solve equations, simultaneously graph and create tables of data for functions, and even generate statistical analysis. It could be the case that a learning-disabled student cannot keep track of sequential steps necessary to solve the problem on paper, but that same student can program a calculator to find the answer. Should we expect students to understand fundamentals such as finding the least common denominator of two fractions, a skill they should have developed before college, or should we be more interested in how the students use the answer found by the calculator to solve the problem? Does this "fundamentally alter" a course? Is this fair to the other students in the class, who may not have been allowed to use any technological help? In light of the outcomes related above, how can this issue best be addressed?

From cases above, it is clear that speed of judgment can be an indicator of qualifications for a program or profession. But does mathematics need to be done quickly? We might ask whether an untimed or extended-time mathematics test really measures the skills or knowledge the exam is intended to measure. When does a mathematics student or a mathematician have to think or calculate quickly?

Research on best practices on teaching K–12 students with learning disabilities in math is available. A meta-analysis from the Center on Instruction in Portsmouth, New Hampshire, listed several pedagogical techniques, including problem-specific step-by-step explicit instruction for finding a solution, student verbalization of steps employed during problem solving, visual representation of math concepts presented in conjunction with traditional problem-solving techniques, and a wide array of examples, to name a few [16]. At the collegiate level the classroom pace is faster, more material is covered, and generalized approaches to solving a problem tend to be presented. With in-class examples to guide them, students are expected to go and explore the concepts outside of class independently. It may be the case that the majority of these techniques can be best applied while working with individual students outside the classroom.[27]

Ideally, the expertise in learning disabilities of a specialist ought to be combined with the subject matter knowledge of a classroom instructor in order to best serve all students. If the use of a calculator is mandated by a learning disabilities professional on placement exams, might that mask an inability to deal with fractions that will later pose a handicap if the student is placed at a higher level than might be mandated were the calculator not used? More fundamentally, is an inability to do well in a college mathematics course without accommodations a disability for purposes of the ADA? Could the "average person" do better? If not everyone is guaranteed a university education that requires a college mathematics course, should a university nonetheless seek to accommodate students with difficulties not legally classifiable as disabilities, especially if the accommodation disadvantages other students?

Conclusion

Instructors faced with requests or demands for accommodations for the learning disabled should not necessarily passively comply. Although they cannot make judgments as to whether the accommodations may be legally required, faculty can and should ask whether the proposed accommodations are actually reasonable and appropriate. Do the accommodations ameliorate the disability, do they fundamentally alter the course or program requirements or lower the academic standards, and are they fair to other students?

Notes

1. See, e.g., OCR re Golden Gate University (CA) (9 NDR 182), July 10, 1996.

2. 42 U.S.C.S §§12101 et seq. (2009).

3. However, under the 2009 amendments one whose vision is corrected merely by ordinary eyeglasses or contact lenses is not considered disabled.

4. See, e.g., Singh [44].

5. Public Law 104–191, 104th Congress.

6. 20 U.S.C. section §1232g.

7. In any case, the sisters would not have been considered disabled and thus entitled to protection if only ordinary glasses were needed to correct their vision to 20-20.

8. Courts have generally held that academic judgments about academic requirements should be granted substantial deference [34], [38].

9. In a case that achieved much publicity, the Supreme Court decided that allowing a disabled golfer to use a cart did not alter the fundamental nature of the competition nor was it unfair to other competitors, as it preserved fatigue as a component of the game [33].

10. The committee later decided that it would and the requirement was retained and not subject to waiver. However, the point made was that a decision as to graduation requirements required a process of academic deliberation.

11. http://www.galvin-group.com/dspsresources/assets/CA_OCR_Letter_Mt_San_ Antonio.pdf (accessed 02/15/2009).

12. http://www.american.edu/american/registrar/schedule.cfm (accessed 03/10/ 2009).

13. In the Boston University case, however, apparently the mathematics department had been consulted by those in learning services about the accommodations to be offered.

14. http://dsp.berkeley.edu/learningdisability.html (accessed 03/09/2009).

15. OCR re Golden Gate University (CA) (9 NDR 182), July 10, 1996.

16. Fuchs and Fuchs [14] have discussed this in a K–12 context.

17. See, e.g., Wong [51].

18. http://www.baruch.cuny.edu/counsel/documents/BM-2-92.pdf (accessed 03/ 10/2009).

19. http://www.parkland.edu/ods/handbook/AudioTapedLectures.pdf (accessed 03/ 10/2009).

20. http://www.copyright.gov/circs/circl.pdf (accessed 02/14/2009).

21. http://enrollment.csusb.edu/~ssd/Documents/Faculty%20Handbook%20–%20 Accessible.pdf (accessed 03/10/2009).

22. http://www.wallace.edu/student_resources/dss/policies.php (accessed 03/10/ 2009).

23. The National Collegiate Athletics Association has rules regarding minimum SAT scores for eligibility for participation in college athletics and for scholarships. They also require a certain number of acceptable high school courses, with courses designated as special education not qualifying unless they can be certified as equivalent to regular courses. Clearly this provides scope for controversy (see e.g., [5], [36]).

24. The ADA prohibits discrimination on the basis of perceived disability, whether or not the person is actually disabled.

25. It has been shown, for example, that accommodated SAT scores overpredict first-year college GPAs [37].

26. If an exam consists primarily of word problems, dyslexia or other difficulties in reading may indeed be a disability.

27. Some institutions establish a special section of required courses specially designed for learning-disabled students, the purpose being to assure maximum assistance without altering the nature of the course or disadvantaging other students. http://www.american .edu/american/registrar/schedule.cfm (accessed 3/10/2009).

References

[1] http://www.american.edu/american/registrar/schedule.cfm (accessed 03/10/2009).
[2] Americans with Disabilities Act, 42 U.S.C. §§1201 et seq. (1990 as amended 2009).
[3] http://www.baruch.cuny.edu/counsel/documents/BM-2-92.pdf (accessed 03/10/ 2009).

[4] Betts v. The Rector and Visitors of the University of Virginia, 18 Fed. Appx. 114 (4th Cir. 2001), 145 Fed. Appx. 7 (4th Cir. 2005).

[5] Bowers v. NCAA, 475 F.3d 524 (3d Cir. 2007).

[6] Carten v. Kent State University, 78 Fed. Appx. 499 (6th Cir. 2003).

[7] http://www.copyright.gov/circs/circl.pdf (accessed 02/14/2009).

[8] H. A. CURRIER, The ADA reasonable accommodations requirement and the development of university service policies: Helping or hindering students with learning disabilities? *U. Baltimore Law Forum* **30** (2000), 42–58.

[9] http://enrollment.csusb.edu/~ssd/Documents/Faculty%20Handbook%20-%20 Accessible.pdf (accessed 03/10/2009).

[10] Falchenberg v. New York State Department of Education and National Evaluation Systems, Ind., 567 F. Supp.2d 513 (S.D.N.Y. 2008).

[11] Falcone v. University of Minnesota, 388 F.3d 656 (8th Cir. 2004).

[12] Family Education Rights and Privacy Act, 20 U.S.C. §§1232g.

[13] E. FENNEMA and T. A. ROMBERG, *Mathematics Classrooms That Promote Understanding,* Erlbaum, Mahwah, NJ, 1999.

[14] L. S. FUCHS and D. FUCHS, Fair and unfair testing accommodations, *School Administrator* **56**(10), (1999), 24–28.

[15] http://www.galvin-group.com/dspresources/assets/CA_OCR_Letter_Mt_San_ Antonio.pdf (accessed 2/15/2009).

[16] R. GERSTEN, D. CHARD, M. JAYANTHI, S. BAKER, P. MORPHY, and J. FLOJO, *Mathematics Instruction for Students with Learning Disabilities or Difficulty Learning Mathematics: A Synthesis of the Intervention Research,* RMC Research Corporation, Center on Instruction, Portsmouth, NH, 2008.

[17] Guckenberger v. Boston University, 974 F. Supp. 106 (D. Mass. 1997).

[18] Health Insurance Portability and Accountability Act of 1996, Public Law 104–191, 104th Congress.

[19] Individuals with Disabilities Education Act, 20 U.S.C. §§1400 et seq. (1990).

[20] J. I. KATZ, Learning disabilities at universities, http://wuphys.wustl.edu/~katz/ LD.html (accessed 12/1/2008).

[21] K. A. KAVALE and S. R. FORNESS, The politics of learning disabilities, *Learning Disability Quarterly* **21**(4) (1998), 245–273.

[22] Kelly v. West Virginia Board of Law Examiners, 2008 U.S. Dist. LEXIS 56840 (July 24, 2008, S.D.W.Va.).

[23] C. S. LERNER, "Accommodations" for the learning disabled: A level playing field or affirmative action for elites? *Vanderbilt Law Review* **57** (2004), 1043–1124.

[24] http://www.mathcurriculumcenter.org/PDFS/HSreport.pdf (accessed 02/15/2008).

[25] McGuiness v. University of New Mexico School of Medicine, 170 F.3d 974 (1998).

[26] National Joint Committee on Learning Disabilities, Secondary to postsecondary education transition planning for students with learning disabilities: A position paper of the National Joint Committee on Learning Disabilities, *Learning Disability Quarterly* **19**(1) (Winter, 1996), 62–64.

[27] No Child Left Behind, 20 U.S.C. §§6301 et seq. (2002).

[28] OCR re Golden Gate University (CA) (9 NDLR 182), July 10, 1996.

[29] Pandazides v. Virginia Board of Education, 946 F.2d 345 (4th Cir. 1991).

[30] http://www.parkland.edu/ods/handbook/AudioTapedLectures.pdf (accessed 03/10/2009).

[31] Pazer v. New York State Board of Examiners, 849 F. Supp. 284 (S.D.N.Y. 1994).

[32] P. L. PETERSON, T. CARPENTER, and E. FENNEMA, Teachers' knowledge of students' knowledge in mathematics problem solving: Correlational and case analyses. *Journal of Educational Psychology* **81** (1989), 558–569.

[33] PGA Tour, Inc. v. Martin, 532 U.S. 661 (2001).

[34] Powell v. National Board of Medical Examiners, University of Connecticut School of Medicine, 364 F.3d 79 (2d Cir. 2004).

[35] Price v. National Board of Medical Examiners, 966 F. Supp. 419 (S.D.W.Va. 1997).

[36] Pryor v. NCAA, 288 F.3d 548 (3d Cir. 2002).

[37] J. D. RANSEEN and G. S. PARKS, Test accommodations for postsecondary students: The quandary resulting from the ADA's disability definition, *Psychology, Public Policy, and Law* **11**(1) (March 2005), 83–108.

[38] Regents of the University of Michigan v. Ewing, 474 U.S. 214 (1985).

[39] Rehabilitation Act, Section 504, 29 U.S.C. §§2701 et seq. (1973).

[40] L. F. ROTHSTEIN, Introduction to disability issues in legal education: A symposium, *Journal of Legal Education* **41** (1991), 301–316.

[41] ———, Disability law and higher education: A road map for where we've been and where we may be heading, *Maryland Law Review* **63** (2004), 122–161.

[42] R. SHALIT, Defining disability down, *The New Republic*, August 25, 1997, at 16, 22.

[43] J. SEARLE and S. SIVALINGAM, Dyslexia and mathematics at university. *Mathematics in School* **33**(2) (March 2004 supp), 3–5.

[44] Singh v. George Washington University School of Medicine, 508 F.3d 1097 (D.C.Cir. 2007).

[45] Sutton v. United Air Lines, Inc., 527 U.S. 471 (1999).

[46] Toyota Motor Manufacturing, Kentucky, Inc. v. Williams, 534 U.S. 184 (2002).

[47] N. L. TOWNSEND, Framing a ceiling as a floor: The changing definition of learning disabilities and the conflicting trends in legislation affecting learning disabled students, *Creighton Law Review* **40** (2004), 229–270.

[48] University of California, Practices for the documentation and accommodation of students with learning disabilities, http://dsp.berkeley.edu/learningdisability.html (accessed 01/20/2009).

[49] S. R. Vaidya, Understanding dyscalculia for teaching. *Education* (Chula Vista, Calif.) **124**(4) (Summer 2004), 717–720.

[50] http://www.wallace.edu/student_resources/dss/policies.php (accessed 03/10/2009).

[51] Wong v. Regents of the University of California, 379 F.3d 1097 (9th Cir. 2004).

[52] Wynne v. Tufts University School of Medicine, 976 F.2d 791 (1st Cir. 1992).

[53] Zukle v. Regents of the University of California, 166 F. 2d 1041, 1048 (9th Cir. 1999).

Audience, Style and Criticism

DAVID PIMM AND NATHALIE SINCLAIR

*We make out of the quarrel with others, rhetoric, but of the
quarrel with ourselves, poetry. (Yeats, 1918, p. 21)*

Throughout its near thirty-year existence, this journal has published articles on a very wide range of topics deemed pertinent 'for the learning of
mathematics'. Among them, the theme of the practices of adult mathematicians themselves has recurred, whether insider accounts by professionals
(e.g. Henderson, 1981; Leron, 1985; Mazur, 2004; Thurston, 1995) or
pieces by others about the views of mathematicians (e.g. Burton, 1999a,
1999b) and their spoken and written cultural practices (e.g. Agassi, 1981;
Morgan, 1996; Smith and Hungwe, 1998). And this is quite aside from
significant attention to explorations concerned with the history of mathematics and ethnomathematics.

In this article, we first attend to questions of style in written mathematics,
particularly from a discursive point of view, expanding on Morgan's (1998)
linguistic feature analysis to include attention to rhetorical style and the potential influence of audience. Next we turn to some aesthetic considerations
of mathematical style, as well as examining published guides to 'good' mathematical writing. We then address questions of taste and criticism as well as
the varied public settings in which mathematical writing might be undertaken, before finally returning to explore the above quotation.

Discursive Aspects of Mathematical Style

Morgan (1998) has made an attempt to characterise certain discourse features of a modern research article in mathematics. Her list includes:

- widespread use of nominalizations rather than verbal forms, which transforms processes into objects and also serves to obscure agency;
- use of non-active verb forms;
- absence of reference to human activity;
- distant authorial voice and lack of direct address;
- continuous present tense throughout;
- preference for imperatives over pronoun use (*let, define, consider, suppose,* . . .);
- prevalent use of connectives, marking the relation of each sentence to antecedent and subsequent sentences (e.g. but, hence, so, then, therefore, . . .).

In addition, these various features interact to support the discursive effects of others. Subsequent work (e.g. Burton and Morgan, 2000; Nardi and Iannone, 2005) has explored the views of practicing mathematicians about facets of written mathematical style, not least in relation to novice university student attempts to acquire a greater command of it. These are some of the surface features that identify a mathematics journal article as a paradigmatic text (in the sense of Bruner, 1986—see [1]), one that speaks its own truth, a truth created in part by the very textual means by which it is asserted.

Part of Foucault's (1980) work reflects centrally the way disciplinary regimes become installed in institutions, including the creation of 'régimes of truth':

> The important thing here, I believe, is that truth isn't outside power, or lacking in power [. . .] truth isn't the reward of free spirits, the child of protracted solitude, nor the privilege of those who have succeeded in liberating themselves. Truth is a thing of this world: it is produced only by virtue of multiple forms of constraint. And it induces regular effects of power. (p. 131)
>
> 'Truth' is to be understood as a system of ordered procedures for the production, regulation, distribution, circulation and operation of statements. 'Truth' is linked in a circular relation with systems of power which produce and sustain it, and to effects of power which it induces and which extend it. A 'régime' of truth. (p. 133)

However, it is not simply a question of *how* it should be done; there are always issues of why *this* rather than *that* form. (See Csiszar, 2003, for an extensive exploration of this question.) What is achieved by this particular style? One such achievement of the writing, Foucault would suggest, is its truth. What would be lost were it altered or hybridized, especially in relation to a more narrative rather than a paradigmatic mode? Part of the declared intent is to render utterly transparent the 'logical structure' of the text. But part of a possibly more covert agenda has to do with creating the very sense of decontextualised authority and certainty that is then claimed as the hallmark of mathematics.

Solomon and O'Neill (1998) firmly contrast the mathematical from the narrative (illustrating their argument on a variety of texts authored by the nineteenth-century Irish mathematician William Rowan Hamilton), arguing the difference lies precisely in this 'glue' of logical versus chronological structuring (and their surface manifestations in terms of verb tense, personal pronoun use, connectives between sentences and other lexical choices). Interestingly, in Hamilton's range of mathematical writing—the focus of their work—the syntactic glue changes depending on whether he was writing diary notes to himself, letters to friends or when he was writing his journal articles or monographs (ostensibly addressed to his colleagues).

> An examination of the letter and the notebook reveals a more complex structure than a simple narrative. The texts contain two distinct component texts: a *mathematical* text is embedded within a *personal narrative*. The difference between the texts is indicated by the tense system, the choice of deictic reference and the forms of textual cohesion employed. (p. 216; *italics in original*)

What might a more narrative style in mathematics look like? How would it differ from what is currently offered in professional mathematical accounts and are such differences significant? In particular, might it provide greater scope for writing about images of mathematics rather than solely its body (Corry, 2001, 2006, also discussed below)? These questions have led us to examine the nature of the semantic/syntactic 'glue' that holds mathematical texts together, the glue that seemed problematically absent in novices' prose. But also, for us, there is an adjacent locus of interest, perhaps of greater interest, and that is the question of audience and its role (both presumptive and actual) in shaping mathematical writing.

A Question of Audience, a Matter of Address

> *Euclid's attitude [towards the reader] is perfectly straightforward:*
> *there is no sign that he notices the existence of readers at all. [. . .]*
> *The reader is never addressed. (Fauvel, 1988, p. 25)*

One of the more taxing questions implicated in the complex interrelationship between language and mathematics has to do with the shaping of form by content and of content by form. One of the less considered aspects of this mutual influence has to do with the nature and influence of the *audience* for the language, especially written mathematical language where the empirical reader (one possible but by no means exclusive audience) is likely not be co-present with the author, either temporally or spatially. Yet as Bakhtin (1952/1986, p. 95) was insistent in claiming, every human utterance is addressed *to* someone, a phenomenon he termed *addressivity* namely an orientation towards an other. One such question of audience signalled by this section's title, then, concerns the addressivity of a mathematical text (see also Pimm, Beatty and Moss, 2007). This is not always a straightforward matter, as the earlier quotation from John Fauvel indicated, especially where a stylistic convention exists that values its denial.

Mathematicians are not unaware of these issues. Norman Steenrod (1973), for instance, in the opening paragraphs of his contribution to the Mathematical Association of America monograph *How to Write Mathematics*, remarked:

> A major objection to laying down criteria for the excellence of an exposition is that the effectiveness of an expository effort depends so heavily on the knowledge and experience of the reader. A clean and exquisitely precise demonstration to one reader is a bore to another who has seen the like elsewhere. The same reader can find one part tediously clear and another part mystifying even though the author believed he gave both parts equally detailed treatment. (p. 1)

And Paul Halmos (1973), in his extensive contribution to the same publication, declared:

> I like to specify my audience not only in some vague, large sense (e.g., professional topologists, or second year graduate students), but also in a very specific, personal sense. It helps me to think of a person, perhaps someone I discussed the subject with two years ago, or perhaps a deliberately obtuse, friendly colleague, and then to keep him in mind as I write. (p. 22)

While this may ease the writer's challenge, such a virtual audience may nevertheless have little impact on the readability of the end result.

Formal mathematics has not always been written in this way. As there have been a variety of written genres used in sophisticated mathematics over the centuries, one need not go back in time very far to find variations. Richard Dedekind (1879/1924), for instance, declared that his considerations regarding the continuity of the straight line:

> are so familiar and well known to all that many will regard their repetition quite superfluous. Still I regarded this recapitulation as necessary to prepare properly for the main question. For, the way in which

the irrational numbers are usually introduced is based directly upon the conception of extensive magnitudes—which itself is nowhere carefully defined—and explains number as the result of measuring such a magnitude by another of the same kind. Instead of this I demand that arithmetic shall be developed out of itself. (pp. 9–10)

In terms of defining continuity, he went on to observe:

As already said I think I shall not err in assuming that everyone will at once grant the truth of this statement; the majority of my readers will be very much disappointed in learning that by this commonplace remark the secret of continuity is to be revealed. To this I may say that I am glad if everyone finds the above principle so obvious and so in harmony with his own ideas of line; for I am utterly unable to adduce a proof of its correctness, nor has anyone the power. (pp. 11–12)

He then reveals the hidden assumption of continuity that had been used in both geometry and analysis for centuries, without explicit attention being paid to it.

Dedekind's account has a number of striking characteristics when read in relation to published late-twentieth-century mathematical writing. It has a strong first-person narrative voice and is addressed to a general, undifferentiated 'everyone' or 'all', who are assumed to have certain knowledge, details of presumption which contribute to what Eco (1979) has termed the 'model reader' of a text. (For discussion of this notion in relation to mathematical text, see Love and Pimm, 1995.)

Addressivity and the linguistic features associated with the paradigmatic mode of mathematical writing, constitute one dimension of style. If these features were the only ones, it might be expected that mathematical style could be fully proceduralised. Yet, as Csiszar (2003) emphasizes, mathematical guides emphasise the importance of "fashioning" mathematical writing so as to make it compelling, understood and appreciated. The mathematician Wolfgang Krull (1930/1987) brings the aesthetic dimension of mathematical style to the fore:

Mathematicians are not concerned merely with finding and proving theorems, they also want to arrange and assemble the theorems so that they appear not only correct but evident and compelling. Such a goal, I feel, is aesthetic rather than epistemological. (p. 49)

Aesthetic Aspects of Mathematical Style

Henderson and Taimina (2006) recount how one of geometer David Henderson's published papers evoked a rash of requests from other mathematicians.

[The paper] has a very concise, simple (half-page) proof. This proof has provoked more questions from other mathematicians than any other of his research papers and most of the questions were of the sort: "Why is it true?," "Where did it come from?," "How did you see it?" They accepted the proof logically, yet were not satisfied. (p. 66)

These questions relate to what Corry (2001, 2006) describes as the *image* of mathematics, which he distinguishes from its *body*; together they form "two interconnected layers of mathematics knowledge" (2006, p. 135). While the body includes "questions directly related to the subject matter of any given mathematical discipline: theorems, proofs, techniques, open problems," the images of mathematics "refer to, and help elucidating, questions arising from the body of knowledge but which in general are not part of, and cannot be settled within, the body of knowledge itself." The images may thus include views about the internal organization of mathematics into fields and sub-fields or even the stated importance of one sub-field over another, or the perceived relationship between mathematics and, say, theoretical physics (see Jaffe and Quinn, 1993, and Csiszar, 2003). Note that the body is marked as singular, whereas there is a plurality of possible images.

Mathematicians simply do not customarily write about their images, in Corry's sense, even if they write from them. Indeed, Corry contends that the black letters and symbols on white pages that constitute formal mathematical texts—that is, the research journal article itself—cannot bring forth forms and colours that constitute the images of mathematics. The same can be said with respect to the arts: questions of categorisation and importance, for instance, are usually settled by outside commentators, most notably art critics who employ aesthetic notions and devices. Indeed, Halmos, in *How to Write Mathematics*, ascribes this situation in mathematics to the desire for efficiency and cumulativeness.

The discoverer of an idea, who may of course be the same as its expositor, stumbled on it helter-skelter, inefficiently, almost at random. If there were no way to trim, to consolidate, and to rearrange the discovery, every student would have to recapitulate it, there would be no advantage to be gained from standing "on the shoulders of giants." (p. 23)

The theme of omission—like Halmos's trimming and consolidation—pervades discussions of style in mathematics. In the *A Manual for Authors of Mathematical Papers* (AMS, 1990), we find, for example, the following advice: "Omit any computation which is routine (i.e. does not depend on unexpected tricks). Merely indicate the starting point, describe the procedure, and state the outcome" (p. 2). Further, we find this distinction between good and bad practice.

> It is good research practice to analyze an argument by breaking it into a succession of lemmas, each stated with maximum generality. It is usually bad practice to try to publish such an analysis, since it is likely to be long and un-interesting. The reader wants to see the path—not examine it with a microscope. (p. 2)

Mathematical writing, according to these guides, should be novel (omitting "routine" parts), interesting (by omitting details and analyses, and providing the central plot of the argument generative (by exposing maximally generative lemmas), and simple (making those lemmas do all the work). This style privileges Krull's "evident and compelling" over Henderson and Taimina's perspicuous and meaningful.

While the Halmos quotation above separates the context of discovery from that of "writing it up," we question the extent to which aesthetic style in mathematical writing can be so summarily disconnected. In other words, do these styles of writing start having an effect on what is considered important and interesting in mathematics? Thurston's (1995) discussion of his early work might well suggest that excessive generalization and simplification (through lemma-writing) may have undesirable effects on the field, and compromise what is taken to be interesting about a specific result. Indeed, Lakatos's (1976) imaginary pupils Alpha and Gamma try to figure out how far they can generalise Euler's formula before it ceases to be *interesting*. Alpha insists that it is a question of taste. So Gamma asks, "Why not have mathematical critics just as you have literary critics, to develop mathematical taste by public criticism?" (p. 98).

Gamma's suggestion bears further consideration, particularly in view of the fact that mathematicians have long seen themselves as artists (rather than scientists), without always following up on the implications of such a designation. For example, speaking of mathematical writing again, Ewing (1984) asserts the aesthetic nature of mathematics production:

> Without taste, mathematics ceases to be Art. Like scientists, we should continue to perform experiments, by any means available. But like artists, sculptors, and composers, we must exercise judgment about what should be placed on public display. (p. 4)

However, he fails to note that unlike the artist, sculptor and composer, the mathematician is not subject to public criticism.

Tymoczko (1993) acknowledges this situation and attempts to provide at least an existence argument for the possibility of engaging in aesthetic criticism. Instead of focusing on whether a particular proof is good or bad (or is beautiful or not), he attempts a form of criticism that resembles that of the art critic, who provides "a way of seeing or hearing a work of art [. . . by means of] drawing attention to certain features of the world and in so doing enhance[s] our appreciation or articulates for us the movement of our own feelings" (p. 71).

As his test case, Tymoczko chooses the standard proof of the Fundamental Theorem of Arithmetic, which states than any natural number greater than 1 is either prime or a composite of powers of primes with a canonical factorisation. He violates the usual uniform pace of mathematical proof by introducing a certain rhythm to his discussion that lingers over some statements while flying over others. He also makes explicit the psychological dimensions of the argument, pointing out that, for example, the ease of the first step (in which one proves existence) only leads [to] greater frustration when the difficulties of uniqueness are encountered. He hints at some of the satisfying and persuasive components of the proof, which have much to do with aesthetics (for example, the proof is graced by the presence of the Euclidean GCD algorithm, which pops up rather unexpectedly—and it is always satisfying to be able to call upon a good technique).

Tymoczko refers to his attempt at criticism as a "performance" through which he seeks to describe a "lived work." Indeed, through his use of fore-shadowing (the difficulties yet to come) and flashbacks (the reflection on a particularly useful technique), he re-inscribes the proof in time. In so doing, he also changes the nature of his authorial voice, as well as the manner of address to the reader: switching from 'we' and 'our' to 'me' and 'my' in such asides, as well as directly addressing the reader as 'you'. Tymoczko claims that his performance of the proof will be more memorable.

A problem of audience arises here too. For whom is such criticism written? Knowing that he cannot expect too many people to understand the mathematics in question, Tymoczko marks as his audience professional mathematicians themselves. Thus, the professional mathematician is not only the one to create new proofs, but also the one who is to *appreciate* the proofs of others—and is sometimes also the one who will "perform" the proof for colleagues and students (see Sinclair, 2005, and Pimm, 2007, for more on this implicit theatrical analogy). We note that this sudden shrinking of potential participants in Tymoczko's attempt at mathematical criticism feels slightly nepotistic, keeping things in-house. We further note that criticism, in artistic circles, may also be critical—pointing to ways in which artists have failed to communicate their work, to achieve coherence, or to transform their audiences. Gamma was alluding to this more critical role of criticism.

If criticism is important for mathematics, as people like Corry, Lakatos, and Tymoczko have argued, then we propose that it is even more important for mathematics education, which of necessity works on the borders between the image and body of mathematics. The aesthetic dimension of mathematical style (which is the province of criticism) is centrally involved in the elucidating, explaining, and challenging questions that arise about the body of mathematics. What it means to do mathematics, or to be a mathematician, changes; and such changes may either transpire under the influence of eminent mathematicians or it may undergo public scrutiny.

Between Rhetoric and Poetry

The opening observation of Yeats's identifies two different antagonists with regard to arguments (or 'quarrels' to use his term), namely others and himself. The question of audience when someone is actually *doing* mathematics may well not be the same as when someone is writing mathematics intended for publication. Both have interesting demands and specificities. Both raise questions with regard to mathematical style.

This article has been focused on the act of writing mathematics for others. However, as Foucault indicated in his comments about truth being produced, the doing of mathematics cannot be cleanly separated from the writing of mathematics (in the same way that the context of discovery cannot be separated from the context of justification). We end this paper with two outline examples of this non-distinction at work.

The first comes from on-going work undertaken by Sinclair to study mathematicians' ways of working on problems. In one interview with a combinatorist, where the interviewer has asked the mathematician to describe the process of collaboration, the following description is given: "If it is somebody who is already familiar with these objects then you talk at what we say a high level. But really what it means is that you can provide broad brushstrokes and the person who has thought about this object would know the aspects that you are talking about and you can think together at a high level." Using "broad brushstrokes" to communicate seems very akin to the style of mathematical writing described above, in which the desire to "see the path" is of utmost concern.

In the interview, the mathematician continued on to describe how mathematical collaboration might look somewhat different amongst mathematicians not at the same level: "But if you bring somebody in who is not familiar with or only has cursory familiarity [. . .] they have to work [. . .]. I show them a paper that explains and then I leave them. They have to go a couple of hours or overnight and go through the process and try to get up to the point of familiarity." Thus, in collaboration too, mathematicians talk to each other much as they write to each other, many adopting a similar style in which the rearranged, trimmed, recapitulated mathematical idea is not only preferred, but assumed to be the best means through which to gain "familiarity" with the idea.

As a second, somewhat related example, we have briefly mentioned the issue of mathematical addressivity. Is there always an audience for mathematics, even when it is a lone mathematician at work alone? Is there always an 'other' for mathematics, an addressee being addressed?

Jacques Nimier (1976) cites a high school student: *Avec les mathématiques, il n'y a personne, on est seul.* (With mathematics, there is no one, one is alone.) (p. 56). Yet Jacques-Alain Miller (2004), the contested inheritor of the Lacanian mantle (according to Sherry Turkle, 1992), has made a challenging contrary claim in this regard. Writing in a collection of essays about mathematics and psychoanalysis, Miller draws on a line of Apollinaire's from *The Mouldering Enchanter*, to the effect that *'Celui qui mange n'est plus seul'* ("The one who eats is no longer alone"). Miller's claim is—to paraphrase him—that the mathematician never dines alone. In other words, when the mathematician is feeding on the sweet fruit of mathematics, there is always an other present.

We like the temporal suggestion ('no longer') of the original Apollinaire, namely that there is a loss of aloneness when engaged in mathematics and that this may actually be one reason for engaging in it. This leaves us with the question posed by Pimm (2007, p. 25), "Who is the mathematician's 'whom', the whomother toward whom the mathematics is oriented?," in response to the inquiry 'Whom have I the pleasure of addressing?'

Most of this article has been about issues of criticism, style and audience. The title of this closing section identifies our sense of mathematics situating itself in *between* rhetoric and poetry, in Yeats's sense of the two different audiences for a quarrel. The poetry of mathematics arises from an often-repeated sense of solitariness in communion with an apparently pre-existing entity. The rhetoric of mathematics draws on a more clearly identified sense of 'other' being addressed. The claim we have just made about these two not being clearly distinguishable suggests that mathematics partakes of both poetry and rhetoric and we lose sight of one or the other at our peril.

Notes

[1] Bruner (1986) makes this general distinction between what he terms narrative and paradigmatic modes of thought. Narratives involve the recounting of sequences of events: "The sequence carries the meaning: contrast the stock market collapsed, the government resigned with the government resigned, the stock market collapsed. But not every sequence is worth recounting" (p. 121). In opposition, the paradigmatic mode "seeks to transcend the particular by higher and higher reaching for abstraction" (p. 13).

As Healy and Sinclair (2007) write: "Paradigmatic thinking is an explicit form of reasoning about the world of facts whereas the narrative mode employs tacit knowledge implied in the telling (and often encourages reading between the lines) and while the paradigmatic favours the indicative mode of speech, the narrative mode is often expressed using the subjunctive verbal mood, or at least through linguistic markers that express possibility, wishes, emotion, judgments or statements that may be contrary to facts in hand." (p. 6).

References

Agassi, J. (1981) 'On mathematics education: the Lakatosian revolution', *For the Learning of Mathematics* **1**(1), 27–31.

AMS (1990) *A manual for authors of mathematical papers*, Providence, RI, American Mathematical Society.

Bakhtin, M. (1952/1986) 'The problem of speech genres', in Emerson, C. and Holquist, M. (eds.), *Speech genres and other late essays*, Austin, TX, University of Texas Press, pp. 60–102.

Bruner, J. (1986) *Actual minds, possible worlds*, Cambridge, MA, Harvard University Press.

Burton, L. (1999a) 'Exploring and reporting upon the content and diversity of mathematicians' views and practices', *For the Learning of Mathematics* **19**(2), 36–38.

Burton, L. (1999b) 'Why is intuition so important to mathematicians but missing from mathematics education?', *For the Learning of Mathematics* **19**(3), 27–32.

Burton, L. and Morgan, C. (2000) 'Mathematicians writing', *Journal for Research in Mathematics Education* **31**(4), 429–453.

Corry, L. (2001) 'Mathematical structures from Hilbert to Bourbaki: the evolution of an image of mathematics', in Bottazzini, U. and Dahan Dalmedico, A. (eds.), *Changing images of mathematics: from the French Revolution to the New Millennium*, London, UK, Routledge, pp. 167–185.

Corry, L. (2006) 'Axiomatics, empiricism, and *Anschauung* in Hilbert's conception of geometry: between arithmetic and general relativity', in Ferreirós, J. and Gray, J. (eds.), *The architecture of modern mathematics: essays in history and philosophy*, Oxford, UK, Oxford University Press, pp. 133–156.

Csiszar, A. (2003) 'Stylizing rigor; or, why mathematicians write so well', *Configurations* **11**(2), 239–268.

Dedekind, R. (1879/1924) *Was sind und was sollen die Zahlen?*, Braunschweig, DE, Vieweg. (*Essays on the theory of numbers* (trans. W. Beman), Chicago, IL, Open Court.)

Eco, U. (1979) *The role of the reader: explorations in the semiotics of texts*, Bloomington, IN, Indiana University Press.

Ewing, J. (1984) 'A breach of etiquette', *The Mathematical Intelligencer* **6**(4), 3–4.

Fauvel, J. (1988) 'Cartesian and Euclidean rhetoric', *For the Learning of Mathematics* **8**(1), 25–29.

Foucault, M. (1980) *Power/knowledge: selected interviews and other writings 1972–1977* (ed. and trans. Gordon, C.), New York, NY, Pantheon.

Halmos, P. (1973) in Steenrod, N. *et al.* (eds.), *How to write mathematics*, Providence, RI, American Mathematical Society, pp. 19–48.

Healy, L. and Sinclair, N. (2007) 'If this is our mathematics, what are our stories?', *International Journal of Computers for Mathematics Learning* **12**(1), 3–21.

Henderson, D. (1981) 'Three papers: mathematics and liberation, Sue is a mathematician, mathematics as imagination', *For the Learning of Mathematics* **1**(3), 12–15.

Henderson, D. and Taimina, D. (2006) 'Experiencing meanings in geometry', in Sinclair, N., Pimm, D. and Higginson, W. (eds.), *Mathematics and the aesthetic: new approaches to an ancient affinity*, New York, NY, Springer, pp. 58–83.

Jaffe, A. and Quinn, F. (1993) 'Theoretical mathematics: toward a cultural synthesis of mathematics and theoretical physics', *Bulletin (new series) of the American Mathematical Society* **29**(1), pp. 1–13.

Krull, W. (1930/1987) 'The aesthetic viewpoint in mathematics', *The Mathematical Intelligencer* **9**(1), 48–52.

Lakatos, I. (1976) *Proofs and refutations: the logic of mathematical discovery*, Cambridge, UK, Cambridge University Press.

Leron, U. (1985) 'Heuristic presentations: the role of structuring', *For the Learning of Mathematics* **5**(3), 7–13.

Love, E. and Pimm, D. (1996) '"This is so": a text on texts', in Bishop, A. *et al.* (eds.), *International handbook of mathematics education*, Dordrecht, NL, Kluwer, pp. 371–409.

Mazur, B. (2004) 'On the absence of time in mathematics', *For the Learning of Mathematics* **24**(3), 18–20.

Miller, J.-A. (2004) 'Un rêve de Lacan', in Cartier, P. and Charraud, N. (eds.), *Le réel en mathématiques: psychanalyse et mathématiques*, Paris, FR, Éditions Agalma, pp. 107–133.

Morgan, C. (1996) '"The language of mathematics": towards a critical analysis of mathematics texts', *For the Learning of Mathematics* **16**(3), 2–10.

Morgan, C. (1998) *Writing mathematically: the discourse of investigations*, London, UK, Falmer.

Nardi, E. and Iannone, P. (2005) 'To appear and to be: mathematicians on their students' attempts at acquiring the "genre speech" of university mathematics', in Bosch, M. (ed.), *Proceedings of the Fourth Congress of the European Society for Research in Mathematics Education*, pp. 1800–1810. Available at http://ermeweb.free.fr/CERME4/.

Nimier, J. (1976) *Mathématique et affectivité: une explication des échecs et des réussites*, Paris, FR, Éditions Stock.

Pimm, D. (2007) 'Euclid, overheard: notes on performance, performatives and proof (an intertextual, interrogative monologue)', in Gadanidis, G. and Hoogland, C. (eds.), *Digital mathematical performance*, London, ON, Faculty of Education, University of Western Ontario, pp. 25–35.

Pimm, D., Beatty, R. and Moss, J. (2007) 'A question of audience, a matter of address', in Pitta-Pantazi, D. (ed.), *Proceedings of the Fifth Congress of the European Society for Research in Mathematics Education*, Nicosia, CY. Available at http://ermeweb.free.fr/CERME5/.

Sinclair, N. (2005) 'Chorus, colours, and contrariness in school mathematics', *THEN* **1**. (http://thenjournal.org/feature/80/).

Smith, J. and Hungwe, K. (1998) 'Conjecture and verification in research and teaching: conversations with young mathematicians', *For the Learning of Mathematics* **18**(3), 40–46.

Solomon, Y. and O'Neill, J. (1998) 'Mathematics and narrative', *Language and Education* **12**(3), 210–221.

Steenrod, N. (1973) in Steenrod, N. *et al.* (eds.), *How to write mathematics*, Providence, RI, American Mathematical Society, pp. 1–17.

Thurston, W. (1995) 'On proof and progress in mathematics', *For the Learning of Mathematics* **15**(1), 29–37.

Turkle, S. (1992) *Psychoanalytic politics: Jacques Lacan and Freud's French revolution*, London, UK, Free Association Books.

Tymoczko, T. (1993) 'Value judgments in mathematics: can we treat mathematics as an art?', in White, A. (ed.), *Essays in humanistic mathematics*, Washington, DC, The Mathematical Association of America, pp. 62–77.

Yeats, W. (1918) *Per Amica Silentia Lunae*, London, UK, Macmillan.

Aesthetics as a Liberating Force in Mathematics Education?

NATHALIE SINCLAIR

Introduction

According to the Ancient Greek divisions of philosophy, questions about aesthetics fall into the branch of axiology, which concerns itself with theories of values, including aesthetic values, and also ethical ones. Given the importance that its root term *axios*—which leads to *axioma*—plays in the discipline of mathematics, it may seem strange that axiology has been mostly ignored in the philosophy of mathematics, which has focused almost exclusively on the branches of ontology and epistemology. Axioms, for the Ancient Greeks, were the things that were taken to be self-evident, not needing proof, but used as starting points for a deductive system. *Axios* had the meaning of "being in balance," "having value," "worthy," and "proper." So aesthetics, far from being confined to more modern questions of artistic taste and style, involved theories about what is valued, how it is valued, and why it is valued. Which conception of aesthetics—the classical one, or the more modern one—should we choose to focus on in the context of mathematics education?

In his book on constructive postmodernism, Martin Schiralli (1999) describes the "fixed" view of meaning represented in the question "What do we mean by X?" (p. 57). He then argues for a view of meaning that attends to the genesis of concepts historically, the development of concepts in individuals, and the possibilities of meaning with regard to empirical and theoretical concerns. This view leads to a different question, namely: "What is there for us to mean by X?" (p. 57). This paper will be driven by the latter formulation in an attempt to help liberate aesthetics from its more modernist, fixed use, which has served both mathematics and mathematics education poorly. I base this claim on encounters I have had with a range of people—including teachers, researchers, parents, and others—who believe that it is either elitist or frivolous to focus on aesthetic concerns in mathematics education.

The elitist perspective derives in part from the fact that most contributors to discussions on aesthetics in mathematics (and in the sciences) are eminent mathematicians who talk about beauty, elegance and purity, which only very few people seem to be able to—or want to—appreciate. The frivolous perspective may be linked to common usages of the word aesthetic itself, either to describe hair salons and spas or to describe capricious, fashionable forms of taste. In the next section, I investigate where these understandings come from and what assumptions they lead to about the role of aesthetic awareness in mathematics education. I then consider some more contemporary views, which offer decidedly non-elitist and non-frivolous interpretations of the role of aesthetics in thinking and learning—and construct the aesthetics domain as essential to human meaning-making. Using these broader conceptions of aesthetics, I propose some relevant connections to current issues in mathematics education as well as questions for future study. I will also identify some pervading assumptions and practices, both in mathematics and in education, that will challenge the emergence of aesthetics as a liberating force in mathematics education.

Interpretations of the Mathematical Aesthetic

Conceptions of aesthetics draw on the multiple understandings of the aesthetic itself, as it has evolved over time. It is most often associated with the arts—aesthetics being seen as the philosophy of art—and is used to describe styles or tastes related to masterpieces of artistic products such as paintings, symphonies, and novels. This particular interpretation dates back to Alexander Baumgarten's (1739, 1758) use of the term aesthetics to mean "criticism of taste" and "the science of the beautiful." When applied in contexts outside of the arts, such as phenomena of the natural world, the word aesthetic is often used as a substitute for terms such as "beautiful," "pretty," or "attractive."

The etymology of the word suggests a somewhat different interpretation, relating the aesthetic to the senses and to sensory perception. As such, we can distinguish "the aesthetic", which relates to the nature of perceptually interesting artefacts, and "aesthetics," which relates to the science of human taste or sensory perception. While the former often appears adjectively, to describe artefacts, experiences, sensibilities, and judgements, the latter represents a more general or systematic theory of what might be considered beautiful, artful or tasteful.

In the domain of mathematics, no overall theories of aesthetics have been proposed. However, scholars have discussed varying ways in which aesthetic judgements arise in mathematics. There is a long tradition in mathematics of describing proofs and theorems in aesthetic terms, often using words such as "elegance" and "depth." Further, mathematicians have also argued that their subject is more akin to an art than it is to a science (see Hardy, 1967; Littlewood, 1986; Sullivan 1925/1956), and, like the arts, ascribe to mathematics aesthetic goals. For example, the mathematician W. Krull (1930/1987) writes: "the primary goals of the mathematician are aesthetic, and not epistemological" (p. 49). This statement seems contradictory with the oft-cited concern of mathematics with finding or discovering truths, but it emphasises the fact that the mathematician's interest is in expressing truth, and in doing so in clever, simple, succinct ways.

While Krull focusses on mathematical expression, the mathematician H. Poincaré (1908/1956) concerns himself with the psychology of mathematical invention, but he too underlines the aesthetic dimension of mathematics, arguing that the aesthetic is the defining characteristic of mathematics, not the logical. In Poincaré's theory, a large part of a mathematician's work is done at the subconscious level, where an aesthetic sensibility is responsible for alerting the mathematicians to the most fruitful and interesting of ideas. Other mathematicians have spoken of this special sensibility as well and also in terms of the way it guides mathematicians to choose certain problems. This choice is essential in mathematics given that there exists no external reality against which mathematicians can decide which problems or which branches of mathematics are important (see von Neumann, 1947): the choice involves human values and preferences—and, indeed, these change over time, as exemplified by the dismissal of geometry by some prominent mathematicians in the early 20th century (see Whiteley, 1999).

While many commentators have argued for the importance of aesthetic judgements in mathematics, they have differed in terms of the extent to which such judgements can be made objectively. As I describe in the following section, some consequences of the objective view include the following: aesthetic judgements are true and immutable; criteria can be established that will identify mathematical objects of aesthetic value; these criteria apply to the aesthetic objects themselves; mathematicians will agree on the aesthetic value of different mathematical objects. After elaborating the objectivist view, I will describe more subjective and contextual conceptions of the mathematical aesthetic.

OBJECTIVE VIEWS OF THE MATHEMATICAL AESTHETIC

Consider the following quotation by the textbook writers Holt and Marjoram (1973): "The truth of the matter is that, though mathematics truth may be beautiful, it can be only glimpsed after much hard thinking. Mathematics is difficult for many human minds to grasp because of its hierarchical structure [. . .]". This statement not only suggests that mathematics object of aesthetic value can only be appreciated by a small elite, it also assumes that aesthetic values belong to the objects themselves, to "mathematical truth," and that mathematical beauty is permanent, like a Platonic ideal. Aristotle ascribes a similar objectivity, locating as he does the beauty within the mathematical object itself, and thus independent of the human mathematician: "The mathematical sciences particularly exhibit order, symmetry, and limitation; and these are the greatest forms of the beautiful" (XIII, 3.107b). In other words, mathematical beauty is independent of time and culture.

The objective view of the mathematical aesthetic holds that mathematicians will agree on judgements of mathematical beauty (if they disagreed, that would imply possible subjectivity). The efforts of mathematicians such as G.H. Hardy (1967/1999) to identify the characteristics of mathematical beauty are dependent on this kind of agreement, and its underlying objectivist view. Hardy proposed "depth" and "significance" as primary features of mathematical beauty. Significance is related to the idea's fruitfulness, or its ability to lead to new ideas and its ability to connect different mathematical ideas. Depth, however, has "*something* to do with *difficulty*: the "deeper" ideas are usually the harder to grasp" (p. 109). As with the Holt and Marjoran, Hardy's conception of mathematical beauty severely restricts its accessibility to those who can understand difficult ideas. But this implies that mathematical practices that exploit more visual, accessible forms of representation and reasoning (as exemplified in many cultures, both historical and contemporary) may not produce beautiful or deep mathematical truths.

Several mathematics educators have followed suit in expressing the belief that the mathematical aesthetic is inaccessible to students and perhaps of less importance than acquiring "the basics." Von Glasersfeld (1985) expresses the belief in his claim that children should not be expected to appreciate mathematics like they appreciate rainbows or sunsets. Dreyfus and Eisenberg (1986) argue that the aesthetic is important in mathematics, but that educators should focus their attention on addressing the problems of teaching students "the basics" before any attention should be paid to less pressing concerns. Silver and Metzger (1989) suggest that aesthetic responses in mathematical problem solving require an advanced level of metacognition, and they only gain momentum in advanced mathematical work.

Research on the mathematical aesthetic as conceived from an objectivist point of view focuses on such as: What criteria determine mathematical beauty? What mathematical proofs have aesthetic value? When do novices make the wrong aesthetic judgement? In the realm of education, research has sought to evaluate the extent to which students use the same criteria, or make the same judgements, as research mathematicians. For instance,

Dreyfus and Eisenberg (1986) ask students to compare their own solutions to a problem to those preferred by mathematicians. They conclude that students' aesthetic judgements were poorly developed since the students did not tend to agree with the mathematicians. As I argue below, their finding suggests that students and mathematicians may not have the same aesthetic preferences, but this does not mean that students are incapable of aesthetic appreciation in mathematics.

In summary, the most prevalent view of the mathematical aesthetic conceptualises it as an objective judgement that only very few are capable of making, but on which the best mathematicians agree, and that apply to the prized objects of mathematics including theorems and proofs. Papert (1978) points out that for Poincaré, the aesthetic sense is innate: "Some people happen to be born with the faculty of developing an appreciation of mathematical beauty, and those are the ones who can become creative mathematicians" (p. 191). Papert stresses that the appreciation of mathematical beauty must be *developed,* but ascribes to Poincaré the belief that a sharp lines separates those who possess the aesthetic faculty from those who do not.

CHALLENGES TO THE OBJECTIVE VIEW

In his chapter of *Grands Courants de Mathematiques*, F. Le Lionnais (1948/1986) discusses the issue of beauty in mathematics, offering a broader conception of mathematical aesthetics than those described above. His discussion stretches the view along two different axes. First, he insists on the subjectivity of aesthetic responses that depend on personal taste. He contrasted mathematicians with Dionysian preferences to those with Apollonian one. Whereas Apollonian tastes privilege equilibrium, harmony and order, Dionysians tend to gravitate toward a lack of balance, form obliteration and pathology. Apollonians will look for structures and patterns, while Dionysians will focus on exceptions, counterexamples, and perhaps even strange or baroque concepts. By admitting a Dionysian proclivity, which contrasts with the more typical visions of mathematics expressed in the quotation from Aristotle, Le Lionnais suggests that mathematicians hold differing— and sometimes even opposing—aesthetic preferences. As a human being, the mathematician is bound to have personal inclinations and perspectives that will have an effect on what she values in mathematics.

Le Lionnais also expands the range of artefacts that are subject to aesthetic judgements by drawing attention to the "facts" and "methods" of mathematics that can be seen as either Dionysian or Apollonian in flavour. Facts include magic squares as well as imaginary numbers, while methods include proof by contradiction and inductive techniques. Le Lionnais' classification scheme of facts and methods is not as interesting as the sheer variety of mathematical ideas and objects that he offers as being capable of eliciting aesthetic responses.

The pluralising move of Le Lionnais challenges the intrinsic view of the mathematical aesthetic described above in several ways. In addition to recognising the subjectivity of aesthetic responses, which also challenges the assumption that mathematicians share a common aesthetic judgement, it recognises the way in which aesthetic responses are not confined to the ivory-tower mathematics of theorems and proofs—artefacts that are barely encountered in school mathematics.

Moreover, the mathematical aesthetic is driven not just by "beauty" and "elegance" but also, sometimes, by the "ugly" and the "vulgar." Interestingly, these latter judgements are often lost to history, as can be seen by looking at the way new ideas were initially seen by mathematicians. For example, Fénélon (1697/1845) warns others of the bewitching and diabolic attractions of geometry: "Défiezvous des ensorcellements et des attraits diaboliques de la géométrie" (p. 493). Charles Hermite recoils with "dread" and "horror" from non-differentiable but continuous functions, writing: "Je me détourne avec effroi et horreur de cette plaie lamentable des fonctions continues qui n'ont pas de derives" (Bailland and Bourget 1905, p. 318). Sometimes, the originally more Dionysian responses are retained in the mathematical words themselves, such as *irrational* numbers, *complex* numbers, the *monster* group, or *annihilators*. I will return to the theme of ugliness later in this article, in terms of the very narrow scope of aesthetics that has been studied in mathematics education.

The survey of mathematicians conducted by Wells (1990) provides a more empirically-based challenge to the intrinsic view of the mathematical aesthetic. Wells obtained responses from over 80 mathematicians, who were asked to identify the most beautiful theorem from a given set of twenty-four theorems. (These theorems were chosen because they were "famous," in the sense that Wells judged them to be well-known by most mathematicians, and of interest to the discipline in general, rather than to a particular sub-field.) Wells finds that the mathematicians varied widely in their judgements. More interestingly, in explaining their choices, the mathematicians revealed a wide range of personal responses affecting their aesthetic responses to the theorems. Wells effectively puts to rest the belief that mathematicians have some kind of secret agreement on what counts as beautiful in mathematics.

Wells also sheds light on the changing values, over different times and cultures, which affect judgements of mathematical beauty. Rota (1997) also echoed this view in relating the mathematical aesthetic to different "schools" and eras: ". . . the beauty of a piece of mathematics is dependent upon schools and periods. A theorem that is in one context thought to be beautiful may in a different context appear trivial."(p. 126). Rota's work on umbral calculus provides a compelling example: this mathematical technique offers a notational device (treating subscripts as exponents) for proving similarities between polynomial equations. This technique first emerged as a sort of magic rule in the 19th century, but was later explained, and placed on a firmer foundation by the work of Rota and his students, which led to further applications and generalisations. While the Rota school found the umbral calculus aesthetically pleasing, it has since gone out of fashion. In sum, beauty is not only subjective; it is context-bound and inseparable from emotions and pleasure.

Moving away slightly from Wells's focus on the evaluation of finished products (theorems and proofs), Burton's (2004) work focuses on the practices of mathematicians and their understanding of those practices. Based on extensive interviews with a wide range of mathematicians, she proposes an epistemological model of "mathematician's coming to know," which includes the aesthetic as one of five categories (the others being: its recognition of different approaches, its person- and cultural/social relatedness, its nurturing of insight and intuition, and its connectivities). She points out that mathematicians range on a continuum from unimportant to crucial in terms of their positionings on the role of the aesthetic, with only 3 of the 43 mathematicians dismissing its importance. For example, one said, "Beauty doesn't matter. I have never seen a beautiful mathematical paper in my life" (p. 65). Another mathematician was initially dismissive about mathematical beauty but later, when speaking about the review process, said: "If it was a very elegant way of doing things, I would be inclined to forgive a lot of faults" (p. 65). While the first point of view arises from a question about defining mathematical beauty, the second statement relates to the way in which aesthetic responses affect decisions and judgements of mathematicians at work. The former view coincides with Schiralli's modernist interpretation of meaning (what is mathematical beauty?). In contrast, the latter, more pragmatic view draws on individual meanings in action and experience.

A more pragmatic approach to thinking about the mathematical aesthetic developed in my earlier work (Sinclair, 2004, 2006b) also draws on mathematical practice among research mathematicians, but focuses instead of the process on mathematical inquiry. Based on interviews with mathematicians, and also on an analysis of the structure of mathematical inquiry and the values that characterise the discipline, my tripartite model of the role of the aesthetic describes the way in which aesthetic values are involved in the selection of mathematical problems, in the generation of hypotheses and conjectures, and in the evaluation of mathematical solutions. In contrast with Burton, who relies solely on mathematicians' understandings of their own practices, the tripartite model also takes into account the way in which mathematical knowledge is produced and communicated in the mathematical community. So, for example, while mathematicians may not care about the beauty or elegance of their solutions or proofs, they do have to assess whether their solutions are "good" and "interesting" since these criteria are used to decide whether their work will be published. Similarly, mathematicians must adopt certain stylistic norms in writing their proofs and these norms are highly aesthetic in nature (see Csiszar, 2003). While mathematicians may not be explicitly aware of the values that guide their work, these values play a crucial role in mathematical practice.

This was the view adopted by Poincaré, who focused on the generative role of the aesthetic. However, for Poincaré, mathematicians were the only to possess the "special aesthetic sensibility" that was capable of generating productive ideas in the mathematician's unconscious mind. Papert (1978) challenges Poincaré on his elitist view, while fully endorsing the work of the aesthetic at the subconscious level. For Papert, mathematicians are not the only ones to possess the "special aesthetic sensibility;" instead, he shows how non-mathematicians can be guided toward correct mathematical ideas through appeal to aesthetic considerations. The aesthetic responses exhibited by Papert's non-mathematicians had little to do with Hardy's qualities of depth and significance, or even of surprise. Instead, they involved emotional reactions to an equation's form and structure—a desire to get rid of a square root sign, or to place the important variable in a prominent position. These responses do not provide solutions, now are they evaluated explicitly by the non-mathematicians; instead, they provide tacit guidance.

Sinclair's tripartite model is somewhat limited in the arena of mathematics education, if only because students rarely have the opportunity to engage in mathematical inquiry in the classroom (in particular, the selection of problems is usually made for them, and the evaluation of a solution is restricted to concerns with veracity). However, some empirical work has been conducted and has shown that, when provided with inquiry opportunities in rich environments, middle school children do indeed use aesthetic values in choosing problems, generating conjectures and evaluating their solutions (Lehrer, 2008; Sinclair, 2001; 2006a). These values sometimes, but not always, overlap with canonically mathematical aesthetic values such as fruitfulness, visual appeal, and surprise. Similar work, focusing on the problem-posing phase of inquiry has shown that prospective elementary teachers can use aesthetic values to pose more interesting mathematical problems (Crespo & Sinclair, 2008).

In working with university-level students, Brown (1973) reports on a "genealogical" tendency for students sometimes to prefer their own solutions to those of mathematicians—in this case, Gauss's solution to finding the sum of the first 100 whole numbers. Brown argues that while these solutions may be seen as "messy" they often encode the parts of the problem solving process that contribute to the student's understanding. These solutions are revelatory. It is interesting to compare Brown's approach to aesthetic preferences with Dreyfus and Eisenberg, who conclude that students lack aesthetic sensibility because they do not agree with mathematicians on the preferred solution. For Brown, a difference in aesthetic preference does not entail a lack of aesthetic sensibility.

Discussions of the mathematical aesthetic, even those of Le Lionnais and Wells, interpret the aesthetic as a mode of judgement that is neither epistemological nor ethical, but instead, related to what is considered good, significant or appealing. As a mode of judgement, the aesthetic is thus most commonly viewed as applying to finished products—such as theorems and proofs—but it can also arise in exploration. In addition, as a mode of judgement, the aesthetic is seen as operating distinctly from other modes of human behaviour, such as affect and cognition. In the next section, I describe very different conceptions of aesthetics that are not specific to the domain of mathematics, but that offer new and productive ways of reflecting on the role of aesthetic awareness in mathematical thinking and learning.

Contemporary Views of the Aesthetic

Over the past several decades, there has been a growing interest in the aesthetic,[1] and new conceptualisations of it that follow in part from contemporary views of the human mind as being inseparable from its body and the world around it. Instead of focusing on judgements, recent interpretations of the aesthetic have talked about human actions and meanings, and have sought to expand the range and deepen the influence of aesthetic responses and experiences. These interpretations draw on scholarly work in cognitive science, neuroscience, anthropology and philosophy, and can be categorised into four different themes related to experience, embodied cognition, inquiry, and evolutionary imperatives. In the sections below, I outline the distinguishing features of each theme and the connections to mathematics education research.

THE AESTHETIC AS A CORE COMPONENT OF BEING HUMAN

In his book *Art as Experience*, Dewey (1934) wants to reclaim the aesthetic from the narrow and elitist confines of "museum art" and place it as a theme of human experience. For Dewey, the *aesthetic experience* is of central interest. Moreover, rarefied aesthetic experiences are simple an extreme form of what all humans experience in a wide variety of endeavours. Indeed,

Dewey seeks, first and foremost, to situate the aesthetic squarely in more common, natural settings:

> in order to *understand* the esthetic in its ultimate and approved forms, one must begin with it in the raw; in the events and scenes that hold the attentive eye and ear of man, arousing his interest and affording him enjoyment as he looks and listens (p. 5).

While he traces the aesthetic to everyday human activities, he draws attention away from aesthetic judgements and instead focuses on the integration of thoughts and feelings that occur in experience. For Dewey, an experience has aesthetic quality whenever there is coalescence into an immediately enjoyed qualitative unity of meanings and values drawn from previous experience and present circumstances. In Dewey's conception, the aesthetic does not describe the qualities of perceptual artifacts; rather, it characterizes experiences that are satisfactory and consummatory. Aesthetic experiences can be had while appreciating art, while fixing a car, while having dinner, or while solving a mathematics problem. They are aesthetic in that they combine emotion, satisfaction and understanding. While previous philosophers focused on the form of perceptual objects (colour, structure, etc.), Dewey looks for integration with the human being in interaction with the world.

Dewey's aesthetic experiences can be had in mathematics, of course, and there have been several descriptions of the kind of overwhelming, satisfying and fulfilling experiences by mathematicians themselves, including the very moving testimony of Andrew Wiles (see Singh, 1997), but also claims that such experiences are the ultimate goal of mathematicians. Interestingly, the mathematician Gian-Carlo Rota (1997) has made a similar move to that of Dewey's in claiming that the notion of mathematical beauty or elegance is nothing more but a safe—and seemingly objective—way for mathematicians to communicate about their own emotionally charged experiences:

> Mathematical beauty is the expression mathematicians have invented in order to obliquely admit the phenomenon of enlightenment while avoiding acknowledgement of the fuzziness of this phenomenon [. . .] (pp. 132–133).

The psychologist Csikszentmihalyi (1990) has also focused on qualities of human experience, and proposes the concept of "flow" to designate optimal experiences that are characterised by states of engagement, satisfaction, and goal-directedness, among others. There are some similarities to Dewey's notion of an aesthetic experience, even though Csikszentmihalyi does not talk explicitly about aesthetics. However, a central difference lies in Dewey's insistence on the location of aesthetic experience within the logic of inquiry, thus integrating cognition with the emotional aspects of experience that "flow" describes.

In terms of mathematics education research, the notion of an aesthetic experience has rarely been used, though similar ideas have been expressed in different ways. For example, in relating the cognitive and motivational dimension of learning mathematics, von Glasersfeld (1985) writes "if students are to taste something of the mathematician's satisfaction in doing mathematics, they cannot be expected to find it in whatever rewards they might be given for their performances but only through becoming aware of the neatness of fit they have achieved in their own conceptual construction" (pp. 16–17). The notion of "neatness of fit" aligns closely with Dewey's qualitative unity; also, in speaking of satisfaction, and in integrating the cognitive and affective, von Glasersfeld describes something very close to Dewey's notion of aesthetic experience.

Neither Dewey nor von Glasersfled offer useful ways of describing what such experiences might look like for mathematical learners. This motivated Sinclair's (2002) work, which applied Beardsley's (1982) list of the defining features of aesthetic experience to the domain of mathematics. By analysing an example of mathematical problem solving, in terms of Beardsley's list (object directedness, felt freedom, detached affect, active discovery and wholeness), she finds that the features apply unevenly to the context of mathematics. However, the feature of object directedness seemed to be necessary to the aesthetic experience, and act as a precursor to active discovery and felt freedom. This feature refers to "a feeling that things are working or have worked themselves out fittingly" (p. 288) as one is fixed on the qualities or relations of a phenomenon. This finding might be useful in guiding the design of situations that can lead to aesthetic experiences.

While Dewey's notion of aesthetic experience might be challenging to operationalise in the mathematics learning context, it proposes two powerful, and distinct, commitments that have still not found sufficient expression in mathematics education research: (1) the refusal to separate emotion from cognition within the process of inquiry, (2) the view of the aesthetic as a continuous, unifying quality that underlies experience—not as a separate mode of judgment exercised after inquiry is complete.

Art as Experience was Dewey's final book. However, it has been argued that his intention was to publish a follow-up in which he linked his philosophical idea about aesthetics, experience and inquiry with his influential work on what constituted *educative experiences* (see Dewey, 1938). Certainly, Dewey would have argued that aesthetic experiences are ones that promote growth—this goal being central to his conception of the goals of education. Jackson (1998) uses Dewey's work to argue for the increased role of arts learning in schools, drawing on Dewey's privileging of artistic experiences as highest expression of the aesthetic dimension of human experience. However, as Dewey himself argued, the arts are not the only enterprise in which aesthetic experience arise. This interpretation of aesthetic experiences in an educational context follows from a longstanding tradition in schooling in which the burden of a child's aesthetic development falls on the art or the music teacher, whereas the burden of that child's logical development gets conferred to the mathematics teacher. As I will argue later, this positioning of aesthetic development in the curriculum represents a common discourse that contributes to the relatively marginal role of aesthetics in mathematics education.

THE AESTHETIC AS A CONSEQUENCE OF EMBODIED COGNITION

More recently, Mark Johnson (2007) uses some of the Dewey's ideas around the notion of aesthetic experience, but adapts them more specifically to contemporary research in embodied cognition. Instead of using experience as the primary locus of the aesthetic, Johnson argues that human meaning-making itself is fundamentally aesthetic, using the word aesthetic now to describe all our physical encounters with the world. Johnson sees human understanding, including images, emotions and metaphors, as rooted in these bodily encounters. To make his argument, Johnson links recent theories of embodied cognition—namely, that even our most abstract concepts are rooted in our sensorimotor experiences—to the notion of the aesthetic as sensuous perception (or, as Kant defined it, as the science that treats the conditions of sense perception). In other words, since what we know is derived from our senses, then our cognitive capacities cannot be separated from our aesthetic ones—even though we may no longer be consciously aware of our underlying body-based conceptual foundations.

Like Dewey, Johnson also sees the aesthetic as being deeply intertwined with other human capacities such as affect and cognition. However, Johnson stresses the way in which all the things that "go into meaning—form, expression, communication, qualities, emotion, feeling, value, purpose" (p. 212) are also rooted in bodily perceptions. Also like Dewey, Johnson sees the arts as the culmination of the aesthetic dimension of human experience, and thus proposes to study the arts as a way to locate the bodily sources of meaning. However, Johnson wants to use human artistic expression as an opportunity to probe human understanding: "[a]esthetics is not just art theory, but rather should be regarded broadly as the study of how humans make and experience meaning" (p. 209).

Johnson's conception of the aesthetic stretches very broadly (perhaps too broadly, in the sense that one could infer that all cognition is aesthetic).

He seeks to replace the traditional focus of philosophy on language with a focus on the body as the bearer of human meaning. As a result, fully acknowledging the aesthetic involves going beyond linguistic meanings, and accepting embodied meanings as well—and not just in the arts, but also in other disciplines. Johnson's argument is that embodied meanings of art—such as the rhythms of poetry or the textures of paintings—are body-based meanings that underlie more abstract understandings as well. Trying to identify these meanings, and even describing them, presents new challenges to researchers attempting to locate the aesthetic underpinnings of students' mathematical understandings. As I discuss later, some mathematics educators are already doing this, though without using the construct of aesthetics explicitly.

THE AESTHETIC AS A DIMENSION OF INQUIRY

The art historian E. Gombrich (1979), in his study of decorative arts around the world, emphasises the human need to find some kind of order or pattern in the flux of experience. He calls this a drive for "a sense of order." Humans are thus biased in their perception for straight lines, circles, and similarly ordered configurations rather than with the random shapes encountered in the chaotic world. Gombrich emphasizes that the order hypothesis is the condition that makes learning possible, since without some initial system, a first guess, no "sense" could be made of the millions of ambiguous stimuli incoming from the environment.

More recent research by cognitive scientists has also posited a mechanism through which humans look for order and pattern. E.O. Wilson (1998) argues that humans have predictable, innate aesthetic preferences they use in making sense of their environments. He notes that basic functioning in the environment depends on discerning patterns, such as the spatial patterns involved in perceiving surfaces and objects, and the rhythmic patterns involved in detecting temporal change. The continued and improving ability to discern such patterns gives rise to what Wilson calls "epigenetic rules," that is, to inherited regularities of development in anatomy, physiology, cognition and behaviour. He argues that such rules account for many pre-dispositions and preferences. For instance, studies in human facial recognition show that humans are particularly sensitive to looking for right/left symmetry (as opposed to looking for up/down symmetry or not attending to symmetry at all). Finding such symmetry provides the simplest (shortest) descriptions of faces, and even of bodies—and thus makes such stimuli easier to encode and recall.

Wilson provides a concrete example of a universally shared aesthetic preference. He describes a study tracing arousal response to a variety of visual images in which the most arousing are those that cognitive psychologists call "optimally complex." Although researchers have a method for quantifying complexity, a qualitative description will suffice here: "optimally complex" designs are those that contain enough complexity to engage the mind but that do not overwhelm it with incomprehensible irregularity or diversity. If too many variations or distortions are made, such that little or no redundancy and repetition can be detected, the design moves from too simple to too complex to provoke arousal. However, if the stimulus is just complex enough, the perceiver is most aroused since, as Gombrich explains, "delight lies somewhere between boredom and confusion" (p. 9).

Dewey and Peirce both offered more philosophical perspectives of the role of the aesthetic in inquiry. Both saw the aesthetic playing a crucial role at the initial stage of inquiry, and as providing the guiding impetus for understanding and solving problems. Both also saw the aesthetic as being imaginative, intuitive and non-propositional. For Peirce (1908/1960), aesthetic responses feature strongly in the free exploration of ideas that gives rise to abductions, which were the only kind of inference to produce new ideas in scientific inquiry. Anticipating Dewey's account of the architecture of inquiry, he elaborates that inquiry begins with "some surprising phenomenon, some experience which either disappoints an expectation, or breaks in upon some habit of expectation of the *inquisiturus*" (6.469).

Dewey's (1938) logic of inquiry offers a similar, but more compelling account of the fundamentally aesthetic nature of inquiry. He claims that there is an aesthetic quality that belongs to any inquiry, be it scientific or artistic: "The most elaborate philosophical or scientific inquiry and the most ambitious industrial or political enterprise has, when its different ingredients constitute an integral experience, esthetic quality" (p. 55). What is this aesthetic quality? Dewey maintains that it relates to the human's inevitable tendency to arrange events and objects with reference to the demands of complete and unified perception.[2] For Dewey, inquiry also starts with surprise, or the feeling of something being problematic. He maintains that a problem must be "felt" before it can be stated; the problematic quality is felt or "had" rather than thought. It cannot be expressed in words. An inquirer is aware of quality not by itself but as the background, thread, and the directive clue in which she acts. Dewey suggests that the types of exclamations and interjections such as "Oh!" "Yes," or "Alas" that open most every scientific investigation supply perhaps the simplest examples of qualitative thought.

For both Dewey and Peirce, the aesthetic of inquiry is linked to the non-propositional, qualitative and felt experience of a situation, which provides the basis for further distinction, conceptualization, and articulation. Dewey saw the aesthetic quality as pervading the whole process of inquiry, and providing the basis for the evaluative judgement made by the inquirer at its close.

Despite the fact that both philosophers believed that the process of inquiry could be studied empirically, and was not simply a succession of mental states that were somehow unobservable or transcendental (as Poincaré might have argued), it has been challenging for researchers to operationalise concepts such as qualitative unity, or even abduction. The most important consequence of their theories has been to underline the important role that the initial stage of inquiry plays, in either providing new ideas or formulating a persistent quality.

In mathematics, there have been few studies of the process of inquiry, with the book *Thinking Mathematically* by Mason, Burton and Stacey (1982) being a notable exception. These authors dwell on the initial part of inquiry, and on the qualitative responses it will give rise to, as exemplified by the importance they accord to "recognizing and harnessing to your advantage the feelings and psychological states that accompany [mathematical enquiry]." Interviews with mathematicians, as well as their autobiographies, have confirmed the way in which qualitative responses, and the exploitation of feelings contribute significantly to the posing and solving of problem (see Albers, Alexanderson & Reid 1990; Davis, 1997; Hofstadter, 1997; Sinclair, 2002; Weil, 1992).

Dewey's and Peirce's ideas have not generated much interest in the mathematics education community. One may hypothesise that aesthetic engagement in mathematics may be dependent on opportunities to engage in the full process of inquiry, which includes exploring situations without specific goals in mind and posing problems that arise out of this exploration (see Hawkins, 2000). Under this assumption, it is not that students are incapable of aesthetic engagement in mathematics, but, rather, that school mathematics offers few opportunities for the kind of mathematical inquiry described by Dewey and Peirce. While some may argue that engaging in mathematical inquiry would be the essence of 'acting like a mathematician,' and therefore strongly desirable, Dewey might instead argue that manufacturing situations in which learners can have aesthetic experiences, the most valuable and satisfying kind of experience possible, should be the driving goal of mathematics education.

In contrast to the focus on inquiry, Gombrich and Wilson both point to possibilities of aesthetic engagement in the more common activities of the mathematical classroom such as solving problems or making sense of ideas. Interestingly, despite the previous emphasis on the subjective, contextual nature of aesthetic judgements, their work may support the conjecture that human beings share many significant penchants and preferences. For example, they often seem to organise their perceptions around symmetry or balance, either because it reduces complexity, as White would argue, or because of its connection to our own bodily symmetries, as Johnson would argue. However, while symmetry acts as an organising principle, Gombrich draws attention to a range of preferences humans have expressed around symmetry: while western decorative art tends to value the presence of symmetric configurations, much of the decorative arts of the east prefer breaking symmetry.[3] Similarly, perceptions of confusion and boredom are highly personal and contextual.

THE AESTHETIC AS AN EVOLUTIONARY IMPERATIVE

The anthropologist Ellen Dissanayake (1992) takes a unique approach to conceptualising the aesthetic in her book *Homo Aestheticus:Where Art Comes From and Why*. Instead of rooting the aesthetic in the human body and the sensory organs, Dissanayake links the aesthetic to more evolutionary concerns. She is concerned with understanding why people everywhere, in different cultures and historical time periods, spend so much time decorating and adorning themselves and their surroundings. The amount of time spent on these activities seems to contradict evolutionary assumptions about survival—no tattooed arm, elaborate dance ritual, or decorated door mat can answer the need for food and shelter. Dissanayake thus sees these aesthetic productions as ways of "making special." The human aesthetic capacity—which she sees as being on par with other capacities such as the emotional, the cognitive and the practical—is nothing more than the need to identify things in the flow of experience as worthy of attention and embellishment.

Dissanayake's approach offers some rather different insights for mathematics education, focusing as it does on the need to highlight and embellish as a means of avoiding either monotony or chaos. There are several ways to see how this kind of need plays out in the mathematics classroom—not all of them conceptually relevant!—whether it's doodling in the notebook to break the monotony of a lecture or seeking repeatable rules to overcome the perceived chaos of algebraic manipulation.

Pinker (1997), another scholar interested in the aesthetic dimension of human behaviour, explains how human emotions become so deeply implicated in aesthetic responses. He focuses on the adaptive responses of human beings to selective pressures in an evolutionary context and, in particular, on responses to the set of "enabling acts" which increase their ability to survive within environmental and social constraints. Some subconscious part of the mind, he argues, registers those highly enabling acts—such as using symmetry to perceive and gather information on family members or hunted animals—through a sensation of pleasure. This pleasure in turn alerts us to, or brings to our consciousness, the advantages of such acts. Enabling acts occur through obtaining information about the improbable, information-rich, consequential objects and forces that dominate everyday lives. Whereas these may have once been acts of predicting rains, fertile hunting grounds, or generosity in other humans, modern humans face very different situations. Nevertheless, Pinker argues that the pleasure-alerting mechanisms function in the same way. When confronted by information-rich and potentially consequential stimuli—the ominous foreign subway map separating me from my hotel, for example—I derive pleasure from being able to discern its underlying pattern.

In contrast with Wilson and Gombrich, who are concerned with the causal dimension of human behaviour, Pinker and Dissanayake examine the consequential dimension of human behaviour, namely, the way in which "making special" or "registering enabling acts" forms the basis for aesthetic sensibility (similar to the way in which Johnson, in his focus on embodiment, attempts to define the basis for aesthetic sensibility). While their approach may help provide persuasive arguments about the centrality and importance of the aesthetic in human thinking and behaviour, it also offers different interpretations of the aesthetic that broaden its relevance to learning, and to mathematics education.

From a theoretical point of view, the work of Pinker suggests a strong connection between affect and aesthetics, and, in particular, the possibility of identifying "enabling acts" through the cue of pleasure responses. More empirically, instead of investigating what students find beautiful (or not), researchers might study the range of "enabling acts" that can occur in problem solving. What sets of actions or transformations can give rise to the kind of pleasure response described by Pinker? To what extent are these acts shared across different learners? From a pedagogical level, it reinforces Brown's critique about "false aesthetic unity" of the mathematics classroom, in which things always work out, in whole numbers, or in orderly patterns. One of the many reasons for the negative affect that is so preponderant amongst students may be linked to the lack of opportunities they have to register such enabling acts. In this respect, Pinker's work links strongly with issues of motivation in mathematics education.

Dissanayake's work leads in a different direction, in that her conception of aesthetics does not have the same close links to cognition. However, it does emphasise the interplay between the aesthetic and the affective in emphasizing the satisfaction that comes from the successful manifestation of the basic "making special." What might constitute an act of making special in mathematics, or in the mathematical classroom? Might researchers be able to use the construct of "making special" in order to assess the engagement of the aesthetic? In the discipline itself, are there characteristic ways in which mathematicians "make special" and are these relevant to mathematics learning?

What the Aesthetic Can Mean in Mathematics Education

Stepping back now, it is clear that the conceptions of the aesthetic used by the scholars cited above vary quite widely. Dewey aims to use the aesthetic as a way of challenging classical distinctions between the cognitive, the affective and the artistic, and to locate the aesthetic as a theme in human experience. Johnson concerns himself with the link between embodied cognition and the aesthetic, also attempting to challenge traditional conceptions of cognition as disembodied and emotion-free. Dissanayake and Pinker want to understand why people engage in activities. Gombrich and White are interested in empirically-derived tendencies in human perception.

In terms of mathematics education, all points of view insist on conceiving the learner as being in possession of aesthetic sensibilities and values needing to be exercised. This is radically divergent from current trends in mathematics education research, which most often ignore the aesthetic dimension of mathematics teaching and learning. In the following sections, I explore some of the reasons for the prevailing gulf between mathematics education research and current theories related to the aesthetic. In particular, I consider the following three factors: (1) the cognitivist orientation to research in mathematics education; (2) the lack of connection between the current theories of aesthetics and the fundamentally social nature of the mathematics classroom; (3) the power dynamics around mathematics education and the accompanying elitist views of the mathematical aesthetic.

INTEGRATING THE COGNITIVE, AFFECTIVE AND AESTHETIC

A major theme of the scholarship discussed above involves the extremely close connection between cognition, affect and aesthetics. Some see the aesthetic as the unifying principle of meaning and experience (see also Schiralli, 2006), but all agree that the human aesthetic plays a role in learning about the world and is intimately related to pleasure and satisfaction. Despite this, the predominantly cognitive approaches in mathematics education acknowledge the existence of affect and aesthetics, but ascribe them both a rather epiphenomenal role in cognitive processing. Moreover, these approaches do not generally take into consideration the cultural and historical aspects of human meaning-making that I have argued are central to understanding the role of aesthetics in mathematical thinking.

Goldin's (2000) work on affective representational systems stands out from other approaches, be they cognitive, affective or sociocultural, in that he explicitly links affective and cognitive representations in his model of problem-solving competence. In particular, he analyses relationships between affective states and heuristic configurations and posits certain pathways through affective states that different problem solvers might take, and that might lead to different types of heuristics. For Goldin, affect comprises a tetrahedral construct which includes (1) beliefs, (2) attitudes, (3) emotional states and (4) values, ethics, and morals. As such, both aesthetics and ethics are subsumed within the affective domain.

The conflation of affect and aesthetics, which defies long-standing distinctions in philosophy, would also be refuted by each of the contemporary interpretations developed above. Most fervent opposition would come from the embodiment viewpoint, which might instead subsume affect under the aesthetic, given that feelings and emotions rely on sensory perception. From the evolutionary viewpoint, the aesthetic, as a form of "making special" or as expressed through enabling acts, involves an attention to values—to what is worthwhile in experience and action. From the inquiry viewpoint, the aesthetic functions as a non-logical form of knowing, which aligns itself much more with cognition (broadly viewed) than with affect. Dewey points to the way in which affective responses might alert the inquirer to the presence of certain perceptions and inferences, but those perceptions and inferences cannot be reduced to feelings, or even beliefs.

Drawing on the last perspective, I would also challenge Goldin's subsumation of the aesthetic. As Dewey would argue, the problem-solver becomes alert to aesthetic responses *through* affective states. Silver and Metzger (1989), in their study of research mathematicians, also support this view: "decisions or evaluations based on aesthetic considerations are often made because the problem solver "feels" he or she should do so because he or she is satisfied or dissatisfied with a method or result" (p. 70). Positive or negative feelings can arise from a perception, or an awareness, of something being worthwhile, important or interesting. In other words, the aesthetic and the affective domains each *function* differently in the problem-solving process: the aesthetic draws the attention of the perceiver to a phenomenon (a pattern, a relationship, a contradiction), while the affective can bring these perceptions to the conscious attention of the perceiver.

From a pragmatic point of view then, in terms of describing and explaining mathematical problem solving, the aesthetic and the affective should retain conceptual distinctiveness, despite their obvious interconnections. Further, theories of affect in mathematics education cannot explain the derivation of aesthetic values, their propagation within different cultures, and their role in guiding the growth of the discipline. Aesthetic responses and values do not exist as biological configurations, which is how Goldin describes affective states. They are socially and historically evolved, contextualized by shared practices within a community, and they exert themselves by determining what should be considered worthwhile, important and useful.

The above discussion suggests some intermingling between aesthetics and affect (feelings arising in relation to perceptions of pleasure, beauty, worthiness, and so on), but the aesthetic can also be tightly coupled with the cognitive. At one level, and perhaps a rather cerebral one, we might talk about perceptions of simplicity, structure, conciseness and lucidity. However, Johnson's view of aesthetics offers a more visceral link. Drawing on Lakoff and Núñez (2000) *Where Mathematics Comes From*,[4] researchers have studied the way in which more body-based experiences can help support the development of abstract mathematical ideas. For example, the work of Radford (2003), which focuses on the *direct* connection between bodily movement and abstract mathematical conceptualisations, has a strong affinity with this aesthetic approach. In particular, Radford shows how the rhythmic utterances of students are used to construct meaning for algebraic patterns they are studying—the rhythm, and not the actual words, act as semiotic markers of generalisation. Recall that for Johnson, rhythm is an aesthetic form of meaning-making, so that the link to cognition occurs through the body.

In fact, several mathematics educators have become interested in the role of bodily actions and gestures in the elaboration of concepts (see Arzarello and Robutti, 2001; Nemirovsky, 2003; Núñez, 2004). This research, while acknowledging the role of body-based meanings, has tended to privilege meanings that are highly cognitive in nature. This stands to reason, given the ultimate interest mathematics educators have in coordinating body-based meanings with abstract mathematical symbolism. However, it does compromise both Johnson's and Dewey's more comprehensive approaches to aesthetics since it usually overlooks both affective and axiological dimensions of meaning.

It may well be close to impossible to coordinate the range of meanings arising from episodes of student mathematical work. Nonetheless, taking Radford's example above, it might be fruitful to examine how the perception and construction of rhythm also relates to affective meanings of comfort and security. Alternatively, might the perception and production of rhythm relate to a heightened sense of interest in, or perceived worthiness of, algebraic patterns? While these considerations may not be immediately germane to the cognitive concerns of researchers, they seem extremely relevant to understanding the full range of meanings that learners attach to mathematical ideas.

In sum, while categories such as cognition, affect and aesthetics provide useful and fruitful analytical tools when considered separately, they clearly intermingle in important ways in mathematical thinking and learning. However, even if assumptions about the primarily cognitive nature of mathematics were to be successfully challenged, the aesthetic dimension of human experience will always be more challenging to study, given their fuzzy, implicit, and ephemeral nature. An important first step would involve forging connections between aesthetics and existing theories in mathematics education. As I argue in the next section, this will require further consideration of the aesthetic dimension of mathematics enculturation.

SOCIAL CONSIDERATIONS

A second reason for the lack of attention to aesthetics relates to considerations that are much more social in nature. While the contemporary perspectives described above draw attention to the importance of the aesthetic in human experience and perception, they focus almost exclusively on individual capacities and tendencies than on the way in which aesthetic values and sensibilities are developed, shared, communicated, and disputed in human interaction. This more social perspective cannot be ignored in mathematics education, where issues of communication and enculturation are central.

The issue of enculturation is especially interesting in mathematics, where the sharing of aesthetic values has traditionally been rather secretive—or at least implicit—and elitist: the general practice is to begin aesthetic enculturation at the PhD level, when students are, often for the first time, having to choose a novel dissertation-level problem. Further, unlike other aesthetically-driven disciplines such as the visual arts, literature, or music, mathematics has no practice of public criticism and thus no mechanism through which aesthetic values might be articulated, defended, or socially mediated. While some philosophers have pointed to the dangers of this for the discipline itself (see Corry, 2001; Csiszar, 2003; Tymoczko 1993), the repercussions for mathematics education may be even worse because they lead to the belief that aesthetic values are either intrinsic to mathematics, or to the mathematicians who control them (see Sinclair and Pimm, 2010).

Bishop (1991) argues that mathematics education should go beyond developing students' conceptual understanding, and should include the teaching of the history and values of the discipline. He links the educational importance of making these values explicit in the classroom to improving the affective environment of the classroom. However, these values also play an important part in determining what mathematicians count as important or interesting in mathematics—these questions being aesthetic in nature (see Sinclair, 2006a). The impact of values extends beyond the emotional, and to the broader activities of mathematics such as inquiry and communication.

The notion of mathematical enculturation, which involves immersion in and reflection on the values of the mathematics culture, offers a more socially oriented opportunity for aesthetics in mathematics education. In particular, it suggests that aesthetic values should be explained and shared at the classroom level, and that the process of doing so may require longer periods of discussion and negotiation. Note that Bishop's perspective of mathematics enculturation is a critical one in that it is meant to expose students to the underlying values that not only drive the conceptual development of the discipline, but that also may interact with other social goals and discourses.

By taking the mathematics classroom as the unit of analysis—rather than the individual learner—researchers such as Yackel and Cobb (1996) have been able to study the way in which various normative values become established in a classroom. They have been particularly interested in mathematical norms that involve decisions about what counts as different when students discuss and offer solutions to problems (although they also acknowledge other mathematical values such as efficiency, elegance and sophistication). They show that it takes time, as well as strong guidance from the teacher, in order for students to understand what it means to be *mathematically* different. In the same way, the normative understandings of mathematical efficiency, elegance, sophistication, and other aesthetic values, would require a certain period of enculturation.

Also taking a classroom-based unit of analysis, Sinclair (2008) studies the ways in which aesthetic values are being communicated in a classroom where the teacher is not necessarily focussed on aesthetic enculturation. The research was based on the assumption that such values would be at least implicitly communicated, as hypothesised by the contemporary research described above. As predicted, while the teacher hardly ever used words such as "beauty" or "elegance" in reference to mathematical ideas, he did appeal to aesthetic values quite frequently, in terms of drawing students' attention to what counts as interesting in mathematics, or what kinds of things mathematicians like to do. For example, in presenting different ways to solve an algebraic equation involving fractions, the teacher talked about how they could "defeat the algebra beast" and turn the fractions into whole numbers. On one hand, the teacher is communicating the fact that techniques that can reduce complexity or "beastliness," are valued in mathematics. On the other, the teacher offers an image of some of the negative aesthetic responses that can be had in mathematics, namely, that fractions are ugly and beastly.

Sinclair also finds a range of responses from the students to the teacher's appeal to aesthetic values. Unless the teacher was able to elicit emotional responses from the students, his appeals to the aesthetic seemed to go unnoticed by the students, and were even interpreted as further supporting the anaesthetic vision of mathematics held by many people. This analysis thus provides some insight into the way in which aesthetic values are negotiated at the classroom level, and how easily unquestioned values from the teacher's perspective might override the aesthetic sensibilities and capacities of learners. Further research might focus more explicitly on students' interpretations of the aesthetic values evinced by teachers (as well as by textbooks and other materials).

CRITICAL PERSPECTIVES: POWER AND AESTHETICS

If one adopts the viewpoint of mathematicians such as Krull and Poincaré, namely, that the discipline of mathematics is fundamentally and crucially an aesthetic enterprise, then one must concede that most learners do not currently have the opportunity to do mathematics. The current elitist (or frivolous) positioning of the aesthetic in mathematics education has important repercussions when it comes to access and power. In fact, Sinclair and Pimm, (forthcoming) propose that the recent emphasis on "mathematics for all" may in fact be compromised by this very positioning.

In discussing the role of power in mathematics education, Valero (2005) points out that power cannot lie in the discipline of mathematics itself, nor in its practitioners, but must, instead, be seen as "a relational capacity of social actors to position themselves in different situations, though the use of various discourses" (p. 10). In terms of aesthetic considerations then, one cannot blame mathematics itself for its inaccessibility: mathematical objects are not beautiful in and of themselves, and they cannot transfer their beauty to potential learners. Nor can one blame mathematicians, even the ivory-tower, eminent mathematicians who are often seen as controlling the exchange of power. Instead, Valero proposes that power transactions evolve out of ever-changing and often subtle practices and discourses.

What might these subtle practices and discourses, which are disempowering learners in their aesthetic engagement with mathematics, look like? As I will show below, they are wide-ranging and surprisingly disparate. For example, consider the current school practice of assigning the aesthetic development of students to arts education—a practice that has developed over many centuries, and differs from schooling practices of the ancient Greeks. The allotment of aesthetic development to the arts makes it acceptable and reasonable to ignore aesthetic development in mathematics (and, of course, art teachers are not supposed to concern themselves with the logical development of their students). This practice seems deeply entrenched (as does the discourse around the purposes of mathematics and arts education more generally), and influential not only in mathematics, but also across the whole schooling system.

A more localised and recent example of the kind of practice that relates to power issues around the mathematical aesthetic involves the decline of geometry in the mathematics curriculum (and the corresponding ascendance of "numeracy," and the attention to fractions and algebra) over the past half-century. Limited experiences with the mathematics of shape and space reduce the range of sensory-based interactions that learners have with mathematical ideas, representations and phenomena, and thus limits learners' aesthetic engagement. This particular practice grows out of a more general turn toward the numeric, the analytic and the algebraic, or the tendency to talk in numbers about almost anything, and the underlying desire for generality, rigor and objectivity (see Sfard, 2008, for a discussion of how these desired properties are actually misleading). The privileging of number not only pushes out the spatial, visual and continuous, thus compromising embodied meanings, it also leads to a certain discourse about what counts are more valuable knowledge, thus affecting judgements about what is interesting, worthy, or even true.

Indeed, there are a wide range of discourses and practices that contribute to current power dynamics underlying the elitist view of the mathematical aesthetic, but I would like to highlight one in particular that is especially germane given its direct relation to aesthetic considerations. It involves the positioning of mathematics (especially by mathematicians) as an artistic discipline rather than a scientific one. For example, the mathematician Sullivan (1925/1956) claims that mathematicians are impelled by the same incentives as artists, citing as evidence the fact that the "literature of mathematics is full of aesthetic terms" and that many mathematicians are "less interested in results than in the beauty of the methods" (p. 2020) by which those results are found. There is also the argument that, unlike with the sciences, mathematics does not have to compare itself against an outside reality—thus, the implication being mathematicians have choice and freedom when it comes to selecting their objects of interest. Sullivan described mathematics as the product of a free, creative imagination and argued that it is just as "subjective" as the other arts. Related to this creative aspect, the mathematician G.H. Hardy also viewed mathematics (the kind he liked anyway) as an art: "I am interested in mathematics only as a creative art" (1967/1999, p. 115).

While one might expect this comparison of mathematics to the arts to enhance its accessibility, I propose that is actually has the opposite effect. The characteristics that mathematics supposedly share with the arts—creativity and free choice, as well as the use of "aesthetic terms"—may sound alluring to [a] non-mathematician, who can recognise them as familiar in other (less exclusive) experiences. Tell a mathematics-fearing artist that the discipline is really about ambiguity (see Byers, 2007), creativity and freedom, and their ears will likely perk up. However, these very characteristics only serve to remove the accessibility of mathematics from the non-mathematician further since, like the aesthetic sensibilities of Poincaré, they only belong to a privileged few.

It may be more fruitful to consider the differences between mathematics and the arts in understanding the power dynamics involved. Indeed, the philosopher Thomas Tymoczko (1993) may well have pointed out the most operative difference between aesthetic judgments in mathematics and those at work in the arts: the mathematics community does not have many "mathematics critics" to parallel the strong role played by art critics in appreciating, interpreting and arguing about the aesthetic merit of artistic products.[5] Lakatos also alludes to this in his *Proofs and Refutations*, when Gamma, exasperated by the never-ending complexities added to a simple equation in order to deal with "monsters", asks "Why not have mathematical critics just as you have literary critics, to develop mathematical taste by public criticism?" (Lakatos, 1976, p. 98). Gamma realises that truth cannot operate separately from taste when it comes to mathematical discovery: not every fact is worth proving.

Mathematics may well be a discipline of freedom and creativity, but in other disciplines that are driven by aesthetics there are critics to interpret and negotiate the meaning and place of creative new products. In mathematics, however, virtually no one stands on that border between the productive and interpretive aspects of creative work for mathematics (see, for instance, Corfield, 2002, on Lakatos's legacy in this regard). This is not just a problem for non-mathematicians, who have little help in assessing the importance of new developments in mathematics; it has been problematic within mathematics itself.

Rota's claim that mathematicians use language full of aesthetic terms to hide the fuzziness of their experiences (and perhaps of their truths) deserves further consideration. This will likely require the introduction of new discourses, ones like Thurston (1994) offers, gently expose rather than hide fuzziness.

Some Final Remarks

The philosophical tides are changing, as scholars become increasingly interested in the axiological dimension of philosophy and, in addition, in articulating the more porous borders between knowledge, feelings and values. The question for mathematics education research is whether these new directions in philosophy help solve any of the perennial problems of mathematics education. The most obvious relevance of aesthetic considerations in mathematics education research relates to student motivation, which persists as one of the greatest problems faced in mathematics education. Many researchers have proposed ways in which to address this problem, ranging from a focus on providing students with a better rationale for why they should study mathematics to finding ways in which to promote students' confidence. By and large, these proposals minimise aesthetics, and in cases where they do not, non-mathematical ideas or activities are frequently used to provide aesthetic values. Further research on the role that aesthetics can play in student motivation deserves urgent attention. This may involve investigating the extent to which students' aesthetic engagement leads to increased interest or intrinsic motivation. The notion of aesthetic engagement may vary depending on which of the contemporary approaches is adopted: from an embodied cognition point of view, for example, researchers might determine whether embodied experiences with mathematical ideas motivate students. In contrast, from a Deweyian perspective, promoting students' aesthetic engagement would involve a very different orientation toward the goals and purposes of mathematics education.

Using the embodied perspective described above, aesthetic considerations in research may also help solve more specific problems such as understanding how students can learn fractions better. Such problems have been studied through mostly cognitive approaches, and they may never be resolved without a broader concept of human understanding to guide the research questions and methods. The fact that students tend to want to avoid fractions probably has some non-cognitive origins, if we accept the theories outlined above. How does the belief that "fractions are hard" relate to the strange way in which they are written, the lack of embodied experiences they give rise to, or the lack of opportunities students have to encounter them in the context of satisfying experiences?

In addition to addressing recognised problems in mathematics education, the attention to aesthetic considerations in research—especially in terms of power dynamics—may also have more transformative influence, helping to suggest new possibilities, draw attention to problems that have not been recognised, or whose solutions have been taken for granted. In particular, it may suggest new ways of positioning school mathematics with respect to research mathematics—not as a mere servant, or as a separate discipline, but as an explicator, mediator and critic.

Notes

1. Higginson (2006) documents some of this in relation to mathematics. Additionally, several books can now be found on the aesthetics of science *It Must be Beautiful: Great Equations of Modern Science* (Farmelo, 2002), *Beauty and the Beast: The Aesthetic Moment in Science* (Fischer, 1999) and of computing, *Aesthetic Computing* Fishwick (2006).

2. Langer (1957) emphasizes this fact by describing how the merest sense-experience is a process of formulation; human beings have a tendency to organise the sensory field into groups and patterns of sense-data, to perceive forms rather than a flux of light-impressions.

They promptly and unconsciously "abstract a form from each sensory experience, and use this form to conceive the experience as a whole, as a thing" p. 90. For Langer, this unconscious appreciation of forms is the primitive root of all abstraction, which in turn is the keynote of rationality; so it appears that the conditions for rationality lie deep in pure animal experience—in the human power of perceiving, in the elementary functions of eyes and ears and fingers.

3. The physicist Freeman Dyson (1982) also distinguishes between two types of scientists, namely, the 'unifiers' and the 'diversifiers,' the former finding and cherishing symmetries, the latter enjoying the breaking of them.

4. In this book, Lakoff and Núñez offer a very stimulating perspective on the genesis of mathematical ideas, based on their theories of embodied cognition. While many scholars (including cognitive science, mathematicians and educators) have expressed reservations about their specific claims (see, for example, Schiralli and Sinclair, 2002), variations of the ideas expressed in this book have motivated many studies in mathematics education that are relevant to an embodied perspective on aesthetics.

5. It might be argued that journal editors play the role of the art critic in mathematics, but their work is done within a small community of mathematicians, rather than being available or addressed to those outside that community. Textbook authors also play a role similar to that of the art critic, in that they seek to organize, explain and even interpret mathematical products for an outside audience; however, they rarely actually criticize or question these products.

References

Albers, D., Alexanderson, G., & Reid, C. (1990). *More mathematical people.* New York: Harcourt Brace Jovanovic, Inc.

Aristotle. (2000). *Metaphysics.* Retrieved October 22, 2000, from http://classics.mit.edu/Aristotle/metaphysics.1.i.html.

Arzarello, F., & Robutti, O. (2001). From body motion to algebra through graphing. In H., Chick, K., Stacey, J., Vincent, & J. Vincent (Eds.), *Proceedings of the 12th ICMI study conference* (pp. 33–40). Vol. *1,* Australia: The University of Melbourne.

Bailland, B., & Bourget, H. (Eds.). (1905). *Correspondance d'Hermite et de Stieltjer. 2 vols.* Paris: Gauthier-Villars.

Baumgarten, A. G. (1739, 1758). *Texte zur Grundlegung der Ästhetik,* trans. H.R. Schweizer, Hamburg: Felix Meiner Verlag, 1983.

Beardsley, M. C. (1982). *The aesthetic point of view. Selected essays.* Ithaca: Cornell University Press.

Bishop, A. (1991). *Mathematics enculturation: a cultural perspective on mathematics education.* Dordrecht: Kluwer Academic Publishing.

Brown, S. (1973). Mathematics and humanistic themes: sum considerations. *Educational Theory, 23*(3), 191–214. doi:10.1111/j.1741-5446.1973.tb00602.x.

Burton, L. (2004). *Mathematicians as enquirers: Learning about learning mathematics.* Dordrecht: Kluwer Academic Publishers.

Byers, V. (2007). *How mathematicians think: Using ambiguity, contradiction, and paradoxes to create mathematics.* Princeton: Princeton University Press.

Corfield, D. (2002). Argumentation and the mathematical process. In G. Kampis, L. Kvasz, & M. Stöltzner (Eds.), *Appraising lakatos: Mathematics, methodology and the man* (pp. 115–138). Dordrecht: Kluwer Academic Publishers.

Corry, L. (2001). Mathematical structures from Hilbert to Bourbaki: the evolution of an image of mathematics. In U. Bottazzini, & A. Dahan Dalmedico (Eds.), *Changing images of mathematics: from the French revolution to the new millennium* (pp. 167–185). London: Routledge.

Crespo, S., & Sinclair, N. (2008). What can it mean to pose a 'good' problem? Inviting prospective teachers to pose better problems. *Journal of Mathematics Teacher Education, 11*, 395–415.

Csikszentmihalyi, M. (1990). *Flow: The psychology of optimal experience.* New York: HarperPerennial.

Csiszar, A. (2003). Stylizing rigor; or, why mathematicians write so well. *Configurations, 11*(2), 239–268. doi:10.1353/con.2004.0018.

Davis, P. (1997). *Mathematical encounters of the 2nd kind.* Boston: Birkhäuser.

de Fénélon, F. (1697/1845). *Oeuvres de Fénélon*, Vol. *1*, Paris, Firmin-Didot frères, fils et cie.

Dewey, J. (1934). *Art as experience.* New York: Perigree.

Dewey, J. (1938). *Logic: The theory of inquiry.* New York: Holt, Rinehart and Winston.

Dissan[ayake], E. (1992). *Homo aestheticus.* New York: Free Press.

Dreyfus, T., & Eisenberg, T. (1986). On the aesthetics of mathematical thought. *For the Learning of Mathematics, 6*(1), 2–10.

Farmelo, G. (2002). *It must be beautiful: Great equations of modern science.* London: Granta Publications.

Fischer, E. (1999). *Beauty and the beast: the aesthetic moment in science.* New York: Plenum.

Fishwick, P. (2006). *Aesthetic computing.* Cambridge: The MIT Press.

Goldin, G. A. (2000). Affective pathways and representations in mathematical problem solving. *Mathematical Thinking and Learning, 17*, 209–219. doi:10.1207/S15327833MTL0203_3.

Gombrich, E. (1979). *The sense of order: A study in the psychology of decorative arts.* Oxford: Phaidon Press.

Hardy, G. H. (1967/1999). *A Mathematician's apology (With a Foreword by C. P. Snow).* New York: Cambridge University Press.

Hawkins, D. (2000). *The roots of literacy.* Boulder: University Press of Colorado.

Higginson, W. (2006). Mathematics, aesthetics and being human. In N. Sinclair, D. Pimm, & W. Higginson (Eds.), *Mathematics and the aesthetic: New approaches to an ancient affinity* (pp. 105–125). New York: Springer.

Hofstatder, D. (1997). From Euler to Ulam: Discovery and dissection of a geometric gem. In J. King & D. Schattschneider (Eds.), *Geometry turned on: dynamic software in learning, teaching and research* (pp. 3–14). Washington: MAA.

Jackson, P. (1998). *John Dewey and the lessons of art.* New Haven: Yale University Press.

Johnson, N. (2007). *The meaning of the body: Aesthetics of human understanding.* Chicago: The University of Chicago Press.

Krull, W. (1930/1987). The aesthetic viewpoint in mathematics. *The mathematical intelligencer, 9*(1), 48–52.

Lakatos, I. (1976). *Proofs and refutations: the logic of mathematical discovery.* Cambridge: Cambridge University Press.

Lakoff, G., & Núñez, R. (2000). *Where mathematics comes from: How the embodied mind brings mathematics into being.* New York: Basic Books.

Langer, S. (1957). *Philosophy in a new key*, 3rd edn. Cambridge: Harvard University Press.

Lehrer, R. (2008). Developing a culture of inquiry in an urban sixth grade classroom. Presentation at the American education research association conference. 27 March, 2008, New York.

Le Lionnais, F. (1948/1986). *Les Grands Courants de la Pensée Mathématique.* Marseille: Rivages.

Littlewood, J. E. (1986). The mathematician's art at work. In B. Bollobás (Ed.), *Littlewood's miscellany*. Cambridge: Cambridge University Press.

Mason, J., Burton, L., & Stacey, K. 1982: *Thinking mathematically*. Reading: Addison-Wesley.

Nemirovsky, R. (2003). Three Conjectures concerning the relationship between body activity and understanding mathematics. In N. A. Pateman, B. J. Dougherty, & J. T. Zilliox (Eds.), *Proceedings of PME 27*, Vol. 1, pp. 103–135.

Núñez, R. (2004). Do real numbers really move? Language, thought, and gesture: The embodied cognitive foundations of mathematics. In F. Iida, R. Pfeifer, L. Steels, & Y. Kuniyoshi (Eds.), *Embodied artificial litelligence* (pp. 54–73). Berlin: Springer.

Papert, S. (1978). The mathematical unconscious. In J. Wechsler (Ed.), *On aesthetics and science* (pp. 105–120). Boston: Birkhäuser.

Peirce, C.S. (1908/1960). A neglected argument for the reality of God. In C. Hartshorne, & P. Weiss (Eds.), *Collected papers of Charles Sanders Peirce (Vol. 6: Scientific metaphysics)*. Cambridge: Harvard University Press.

Pinker, S. (1997). *How the mind works*. New York: W·W. Norton & Company.

Poincaré, H. (1908/1956). Mathematical creation. In J. Newman (Ed.), *The world of mathematics*, Vol. 4 (pp. 2041–2050), New York: Simon and Schuster.

Radford, L. (2003). Gestures, speech and the sprouting of signs. *Mathematical Thinking and Learning*, 5(1), 37–70. doi:10.1207/S15327833MTL0501_02.

Rota, G.-C. (1997). *Indiscrete thoughts*. Boston: Birkhäuser.

Schiralli, M. (1999). *Constructive postmodernism: Toward renewal in cultural and literary studies*. Westport: Bergin & Garvey.

Silver, E., & Metzger, W. (1989). Aesthetic influences on expert mathematical problem solving. In D. McLeod, & V. Adams (Eds.), *Affect and mathematical problem solving* (pp. 59–74). New York: Springer.

Sinclair, N. (2001). The aesthetic is relevant. *For the Learning of Mathematics*, 21(2), 25–32.

Sinclair, N. (2002). The kissing triangles: The aesthetics of mathematical discovery. *International Journal of Computers for Mathematical Learning*, 7(1), 45–63. doi:10.1023/A:1016021912539.

Sinclair, N. (2004). The roles of the aesthetic in mathematical inquiry. *Mathematical Thinking and Learning*, 6(3), 261–284. doi:10.1207/s15327833mt10603_1.

Sinclair, N. (2006a). *Mathematics and beauty: Aesthetic approaches to teaching children*. Teachers College Press.

Sinclair, N. (2006b). The aesthetic sensibilities of mathematicians. In N. Sinclair, D. Pimm, & W. Higginson (Eds.), *Mathematics and the aesthetic: New approaches to an ancient affinity* (pp. 87–104). New York: Springer.

Sinclair, N. (2008). Attending to the aesthetic in the mathematics classroom. *For the Learning of Mathematics*, 28(1), 29–35.

Sinclair, N., & Pimm, D. (2010). The many and the few: Mathematics, Democracy and the Aesthetic. *Educational Insights*, 13(1).

Singh, S. (1997). *Fermat's enigma: the epic quest to solve the world's greatest mathematical problem*. New York: Viking.

Sfard, A. (2008). *Thinking as communicating: Human development, the growth of discourses, and mathematicizing*. Cambridge: Cambridge University Press.

Sullivan, J. (1925/1956). Mathematics as an art. In J. Newman (Ed.), *The world of mathematics*, Vol. 3 (pp. 2015–2021). New York: Simon and Schuster.

Thurston, W. (1994). On proof and progress in mathematics. *Bulletin of the American Mathematical Society*, 30(2), 161–177. doi:10.1090/S0273-0979-1994-00502-6.

Tymoczko, T. (1993). Value judgments in mathematics: can we treat mathematics as an art? In A. White (Ed.), *Essays in humanistic mathematics* (pp. 62–77). Washington: The Mathematical Association of America.

Valero, P. (2005). What has power got to do with mathematics education? In D. Chassapis (Ed.), *Proceedings of the 4th dialogue on mathematics teaching issues*, Vol. 1, (pp. 25–43). Thessaloniki, University of Thessaloniki.

von Glasersfeld, E. (1985). *Radical cconstructivism: a way of knowing and learning*. London: Falmer Press.

von Neumann, J. (1947). The mathematician. In R. Heywood (Ed.), *The works of the mind* (pp. 180–196). Chicago: The University of Chicago Press.

Weil, A. (1992). *The apprenticeship of a mathematician*. Trans. J. Gage. Berlin: Birkhauser.

Wells, D. (1990). Are these the most beautiful? *The Mathematical Intelligencer*, *12*(3), 37–41.

Whiteley, W. (1999). The decline and rise of geometry in 20th century North America. In J. G. McLoughlin (Ed.), *Canadian mathematics study group conference proceedings* (pp. 7–30). St John's: Memorial University of Newfoundland.

Wilson, E. (1998). *Consilience: the unity of knowledge*. New York: Knopf.

Yackel, E., & Cobb, P. (1996). Sociomathematical norms, argumentations and autonomy in mathematics. *Journal for Research in Mathematics Education*, *27*, 458–477. doi:10.2307/749877.

Mathematics Textbooks and Their Potential Role in Supporting Misconceptions

ANN KAJANDER AND MIROSLAV LOVRIC

Introduction

Mathematics textbooks are integral parts of our daily lives as mathematicians and mathematics teachers. Students use mathematics textbooks to study and to do homework questions, while professors and teachers may use them to prepare classes and to teach. We also use them to look up a formula or a theorem, and to prepare tests and exams for our students. While at the elementary and secondary levels teachers also have curriculum documents from which to work, the reality is that most teachers still use the textbook as their primary resource [1]. It determines both the material that needs to be covered and the way it is presented. 'Textbooks form the backbone as well as the Achilles' heel of the school experience in mathematics. The dominance of the textbook is illustrated by the finding that more than 95% of 12th-grade teachers indicated that the textbook was their most commonly used resource.' [2]. However, while textbooks support our thinking about teaching, we rarely think about studying the textbooks themselves!

There have been attempts at evaluating textbooks, studies exploring the relationship between textbooks and curriculum (mostly K–12), and also global comparisons (for example [3–5]). Some effort has been put into content analysis and exploring the ways in which textbooks are used in classrooms and beyond (for example [6, 7]). However, very few *mathematics education* researchers have taken a really close look at what is in the textbooks, with the focus on *how* the material is presented and *what kind of learning* may be implied. In many cases, 'unscientific market research is chiefly used to determine content and approach' ([8], p. 55). Furthermore, 'Commercially published, traditional textbooks dominate mathematics curriculum materials in US classrooms and to a great extent determine teaching practices' ([8], p. 55).

Project 2061, an initiative of the American Association for the Advancement of Science, is a long-term project aimed at evaluating teaching and learning resources in science and mathematics. Founded in 1985, its goal is to 'help all Americans become literate in science, mathematics, and technology' (Project 2061 [9]). According to one of their studies, '. . . the majority of textbooks used for algebra—considered the gateway to higher mathematics—have some potential to help students learn, but they also have serious weaknesses' [10]. More than half of the 12 middle school and secondary textbooks evaluated were considered adequate, but none was rated highly. Three textbooks that have been widely used were rated '. . . so inadequate that they lack potential for student learning' [10]. The same report claims that '. . . No textbook does a satisfactory job of building on students' existing ideas about algebra or helping them overcome their misconceptions or missing prerequisite knowledge.' For details and complete report, see [11]. Project 2061 used the guidelines for how and what students should learn (referred to as the *Standards*), as defined by the National Council of Teachers of Mathematics [12]. According to the Project 2061 findings, authors of textbooks generally ignore the research on how students acquire ideas and concepts.

Framework and Methodology

The theory of *conceptual change* [13, 14] provides an ideal framework for the study of potential student misconceptions related to learning from textbooks. The theory describes learning processes of adults as well as children and hence is appropriate in addressing high school as well as university students. Of particular interest are situations in which the new knowledge (be it learned by a child or an older student, or discovered by a research scientist) is incompatible with the prior knowledge, and hence might affect the understanding of the material. Based on a certain amount of information (learned from a book, heard in a class, etc.), a learner uses her/his own ideas and current understanding to create an initial explanatory framework. The presuppositions that inhabit this framework, also known as naïve beliefs or alternative conceptions, will form student's current beliefs about the material. Faced with having to absorb the material that is in some way incompatible with the prior knowledge, the student will try to assimilate new information into their existing framework, thus creating a so-called synthetic model. As a mixture of beliefs and scientific facts, this synthetic model represents student's misconceptions about the subject.

Evidence suggests [13, 14] that formed beliefs are very strong and, consequently, synthetic models become quite robust in one's cognitive environment. It is an important purpose of a course instructor to identify these models and to apply adequate means in teaching and learning environments to help students cope with them.

One method of using the textbook to support the use of classroom time in overcoming misconceptions grounded in prior beliefs might be to require that students prepare material in advance of a class by reading a textbook. The course instructor might then transform the way she/he teaches in many productive ways such as focussing on the material that students are unclear about or indicate as challenging, or emphasizing topics that are of particular importance. Discussion replaces lecturing, and there is ample opportunity to go deeper into material and to make connections with other areas of mathematics that, for various reasons, are not covered in the textbook.

Assuming that an instructor will teach *with* the textbook (i.e. students will use it to read and study), the kind of research presented here raises awareness of a certain type of problems and issues that need to be addressed in classes. The fact that (mathematically incorrect) synthetic models can be quite robust means that a serious effort needs to be employed to devise strategies aimed at weakening the beliefs that inhabit them.

We have begun to take a closer look at textbooks commonly used in Ontario at the grade 12 and first year university levels to teach calculus, in order to determine to what extent, and how, mathematics textbooks potentially contribute to the creation and strengthening of students' conceptions and misconceptions about mathematics. More generally, we are investigating to what extent textbooks promote (or not) deep, conceptual understanding of the material that they present. The case study we present here serves to illustrate potential issues, and also suggests a framework for our further analyses.

Tangent Line to a Curve and Related Synthetic Models

Examining calculus textbooks, we have identified a number of misconceptions related to the presentation (both narrative and visual) of the concept of the line tangent to the graph of a function $y = f(x)$. For clarity, we have classified issues potentially leading to synthetic models into several categories, and selected typical cases to illustrate each category (in a few cases, we mention misconceptions arising elsewhere in calculus). Naturally, there is a certain amount of overlap between these categories (i.e. a synthetic model could belong to two, or several categories). Furthermore, we do not claim that our list is comprehensive in any way.

A fairly common synthetic model many entering first-year university students possess includes ideas described by statements such as 'the tangent is the line that touches the graph of $y = f(x)$ at only one point,' or a 'tangent line crosses the graph at one point.' When we examined several grade 12 calculus textbooks currently in use in some Ontario schools, we found evidence that could support (or at best fail to correct) such a misconception. For example, while the initial explanation near the beginning of the chapter on tangents states (with accompanying diagrams)

> In the graphs of the circle and the parabola, a tangent line touches exactly one point of the graph P. For other curves, such as the one in the third diagram [an example of a tangent which also crosses the curve at two other points] a tangent line touches the graph at the point of tangency, P, but may pass through other points on the graph as well ([15], p. 183).

but later on [in] a summary box highlighted in blue, the same textbook contains the unadorned statement

> A tangent is a line that touches exactly one point on the graph of a relation (p. 190).

In a university textbook, we find the statement

> If there is a tangent line to a curve at P—a line that just touches the curve like the tangent to a circle—it would be reasonable to identify the slope of the tangent as the slope of the curve at P ([16], p. 57).

Later, the textbook presents the usual limit-of-the-secant-lines approach to defining the tangent, and shows an example where the tangent crosses the curve at the point of tangency. However, the authors do not adequately address the 'touching' misconception. Another textbook [17] shows the graph of a function and two lines, both of which intersect the graph at the same point. In answer to 'Are these both tangent lines?' it states

> Certainly not; only one of these acts like a tangent line to a circle, by hugging the curve near the point of tangency ([17], p. 150).

The book then continues by building correct conceptions (such as discussing direction as what is common to the tangent and the curve, and local linearity), but, again, does not revisit the 'hugging/touching' metaphor to correct it. Similarly, we find

> The word tangent stems from the Latin word *tangens*, for 'touching.' Thus a line tangent to a curve is one that 'just touches' the curve ([18], p. 45).

However, there are textbooks that recognize the issues related to the 'touching' cognitive model and address it in an adequate way. For instance, in Larson et al. [19], the authors

- recall the concept of the tangent to the circle, and announce its inadequacy in the general case of the tangent to the graph of a function
- address the 'tangent touches the curve' notion directly, by showing examples (with illustrations) where the notion no longer applies
- develop carefully and precisely a correct mathematical definition of the tangent line
- discuss the case of the vertical tangent appropriately (i.e. they say how to define the tangent in the case when the limit of the slopes of secant lines is plus/minus infinity and the general definition for slope of a tangent does not apply)
- do not mention the word 'touching' even once after the precise definition is given, thus suggesting that the inappropriate 'tangent touching' cognitive model be abandoned.

We suggest that such treatment corrects, refines, and builds upon the earlier-held beliefs, and supports deeper understanding.

The textbook 'Calculus and Its Applications' [20], in our opinion, provides the most adequate way of addressing the 'touching' cognitive model. In the introduction to a rigorous treatment of the tangent and the derivative, the authors explicitly state four most common misconceptions related to the tangent line. Using graphs, i.e. constructing visual (counter)examples, they carefully analyse all of them. With all misconceptions now fully exposed, readers are invited to reflect, and to construct their own definition of the tangent line. Next, the authors begin to develop the concepts in a precise mathematical framework. Once this is accomplished, there is no going back to the colloquial language of 'touching' or 'crossing.'

In conclusion to this section, let us mention that there is another reason why the term 'touching' is not appropriate. Imagine touching a book. Although placed very closely next to each other, our finger and the book do not share any part of space (not a single atom!). However, the graph of a function and its tangent actually do share part of the space—the point of tangency.

INCORRECT GENERALIZATIONS, STATEMENTS TAKEN OUT OF PRECISE MATHEMATICAL CONTEXT

Certain misconceptions arise from correct statements, when one forgets the precise (narrow) context in which they originally appeared. For example, the tangent to a circle is usually defined or described as the line that touches (or crosses) the circle at one point only, and the circle lies entirely on one side of it. The concept of the tangent to conic sections, introduced some time later in secondary school, is relatively easily incorporated into students' cognitive model of the tangent, since it is a straightforward generalization of the concept of the tangent to a circle. However, the introduction of tangents to general curves produces new knowledge (which might not surface before first-year university) that is incompatible with the previous knowledge, and fosters the creation of (or reinforces already existing) synthetic models of the tangent that are mathematically incorrect. Many secondary as well as first-year university calculus textbooks that we examined contain material that strengthens these models even further, or, at best, do not go far enough to convincingly dispel the naïve beliefs students might possess.

Perhaps the most common misconception of this type is the one that relates to vertical asymptotes. In high school, students study the case of rational functions with numerator 1, and use the correct fact that such functions have vertical asymptotes at all points where they are not defined (i.e. where the denominator is zero). Later, when studying general rational functions (i.e. with an arbitrary polynomial in the numerator), they use the same model, although it no longer holds (for instance, $(x^2 - 1)/(x - 1)$ does not have a vertical asymptote at $x = 1$).

ILLUSTRATIONS AND DIAGRAMS AS SOURCES OF MISCONCEPTIONS

Standard university calculus texts contain numerous illustrations of tangent lines. However, in a vast majority of cases, the tangent is shown in the 'generic' position where it 'touches' the curve at one point (and does not cross it).

In Ellis and Gulick [21], the authors recall the idea of the tangent to a circle (accompanied by an appropriate illustration), and then also show a tangent to a spiral (page 54). However, the point of tangency on the spiral is chosen in such a way (on the outmost winding) that the tangent does not cross the spiral anywhere else. In fact, an ideal opportunity to show how a tangent can cross a curve at many points has been missed! As a side issue, the choice of the spiral is not appropriate to announce the generalization (the generalization deals with graphs of functions $y = f(x)$ and a spiral (as shown there) is not the graph of a function). Interestingly, the authors do note

> Yet the idea of a tangent line 'touching' a curve does not lend itself well to drawing a tangent line, nor does it give a procedure for deriving an equation of a tangent line, which is important in calculus ([21], p. 55).

but they do not provide a clear alternative to drawing the tangent. Furthermore, the tangent to a circle is used to show (visually) how secant lines approach the tangent, which could also potentially reinforce the 'tangent touches the curve' misconception.

The concept of 'touching' might be further supported by examples in which students are given the graph of a function $f(x)$ and are asked to sketch the graph of its derivative $f'(x)$. Although the tangent is defined as limit of secant lines, the illustrations or accompanying text in these examples do not attempt to encourage drawing the tangent or thinking about it as the limiting position of secant lines.

To address this misconception, a textbook (or instructor) could ask students to create illustrations that show relationships between curves and lines and identify which are (or are not) tangents. These illustrations should include cases such as (a) a tangent which does not 'touch' on one side but instead crosses the graph at the point of tangency, (b) a tangent which crosses the graph at two (or more) points (one of which is the point of tangency), (c) a tangent line which 'touches' the graph at more than one point, (d) a line which 'touches' the graph, does not cross it, but is not a tangent (cusp), etc. By analysing such situations, students will potentially rework some of their misconceptions and expand their cognitive model of the tangent.

OVERSIMPLIFICATION (SUMMARY DEFINITION; INTERPRETATION)
OR OMISSION OF SPECIAL CASES

Related to the issue of over-use of colloquial language, examples of the situation of over-simplification can be found in attempts to verbalize or rephrase a notion that has already been defined, to make it more 'clear' or 'understandable'. Also, they could be found in attempts at summarizing an idea or concept. For example, consider the following description of the secant line:

> In particular, consider the line joining the given point P to the neighboring point Q on the graph of f, as shown in Figure 3.1. This line is called a secant (a line that intersects, but is not tangent to, a curve). ([22], p. 128)

The first sentence is a precise and correct definition of the secant. In an attempt to describe it in words (the statement in the brackets in the second sentence) the authors provided a misleading (and actually wrong) interpretation.

All textbooks we have examined define the tangent line as the limit of secant lines. However, very few mention (as part of the definition, or, say, immediately following the definition), the fact that the tangent line could be vertical (this can be illustrated, for instance, by analysing the tangent to the graph of the cube root function $y = x^{1/3}$ at $x = 0$; the textbook we mentioned earlier ([19]) is an exception). From the definition of the tangent, it is not at all clear why this case deserves special attention—that is usually given to it later, in a section on differentiability, where it is named 'vertical tangent.' The fact that the statements 'function $f(x)$ is differentiable at a' and 'the graph of $f(x)$ has a tangent at $(a, f(a))$' are not equivalent (precisely because of 'vertical tangents') is, very often, inadequately addressed in the presentation of the material.

A common strategy that many students employ when finding vertical asymptotes is to identify points where the given function is not defined. Quite often, they identify the statement 'if $f(a)$ is defined then f cannot have a vertical asymptote at a' as false. (It is easy to construct a counterexample: the function $f(x) = 1$ if $x = 0$, and $f(x) = 1/x$ otherwise is defined at $x = 0$, and has a vertical asymptote there.) The source of this misconception can be traced, again, to attempts at verbalizing a correctly introduced concept in order to 'simplify' it (for instance, because the definition involves numerous mathematical symbols and/or concepts).

DISCUSSING CONCEPTS THAT HAVE NOT BEEN PRECISELY DEFINED

In the introductory part of the section on tangents and derivatives in Bradley and Smith [22], the reader is reminded that

a tangent to a circle is defined as a line in the plane of the circle that intersects the circle in exactly one point (p. 128).

Then, the text states 'however, this is much too narrow a view for our purposes in calculus' (p. 128) without any indication of why this is so. Continuing in the same, somewhat vague manner, the textbook does not precisely define the meaning of the word 'tangent.' Referring to an illustration that shows four secant lines and a tangent line (with no explanation of how the tangent line was obtained), we read

Notice that a secant is a good approximation to the tangent at point P as long as Q is close to P (p. 128).

with no elaboration on meaning of the phrase 'good approximation'. Later (p. 129) the text begins the explanation 'To bring the secant closer to the tangent . . .', i.e. starts discussing properties of the tangent as if the tangent were already defined. In a similar vein, after reading the sentence

> By the definition of the derivative, $f'(a)$ represents the slope of the line tangent to the graph of f at $(a, f(a))$ ([21], p. 106).

we tried to identify where, previously in the text, the authors defined the tangent. Following through the pages, we found the usual material (reference to the tangent of the circle, the mention of the idea of a tangent line 'touching' a curve, etc.) but we could not find a rigorous mathematics statement defining the tangent line.

Suggestion for an Alternative Cognitive Model of the Tangent

As an alternative model, visualize the tangent definition (i.e., the limit of secant lines) as a sequence of magnifications, zooming in on the point of tangency (for example see the grade 12 text by Alexander *et al.*, [23], p. 21). Some textbooks do show this, but do not emphasize strongly enough its importance. If these magnifications tend to flatten the graph (i.e. make it look more and more like a line), then the graph (most likely) has a tangent line at the point in question. Using this approach, it becomes obvious, for example, that $y = |x|$ does not have a tangent at $x = 0$.

The 'magnification' approach extends naturally to functions of two variables (i.e. to the concept of the tangent plane to the graph of $f(x, y)$). Moreover, it reinforces the crucial idea that the tangent (line, or plane) is a local—and also linear—object.

The fact that the 'magnification' cognitive model of the tangent is fundamentally different from the usual 'tangent touches the curve' model potentially prevents the student from building a synthetic model of the tangent. It becomes difficult to assimilate the notion of 'zooming in' on the graph of the curve $f(x)$ to dynamically approximate the tangent more and more precisely, with the notion of a (static) line 'touching' the curve. Instead, it might force the student to completely abandon 'the tangent touches the curve' model and adopt the new, 'magnification' model. Furthermore, the 'magnification' model naturally leads into the discussion of the role of tangent as the best approximation of the graph of a function by a straight line.

Final Remarks

We were surprised that this preliminary analysis uncovered such a breadth of issues related to basically one topic. Situations leading to potential misconceptions occurred consistently in multiple sources. Acknowledging that textbooks remain a fundamental teaching resource, we suggest that more attention be paid to the presentation of mathematics. Furthermore, analyses of textbooks should include developmental as well as subject matter scrutiny.

In our forthcoming work, using the initial exploratory framework suggested here, we plan to investigate further misconceptions related to tangents and derivatives (local linearity, linear approximation and differentiability), as well as those related to other concepts [24].

We believe that as students move through the secondary curriculum and (potentially) develop deeper and more accurate conceptual understandings of fundamental concepts as they enter university, more attention should be paid to textbooks to continue to support rather than marginalize such growth.

References

[1] Kajander, A., & Mason, R. (2007). *Examining teacher growth in professional learning groups for in-service teachers of mathematics.* Canadian Journal of Mathematics, Science and Technology Education, 7(4), 417–438.

[2] C. McKnight, F.J. Crosswhite, and J.A. Dossey, *The Underachieving Curriculum*, Stipes Publishing, Champaign, IL, 1987.

[3] C.H. Schutter and R.L. Spreckelmeyer, *Teaching the Third R: A Comparative Study of American and European Textbooks in Arithmetic*, Council for Basic Education, Washington, DC, 1959.

[4] J.W. Stigler, et al., *An analysis of addition and subtraction word problems in American and Soviet elementary mathematics textbooks*, Cognition Instruct. 3 (1986), pp. 153–171.

[5] Y. Li, *A comparison of problems that follow selected content presentations in American and Chinese mathematics textbooks*, J. Res. Math. Educ. 31 (2000), pp. 234–241.

[6] E. Love and D. Pimm, *'This is so': a text on texts,* in International Handbook of Mathematics Education: Part One, A. Bishop, et al., ed., Kluwer Academic Publishers, Dordrecht, 1996, pp. 371–409.

[7] R. McCrory, *Mathematicians and mathematics textbooks for prospective elementary teachers*, Not. AMS 53(1) (2006), pp. 20–29.

[8] D.H. Clements, *Curriculum research: toward a framework for "research-based curricula"*, J. Res. Math. Educ. 38 (2007), pp. 35–70.

[9] Project 2061 (2000a). Available at http://www.project2061.org/about/.

[10] Project 2061 (2000b). *Algebra for all—not with today's textbooks, says AAAS.* Available at http://www.project2061.org/about/press/pr000426.htm.

[11] Project 2061 (2000c). Available at http://www.project2061.org/publications/textbook/default.htm.

[12] National Council of Teachers of Mathematics. *Principles and Standards for School Mathematics*, NCTM, Reston, VA, 2000.

[13] J. Davis, (2001) Conceptual change, in Emerging Perspectives on Learning, Teaching, and Technology (e-book), M. Orey, ed., Available at http://www.coe.uga.edu/epltt/ConceptualChange.htm

[14] I. Biza, A. Souyoul, and T. Zachariades, *Conceptual Change In Advanced Mathematical Thinking*, Discussion paper, Fourth Congress of ERME, Sant Feliu de Guíxols, Spain, 17–21 February 2005.

[15] C. Kirkpatrick, et al., *Advanced Functions and Introductory Calculus*, Nelson, Scarborough, ON, 2002.

[16] J. Hass, M.D. Weir, and G.B. Thomas Jr, *University Calculus*, Pearson, Boston, 2007.

[17] T. Smith and R.B. Minton, *Calculus*, McGraw-Hill, New York, 2002.

[18] C.H. Edwards and D.E. Penney, *Calculus with Analytic geometry; Early Transcendentals*, 5th ed., Prentice Hall, Upper Saddle River, 1998.

[19] R. Larson, R.P. Hostetler, and B.H. Edwards, *Calculus, Early Transcendental Functions*, 2nd ed., Houghton Mifflin Company, Boston, New York, 1999.

[20] S.J. Farlow and G.M. Haggard, *Calculus and Its Applications*, McGraw-Hill, New York, 1990.

[21] R. Ellis and D. Gulick, *Calculus with Analytic Geometry*, 4th ed., Harcourt Brace Jovanovic, Publishers and its subsidiary, Academic Press, San Diego, 1990.

[22] G.L. Bradley and K.J. Smith, *Calculus*, 2nd ed., Prentice Hall, Upper Saddle River, New Jersey, 1999.

[23] R. Alexander, et al., *Advanced Functions and Introductory Calculus 12*, Addison Wesley, Toronto, 2003.

[24] M. Lovric, *Mathematics textbooks and promotion of conceptual understanding.* Presentation to the Education Forum of the Fields Institute for Mathematical Sciences, University of Toronto, Toronto, 2007.

Exploring Curvature with Paper Models

HOWARD T. ISERI

Curvature is a fundamental concept in modern geometry, and yet the topic is often neglected in the high school and undergraduate curriculum. A concept of curvature dating to Carl Friedrich Gauss (1777–1855) is used by differential geometers to study both Euclidean geometry and the non-Euclidean geometries discovered by Janos Bolyai (1802–1860), Nikolai Lobachevsky (1792–1856), and Bernhard Riemann (1826–1866) as curved surfaces. Although a formal investigation of the curvature of surfaces requires advanced multivariate calculus, the underlying concept can be explored using only basic plane geometry. The origin of this idea can be traced to Rene Descartes (1596–1650.) By measuring angles on polyhedral surfaces, Descartes measured the same quantities that Gauss would measure two centuries later and, at the same time, gave us a simple way to explore non-Euclidean geometry in terms of curvature.

Although curvature originates in mathematics, popular interest in the topic is due mostly to scientists. Einstein's discovery of the general theory of relativity (1915) introduced the notion that the behavior of gravity can be explained by equating the presence of matter in space with curvature. This curvature of space would allow a photon (with no mass, and hence no gravitational attraction) to travel in a straight line but also to appear to change direction when passing a massive object. This phenomenon was observed in the famous experiment that brought general relativity to the public's attention, when in 1919 Arthur Eddington, during a solar eclipse, observed that a distant star appeared to have changed position because the light coming from the star had traveled close to the sun. Today, the curvature and geometry of our universe are at the center of the debate concerning whether the universe contains enough matter to keep it from expanding forever.

The first part of this article examines how the curvature of curves and surfaces is related to the angles of polygons or polygonal curves and "solid" angles at the vertices of polyhedra. We examine the theorem at the center

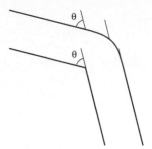

FIGURE 1. These two curves change direction by the same amount.

of Descartes' work in this area to give us a simple measure of the sharpness of a solid angle and then show how Descartes' theorem can be proved from the seemingly unrelated Euler theorem. The second part of the article presents examples of paper models that use Descartes' version of curvature to explore hyperbolic and elliptic figures in a precise and accessible way.

The Curvature of Curves

Figure 1 shows two paths, one polygonal and one smoothly curved. Following each from bottom to top, we see that the beginning direction and the ending direction are the same for both curves. The change in direction, therefore, is measured in both instances by the same angle, θ. For the polygonal curve, we call this angle the *exterior angle*. The *interior angle*, the one we typically measure, is supplementary to the exterior angle, so they sum to 180 degrees, or π radians, and we can easily find one from the other on a polygonal curve.

If we interpret a curve as a path that we travel along, an external angle on a polygonal path measures an abrupt change in direction. On a smooth curve, this change in direction occurs gradually, and we have, in some sense, infinitely many, infinitesimally small external angles. So although we can measure the change in direction between two points on the same curve, another concept is needed to adequately describe how this change in direction takes place. This concept, called *curvature*, is defined to be the rate at which the direction changes with respect to arc length. From point A to point B on a curve, for example, the direction will change by some angle $\Delta\theta$, and the arc length along the curve between these two points will be a distance Δs. This notation gives us an *average* curvature of $\Delta\theta/\Delta s$. Taking shorter sections of the curve and a limit, we arrive at the curvature at each point, which is an instantaneous rate of rotation.

In figure 2, we have a circle of radius r. The arc between points A and B is a quarter of the circle and has length $2\pi r/4 = \pi r/2$. The change in

FIGURE 2. The change in direction from point A to point B is $\pi/2$ radians over a distance of $\pi r/2$.

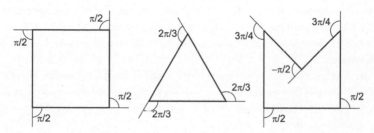

FIGURE 3. The external angles on a polygon measure the rotations necessary to walk around it.

direction between points A and B is $\pi/2$ radians. Dividing the change in direction by the arc length gives an average rate of change in direction (curvature) of $1/r$ radians per unit distance. The average curvature, not surprisingly, is the same for any arc of the circle, so in the limit, the curvature (denoted by the Greek letter κ, "kappa") at every point on a circle of radius r is $\kappa = 1/r$ radians per unit distance. Note that small circles will have more curvature, which should make sense. In general, if a curve has curvature κ at some point, then the part of the curve near that point will look like part of a circle of radius $1/\kappa$.

The significance of curvature to geometry becomes clearer in higher dimensions, and Descartes' theorem allows us to explore that significance. Before discussing surfaces, however, we will be helped by considering a lower-dimensional version of Descartes' theorem.

Imagine an ant walking around each shape in figure 3 in a counter-clockwise direction. To do so it must turn as indicated by the shapes' exterior angles. The first figure is a square, and the four exterior angles each measure $\pi/2$ radians. The sum of the exterior angles in this instance is 2π radians. The sum is the same for the triangle. The figure on the right has one turn to the right, which we define to be negative, and with that, the exterior angles again sum to 2π radians.

One-dimensional Descartes' theorem: Given any simple, closed polygonal curve in the plane, the exterior angles will sum to 2π radians.

This theorem holds true for smooth curves. Imagine a circle painted on the floor. If you walk around it once, your body will make one complete rotation. If you face north initially, you will turn gradually until you face west, then south, and then north again. That is, you will make a 2π radian rotation. On a polygonal curve, the only difference is that the turns happen abruptly, but the net rotation is still 2π radians. For smooth curves, because curvature is essentially a derivative of the direction function with respect to the arc length, integrating the curvature around a simple, closed curve will give the net rotation, 2π radians. For this reason, an accumulated amount of curvature is called *total curvature*, and so total curvature is the same as a net change in direction. That is, the sum of the external angles measures total curvature.

In Euclidean (plane) geometry this sum of angle measures around a closed figure is 2π radians, and we will see that deviations from the number 2π serve as measures of how much a non-Euclidean geometry differs from the Euclidean model.

The Curvature of Surfaces

We have seen that the total curvature around a simple closed curve in the plane is always the same. As we turn to surfaces, the situation is similar, as long as we choose a suitable measure of curvature. *Gaussian curvature* is the standard measure of curvature for a smoothly curving surface, and it nicely ties the standard non-Euclidean geometries together as surfaces with constant Gaussian curvature. In particular, Gaussian curvature is 0 for the Euclidean plane, -1 for the hyperbolic "plane," and 1 for the elliptic and spherical "planes." Furthermore, we can show that integrating Gaussian curvature over any surface that is a smooth deformation of a sphere always yields the number 4π, which should remind us of the fact that the net rotation for a simple closed curve in the Euclidean plane is always 2π. Here, as with curves, an accumulated amount of Gaussian curvature is called *total curvature*. We can say, therefore, that closed curves (which are deformations of a circle) in the Euclidean plane always have total curvature 2π, and closed surfaces in Euclidean space that are deformations of a sphere always have total (Gaussian) curvature 4π. Being a deformation of a sphere here is important, because deformations of a torus, for example, always have total curvature 0.

Unfortunately, the definition of Gaussian curvature is somewhat involved, and calculating it for a typical surface is impractical. As with the curvature of curves, however, Gaussian curvature has an angular analog, which Descartes described in a manuscript with a tenuous existence. This manuscript

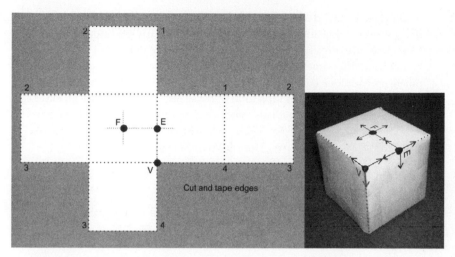

FIGURE 4. The angles around points E and F sum to 2π. The angles around vertex V sum to $3\pi/2$. To construct the model, remove shaded regions and tape them together so that the vertices labeled with 1 come together, the vertices labeled 2 come together, and so on.

was apparently not published during Descartes' lifetime. After Descartes died in 1650, when his papers were collected, it was accidentally dropped in a river, fished out, and hung to dry. Gottfried Leibniz (1646–1716), the coinventor of calculus, was given the opportunity to copy some of these papers including the manuscript containing our theorem. Leibniz' copy was then lost, found, and published in 1860 (Federico 1982; Phillips 1999).

Descartes' work focused on the angles of polyhedra. The *external angle* at a vertex of a polyhedral surface measures how much that vertex is "not flat," just as the external angle on a polygonal curve measures how much the curve is not straight. To see how this measure can be accomplished, consider a point in the plane. If we were to fit a collection of angles around this point, their measures would sum to 2π radians. Compare this result with the surface of a cube, a polyhedral surface with eight vertices. Each point on a face of the cube is surrounded by 2π radians. For example, on the one hand the point F shown in figure 4 has four right angles around it. Each point on the edge of the cube also is surrounded by 2π radians, as illustrated by the point E in figure 4. On the other hand, each vertex point, such as the point V, is surrounded by only three right angles, which total $3\pi/2$ radians.

Compared with a point in the plane, therefore, each vertex comes up short by $\pi/2$ radians. Descartes called this deviation from 2π the *angulum externum*, or *external angle*. Since the eight vertices of the cube each have an external angle of $\pi/2$, the total external angle is 4π, or as Descartes says, "eight right angles" (Federico 1982). This is the same number that differential geometers find for the total amount of Gaussian curvature (i.e., the total curvature) on a sphere or any figure that can be deformed to a sphere.

Let us try this same computation with another simple polyhedron, a tetrahedron with four equilateral triangular faces. Each vertex is surrounded by three $\pi/3$-radian angles for a total of π radians. We are left with an external angle of $2\pi - \pi = \pi$ radians. Since four vertices are involved, we have an external angle sum of 4π. The result will be the same for any convex polyhedron, as we will prove. This outcome is Descartes' theorem.

(Two-dimensional) Descartes' theorem: For any polyhedron the sum of the external angles is 4π.

This theorem is relatively easy to prove from Euler's formula,

$$(1) \qquad\qquad f - e + v = 2,$$

where, for a given polyhedron, f is the number of faces, e the number of edges, and v the number of vertices. For simplicity, each of the faces can be subdivided into triangles by inserting diagonals, so we can assume that all the faces are triangles, although not necessarily on distinct planes. Each face now contributes three edges, each of which are shared by two faces. In other words, $3f = 2e$. Multiplying equation (1) on both sides by 2 and substituting $3f$ for $2e$ gives us

$$(2) \qquad\qquad -f + 2v = 4.$$

Multiplying by π on both sides gives us

$$(3) \qquad\qquad \pi v - \pi f = 4\pi.$$

We can now interpret equation (3). The (interior) angle sum of each of the f triangular faces is π, so the sum of all the angles of all the faces must be πf, which is also the sum of the angles at all the vertices. At each vertex the external angle is equal to 2π minus the angle sum. For all v vertices the sum of the external angles must be $2\pi v$ minus the sum of all the angles, which we already know is πf. Therefore, if we let ΣC be the sum of all the external angles, we get

$$(4) \qquad\qquad \Sigma C = 2\pi v - \pi f = 4\pi.$$

Note that if we were to allow an angle *excess* (i.e., angles surrounding a vertex that sum to *more* than 2π radians) to be a *negative* external angle, no modification to these computations is necessary. As a result, equation (4) applies to nonconvex as well as to convex polyhedra.

As an aside, we should note that Descartes considered only those polyhedra that were deformable to a sphere. However, both Euler's formula and

Descartes' theorem can be extended to other surfaces. The external angles of a polyhedral torus, for example, sum to 0.

For curves in the Euclidean plane, we saw that the change in direction measured by the external angle occurs gradually for a smoothly curving curve, and the amount of change per unit arc length is called *curvature*. The external angle on a surface measures the "sharpness" at a vertex where the inclination of the faces abruptly changes. But again, on a smooth surface, this change occurs gradually. Therefore, to define Gaussian *curvature* as the amount of change per unit surface area makes sense. In fact, as noted earlier, integrating Gaussian curvature over a region on a surface yields a quantity called *total curvature*, and this phenomenon is what the sum of the external angles measures on a polyhedral surface. For both curves and surfaces, therefore, external angles provide a way to measure total curvature. Virtually any theorem about a smooth surface involving total curvature can be transformed into a theorem about polyhedral surfaces by replacing the words *total curvature* with the words *sum of the measures of the external angles*. The paper models discussed subsequently, therefore, are legitimate approximations for the geometry of curved surfaces.

The Geometry of Geodesics

Lines in the plane minimize distance in the sense that given two points, the shortest distance between them is measured along a line. For measurements taken along a surface, those curves that minimize distance are called *geodesics*. The lines on a plane are nothing more than the geodesics of a flat surface, and Euclidean geometry is just a particular example of the geometry of geodesics on surfaces.

We have already seen that the sum of the external angles is 4π for a cube, and coincidentally, the total (Gaussian) curvature for a sphere is also 4π. The geodesics on these two surfaces, as we shall see, also behave in a similar way.

On the sphere, the geodesics are the great circles, and on the sphere, a triangle with geodesic sides has an angle sum greater than 180 degrees, or π radians. The particular angle sum turns out to be directly related to the total curvature contained in the triangle.

For example, on a sphere a triangle can have three right angles. We can put one vertex at the north pole and two on the equator as in figure 5. This triangle contains one-eighth of the surface area of the sphere, and so it also must contain one-eighth of the total curvature, in particular, $C = 4\pi/8 = \pi/2$. The angle sum of the triangle is $3\pi/2$, which is $\pi/2$ greater than the Euclidean angle sum of π radians. The agreement between the angle sum excess and the curvature contained within the triangle is not a coincidence.

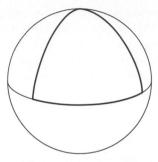

FIGURE 5. A triangle on a sphere can have three right angles.

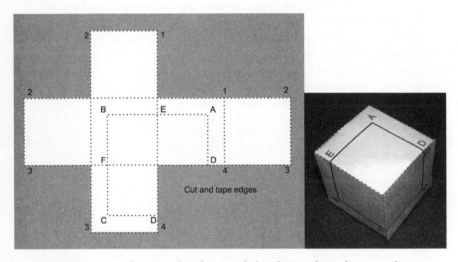

FIGURE 6. A triangle on a cube. The triangle has three right angles, an angle sum of 270 degrees, and an angle-sum excess of 90 degrees, or $\pi/2$ radians.

In fact, the angle sum of a triangle is always $\pi + C$, where C is the curvature contained in the triangle. This outcome is the spherical version of the fundamental theorem of surface geometry, the Gauss-Bonnet theorem.

On the cube, a geodesic must be as straight as possible and, equivalently, must minimize distance. Plans for a paper model of a cube are shown in figure 6. Four straight-line segments are also drawn. Since folding along the edges will not change the length of the segments, nor will it change any small measurements taken along the surface, these segments will become geodesics on the cube after being folded and taped. Note also that after the edges are taped together, the segments AD and DC come together just as AE and EB do. The fact that AB, BC, and CA are each geodesic segments follows, and we have a triangle with three right angles. We again have a triangle with an angle sum

$\pi/2$ more than the Euclidean angle sum of π radians. Furthermore, this tri-angle contains one vertex with an external angle of $C = 2\pi - 3\pi/2 = \pi/2$. In other words, the excess angle sum equals the curvature contained within.

Note, however, that for triangles on a sphere or a cube, the notion of "interior" is somewhat ambiguous. The three edges, in fact, separate the cube's surface into two regions. If we consider the larger region as the "interior" of the triangle, then the triangle's three "interior" angles each measure $3\pi/2$ radians for an angle sum of $9\pi/2$ radians. The angle sum excess, therefore, is $7\pi/2$, which is equal to the sum of the seven external angles of the cube, each measuring $\pi/2$ radians, that lie in the larger region. This result again matches the total curvature contained within this interpretation of the triangle, and so both interpretations of the triangle have this property. In general, for any triangle, we have

(5) $$\Sigma = \pi + C,$$

where Σ is the angle sum of a triangle and C is the total curvature contained in the triangle.

The sphere is an example of an *elliptic* geometry, in which we see triangles with angle sums greater than π radians. A differential geometer would equate elliptic geometry with *positive* Gaussian curvature. In *hyperbolic* geometry, we see triangle angle sums that are less than π radians, and *negative* Gaussian curvatures characterize hyperbolic geometries. As we shall see, the corresponding solid angles have *more* than 2π radians surrounding them, and so extra radian measure beyond 2π can naturally be viewed as a *negative* external angle.

In his book *The Shape of Space*, Jeffrey Weeks (1985) presents a visualization of the hyperbolic plane, and therefore negative curvature, with something he calls *hyperbolic paper*, which he attributes to the well-known geometer William Thurston. The idea is based on the fact that equilateral triangles tile the plane with six triangles coming together at each vertex and each contributing an angle measuring $\pi/3$ radians around that vertex. These six angles total 2π radians around each of these vertex points. Hyperbolic paper, in contrast, squeezes *seven* equilateral triangles around each vertex, creating a very wavy surface and an approximation to the hyperbolic plane.

In figure 7 the seven equilateral triangles, when taped together, create a small piece of hyperbolic paper. As in the cube example, several line segments are drawn in this figure, and after the seven triangles are taped together, the lines come together to form geodesic $\triangle ABC$ with angle sum $2\pi/3$ radians. In constructing geodesics across "cuts," one must make sure that vertical angles formed by the cut and the geodesic are equal. The photograph shows what the paper looks like after being taped together. Note that the paper takes a saddle shape, which is characteristic of a hyperbolic surface

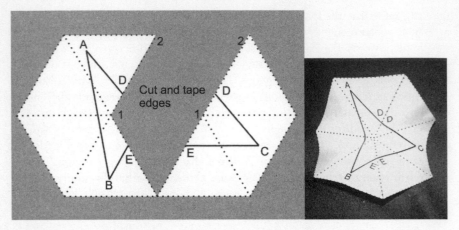

FIGURE 7. A triangle on hyperbolic paper. The angle at the top is 30 degrees, at the bottom is 40 degrees, and on the right is 50 degrees, for an angle sum of 120 degrees and an angle-sum deficit of 60 degrees, or $\pi/3$ radians.

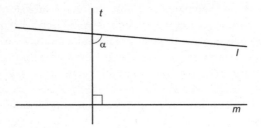

FIGURE 8. Two lines cut by a transversal. In Euclidean geometry, if $\alpha <$ 90 degrees, then l and m intersect.

with negative curvature. The triangle in this figure has an angle sum that is less than π radians; in particular, the angle sum "excess" is $-\pi/3$ radians. This result agrees with the fact that the external angle at the central vertex of the surface is $-\pi/3$ radians. Again the deviation of the angle sum of a triangle from π radians equals the total curvature contained in the triangle.

Exploring Euclidean and Non-Euclidean Geometry

Virtually all the phenomena observed in Euclidean and non-Euclidean geometry can be explored concretely on our paper models. For example, Euclid's fifth postulate states that given a figure like that shown in figure 8 with $\alpha < \pi/2$, the lines l and m must intersect. In hyperbolic geometry, however, the two lines might not intersect. In Euclidean geometry the

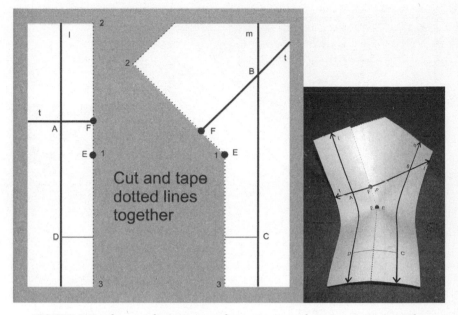

FIGURE 9. Two lines cut by a transversal near a point with negative curvature. The interior angles below the transversal measure 90 degrees and 45 degrees, but the two lines do not intersect.

distance between the lines decreases linearly, but in hyperbolic geometry this rate of decrease may get smaller. Negative curvature between the lines lessens the rate of decrease in the distance between l and m. In fact, total curvature equal to $-\alpha$ between the lines stops this decrease completely. The ability to contain this amount of curvature in a small region helps make this phenomenon easier to see.

If $\alpha = \pi/4$, then curvature totaling $-\pi/4$ can prevent these lines from intersecting, and we can build a model that illustrates this outcome. Adding a $\pi/4$-radian wedge gives us the necessary Cartesian curvature, and we can construct a model as in figure 9. Note that since the edges on a polyhedral surface do not affect the geometry, we have no need to crease edges in the paper models. With the introduction of negative curvature, the lines l and m diverge in one part of the model and remain equidistant in another. Also notice that $ABCD$ has three right angles and one acute angle. A quadrilateral with three right angles is called a *Lambert quadrilateral* after one of the mathematicians whose work paved the way for the development of non-Euclidean geometry. In Euclidean geometry the fourth angle of a Lambert quadrilateral is necessarily a right angle, but in hyperbolic geometry it is always acute. This outcome illustrates how the history of non-Euclidean geometry can be

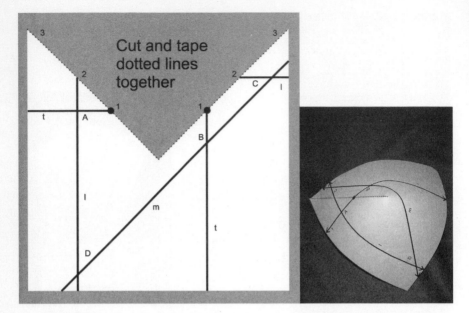

FIGURE 10. Two lines cut by a transversal near an "elliptic" vertex. The interior angles, $\angle A$ and $\angle B$, measure 90 degrees and 135 degrees, respectively, but the two lines still intersect on that side of the transversal.

investigated with these paper models by building counterexamples to Euclidean objects.

Elliptic comparisons with Euclid's fifth postulate can be made just as easily. In figure 8, if $\alpha > \pi/2$, then the lines l and m can still intersect on the right if enough positive curvature occurs between the lines. For example, when $\alpha = 3\pi/4$ radians, in a Euclidean plane l and m diverge on that side of the transversal. Positive curvature of $C = \pi/4$ is sufficient to stop that divergence, and more curvature forces the lines to intersect. In figure 10, positive curvature of $C = \pi/2$ is introduced between l and m by removing a $\pi/2$-radian wedge. This removal results in the two lines' intersecting on the side of the transversal with interior angles summing to more than π radians. In this instance, the lines l and m have two points of intersection. Here we can see why in elliptic geometry we can have "2-gons," which are impossible in Euclidean geometry.

Summary

Through all these examples, we see that a direct relationship exists between the curvature of a surface and such properties as the angle sum of a triangle.

We have seen how total curvature on paper models can be used to complement and contrast with the phenomena encountered in the usual Euclidean and non-Euclidean geometries. These polyhedral models make possible the construction of explicit, concrete models with precise specifications, and with these come the ability to explore isolated properties by comparing and contrasting spaces with and without these properties. This ability to play with and manipulate geometric spaces is crucial to reaching a deep understanding of geometry.

References

Federico, Pasquales J. *Descartes on Polyhedra*. New York: Springer Verlag, 1982.
Phillips, Tony. "Descartes's Lost Theorem." American Mathematical Society Web Site Feature Column, September 1999. www.ams.org/featurecolumn/archive/descartesl.html.
Weeks, Jeffrey. *The Shape of Space*. New York: Marcel Dekker, 1985.

Intuitive vs Analytical Thinking:
Four Perspectives

URI LERON AND ORIT HAZZAN

I think, therefore I err.[1]

Introduction

This article is an attempt to place mathematical thinking in the context of more general theories of human cognition. We describe and compare four perspectives, each routed in a different research community, and each offering a different view on mathematical thinking and learning and, in particular, on the sources of mathematical errors.

1. Mathematics: where the focus is on the rules and norms of the mathematical research community.
2. Mathematics education: where the focus is on the individual and social processes in a community of learners, in and out of the classroom.
3. Cognitive psychology: where the focus is on the universal characteristics of the human mind and behavior, which are shared across individuals, cultures, and different content areas.
4. Evolutionary psychology: where the focus is on the evolutionary origins of human cognition and behavior and their expression in "universal human nature."

The four perspectives represent four levels of explanation, and we see them not as competing but as complementing each other, much as descriptions at different levels of the same phenomena in other sciences are necessary for fuller understanding of the observed phenomena. For example, here is a typical quote on the usefulness of multilevel view in biological psychology.[2] The "explanatory pluralism" discussed in that article seems very relevant to the multilevel character of mathematics education.

> [Theories] at different levels of description, like psychology and neuroscience, can co-evolve, and mutually influence each other, without the higher-level theory being replaced by, or reduced to, the lower-level one. Such ideas seem to tally with the pluralistic character of biological explanation. In biological psychology, explanatory pluralism would lead us to expect many local and non-reductive interactions between biological, neurophysiological, psychological and evolutionary explanations of mind and behavior. (Looren de Jong, 2002, p. 441)

In the classroom or in research data, all four perspectives may be observed. They may differentially account for the behavior of different students on the same task, the same student in different stages of development, or even the same student in different stages of working on a complex task.

The article can also be considered as an attempt to deal with the (admittedly vague) question, "Are mathematical errors good or bad?" (We are referring here not to accidental errors, but to those of the universal recurring kind.) The mathematical perspective typically views errors as bugs, something that went wrong due to faulty knowledge, and needs to be corrected.[3] The mathematics educational perspective typically views errors as partial knowledge, still undesirable, but a necessary intermediate stage on the way towards attaining professional norms, and a base on which new or refined knowledge can be constructed. Cognitive psychologists[4] typically view errors as an undesirable but unavoidable feature of the human mind, analogical to optical illusions, which originate at the interface between intuitive and analytical thinking (Evans, 2008; Kahneman & Frederick, 2005; Stanovich & West, 2000). Evolutionary psychologists, in contrast, view errors as stemming from useful and adaptive features of human cognition (Barkow, Cosmides & Tooby, 1992; Buss, 2005; Cosmides & Tooby, 1996; Gigerenzer, & Todd, the ABC research group, 1999). In this view, mistakes are not really errors but, rather, the expression of an intelligent system that had been adapted by natural selection to Stone-Age ecology, and now has been "tricked" by unexpected conditions of modern civilization or in the psychologist's lab. According to this perspective, people make mistakes (at least of the universal recurring kind) not because of deficiencies in their intelligence or their knowledge but because the requirements of modern mathematics, logic, or statistics clash with their "natural" intelligence. In addition, by the very nature of an intelligent system, it must calculate and make inferences beyond the information given; hence, it must, to a certain extent, be prone to errors in the face of atypical conditions. Hence, the opening quotation, "I think, therefore I err." (Gigerenzer, 2005, p. 1).

The main part of the article consists of the application of the four perspectives to two mathematical tasks, one from statistics (taken from the psychological research literature) and one from abstract algebra (taken from our own research). The two tasks are known to elicit high rates of errors even among students in top universities. The two tasks are introduced in the next section. We then introduce each of the perspectives by reviewing its basic ideas and research base. Finally, we put the four perspectives to work by applying them to the analysis of the two tasks and the errors they engender.

The Two Tasks

The following two tasks will be used later in the paper to illustrate how each of the four perspectives can be used to explain students' misconceptions. The tasks were selected for the prevalence and magnitude of the misconceptions they engender.

STATISTICAL THINKING: "ARE HUMANS GOOD INTUITIVE STATISTICIANS AFTER ALL?"[5]

Many "biases" concerning statistical thinking are discussed in the psychological research literature. One of the most famous is the "medical diagnosis problem" and the related phenomenon of "base-rate neglect." Here is a standard formulation of the task and data, taken from Samuels, Stitch and Tremoulet (1999):

Casscells, Schoenberger, and Grayboys (1978) presented [the following problem] to a group of faculty, staff and fourth-year students at Harvard Medical School.

> If a test to detect a disease whose prevalence is 1/1000 has a false positive rate of 5%,[6] what is the chance that a person found to have a positive result actually has the disease, assuming that you know nothing about the person's symptoms or signs? ___%

Under the most plausible interpretation of the problem, the correct Bayesian answer is 2%. But only eighteen percent of the Harvard audience gave an answer close to 2%. Forty-five percent of this distinguished group completely ignored the base-rate information and said that the answer was 95%. (p. 79–80)

This task is intended to test what is usually called *Bayesian thinking*: how people update their initial statistical estimates (the *base rate*) in the face of new evidence. In this case, the initial estimate (the prevalence of the disease in the general population) is $1/1,000$, the new evidence is that the patient has tested positive (together with the 5% false-positive rate), and the task is intended to uncover how the subjects will update their estimate of the chance that the patient actually has the disease. *Base-rate neglect* reflects the widespread fallacy of ignoring the base rate, instead simply subtracting the false-positive rate (5%) from 100%.

A formal solution to the task is based on Bayes' theorem, but there are many complications and controversies surrounding the interpretation and application of that theorem.[7] This is a deep and fascinating issue involving mathematics, psychology, and philosophy, which is beyond the scope of this paper. See Cosmides and Tooby (1996) and Barbey and Sloman (2007) for extensive discussion.

Interestingly, it is possible to arrive at the Bayesian solution intuitively without using the formal Bayes' theorem: Assume that the population consists of 1,000 people and that all have taken the test. We know that one person will have the disease (because of the base rate) and will test positive (because no false-negative rate has been indicated). In addition, 5% of the remaining healthy 999 people (approximately 50) will test false-positive—a total of 51 positive results. Thus, the probability that a person that tests positive actually has the disease is $1/51$, which is a little less than 2%.

We digress for a classroom-oriented comment. Confronting people who gave the 95% answer with the correct mathematical solution is not, in our view, a satisfactory way to deal with the conflict between their primary intuition and the mathematical solution (in either its intuitive or formal versions). In our experience, students will reluctantly accept the analytical solution but will remain confused and unhappy. This is because, we submit, they still hold in their mind the two solutions: one immediate and intuitive but now declared illegitimate; the other mathematically respectable but clashing with the intuitive one. And we as mathematics educators have not really done much to help them alleviate the tension between the two incompatible entities that simultaneously inhabit their mind.

A good way of helping students deal with this clash is to look at some variations of the original problem, in particular extreme cases. For example, assume there is a country where the disease has been totally eliminated (that is, its base rate is zero). Now if you ask your students what is the probability that someone from that country who tested positive actually has the disease, they will likely say, "you must be kidding, you just told us that no one has the illness, so of course the probability is zero"—no base-rate neglect here! But now assume that the disease in that country is *almost* totally eliminated, say only 1 out of 1,000 inhabitants has it, would you not expect that the answer to the same question would be, "the probability is *nearly* zero"? Similarly, we might consider the same question in an unfortunate country where *everybody* has the disease. In general, we could imagine a computer microworld (or even let the student develop one) that models this problem, with a slider that assigns variable base rates and demonstrates the resulting probabilities. Similarly, we can add a slider to vary the false positive rate between 0 and 100 and follow the resulting change in the intuitive answer.

MATHEMATICAL REASONING: "STUDENTS' USE AND MISUSE OF LAGRANGE'S THEOREM"[8]

The data are drawn from the performance of university students on a group theory task. We first lay out the necessary mathematical background.

The entire task takes place within the group Z_6, consisting of the set $\{0,1,2,3,4,5\}$ and the operation of addition modulo 6, denoted by $+_6$. For example, $2 +_6 3 = 5$, $3 +_6 3 = 0$, $3 +_6 4 = 1$, and, in general, $a +_6 b$ is defined as the remainder of the usual sum $a+b$ on division by 6. Z_6 is a *group* in the sense that it contains 0 and is *closed* under addition mod 6: if a and b are in Z_6, then so is $a +_6 b$.[9] Similarly, we define Z_3 to be the group consisting of the set $\{0,1,2\}$ and the operation $+_3$ of addition modulo 3. A *subgroup* of Z_6 is a subset of $\{0,1,2,3,4,5\}$, which is, in itself, a group under the operation defined in Z_6. For example, it can be checked that the subset $\{0,2,4\}$ is a subgroup of Z_6, since it contains 0 and is closed under $+_6$.

All the groups in this discussion are *finite,* in the sense that they have a finite number of elements; this number is called the *order* of the group. Thus, the order of Z_6 is 6 and the order of Z_3 is 3. Finally, an important theorem of group theory, called *Lagrange's theorem*, states that *if H is a subgroup of G, then the order of H divides the order of G*; for example, if H is a subgroup of Z_6, then the order H divides 6. Thus, the order of H cannot be 4 or 5 but 3 is possible, and indeed, we have seen above an example of a subgroup of Z_6 with three elements. For what follows, it is relevant to mention that the *converse* of Lagrange's theorem is *not* true in general. For example, it is possible to give an example of a group G of order 12 that does not contain a subgroup of order 6 (see, e.g., Gallian, 1990, example 13, p. 151).

THE TASK AND DATA (HAZZAN & LERON, 1996)

The following task was given to 113 computer science majors in a leading Israeli university, who had previously completed courses in calculus and in linear algebra (an abstract approach), and were now in the midst of an abstract algebra course:

A student wrote in an exam, "Z_3 is a subgroup of Z_6."
In your opinion, is this statement true, partially true, or false?
Please explain your answer.

An incorrect answer was given by 73 students, 20 of whom invoked Lagrange's theorem, in essentially the following manner:

Z_3 is a subgroup of Z_6 by Lagrange's theorem, because 3 divides 6.

Mathematical remark 1 The correct answer is that Z_3 is *not* a subgroup of Z_6. The reason is that Z_3 is not closed under the operation $+_6$ (for example, $2 +_6 2 = 4$, and 4 is not in Z_3). The question is tricky because Z_3 is a sub-*set* of Z_6 and is a group (relative to $+_3$), but it is not a sub-*group* of Z_6 (since it is not a group relative to $+_6$).[10] There is a sophisticated sense in which the statement "Z_3 is a subgroup of Z_6" is partially true, namely, that Z_3 is *isomorphic* to the subgroup $\{0, 2, 4\}$ of Z_6. We would, of course, have been thrilled to receive this answer, but none of our 113 students gave it.

Mathematical remark 2 As can be seen from the previous remark, the correct solution does not use Lagrange's theorem. It is relevant to mention that, in spite of superficial resemblance, there is no way Lagrange's theorem could even *help* on this task, since "*H* is a subgroup" is the hypothesis of Lagrange's theorem, not its conclusion. What the students seem to be using is an incorrect version of a nontheorem (namely, the *converse* of Lagrange's theorem).[11]

Reviewing the Four Perspectives

We now introduce each of the four perspectives outlined in the introduction. Obviously, not all members of the relevant communities hold the same views, and sometimes, not even similar views. What we are portraying is what we believe is a typical view held in the relevant community, or at least in a significant subcommunity. Where possible, we support our profiles with appropriate references to the literature.

Since the first two perspectives (mathematics and mathematics education) are well known to readers of this collection, we dwell on them only briefly, as a basis for comparison with the other perspectives (cognitive and evolutionary psychology), which are introduced and discussed more thoroughly.

A MATHEMATICAL PERSPECTIVE: ERRORS AS BUGS

This perspective looks at mismatches between students' work and the norms of the professional mathematician. The source of students' errors is typically assumed to be their faulty mathematical knowledge, and the way to address them contains teaching suggestions, such as explaining a difficult point, giving more examples or exercises, etc.

A MATHEMATICS EDUCATION PERSPECTIVE: ERRORS AS EXPRESSING INTERMEDIATE STATE IN KNOWLEDGE CONSTRUCTION

In the mathematics education literature, especially in research on misconceptions, knowledge is typically viewed as being actively constructed by the learner, rather than being transmitted from teacher to student. The process of knowledge construction may be lengthy, and in the intermediate stages, the learner will have partial knowledge, which may result in errors. However, these errors are viewed as a normal and acceptable part of the learning process rather than expressing faulty performance. Moreover, errors are valued as offering a window into the learner's states of mind in the process of learning. Thus, researchers document recurring errors and develop theoretical interpretations to account for the process that yields these errors (e.g., Davis, Maher, & Noddings, 1990).

A COGNITIVE PSYCHOLOGY PERSPECTIVE: ERRORS AS A CLASH BETWEEN TWO SYSTEMS OF THINKING

Here, we will mainly discuss the *dual-process theory* (introduced below) and the heuristics-and-biases research program in cognitive psychology (led by Kahneman & Tversky over the last 30 years; cf., e.g., Evans 2003, 2008; Gilovich, Griffin, & Kahneman, 2002; Kahneman, 2002; Stanovich & West, 2000, 2003).

Dual-process theory The ancient distinction between intuitive and analytical modes of thinking has achieved a new level of specificity and rigor in what cognitive psychologists call *dual-process theory*. The first application of this theory to mathematics education research, to the best of our knowledge, appears in Leron and Hazzan (2006); the present exposition and analysis is an abridged version of the one given in that paper.

According to dual-process theory, our cognition and behavior operate in parallel in two quite different modes, called *System 1* (S1) and *System 2* (S2), roughly corresponding to our commonsense notions of intuitive and analytical modes of thinking. These modes operate in different ways, are activated by different parts of the brain, and have different evolutionary history (S2 being evolutionarily more recent and, in fact, largely reflecting *cultural* evolution). The distinction between perception and cognition is ancient and well known, but the introduction of S1, which sits midway between perception and (analytical) cognition, is relatively new and has important consequences for how empirical findings in cognitive psychology are interpreted, including the wide-ranging *rationality debate* (Samuels et al., 1999; Stanovich & West, 2000, 2003; Stein, 1996), and the application to mathematics education research.

Like perception, S1 processes are characterized as being fast, automatic, effortless, unconscious, and inflexible (hard to change or overcome); unlike perceptions, S1 processes can be language-mediated and relate to events not in the here-and-now. S2 processes are slow, conscious, effortful, computationally expensive (drawing heavily on working memory resources), and relatively flexible. The two systems differ mainly on the dimension of *accessibility*: how fast and how easily things come to mind. In most situations, S1 and S2 work in concert to produce adaptive responses, but in some cases (such as the ones concocted in the heuristics-and-biases and in the reasoning research), S1 generates quick automatic *nonnormative* responses, while S2 may or may not intervene in its role as monitor and critic to correct or override S1's response.

Many of the nonnormative answers people give in psychological experiments—and in mathematics education tasks, for that matter—can be explained by the quick and automatic responses of S1, and the frequent failure of S2 to intervene in its role as critic of S1.

Here is a striking example (Kahneman, 2002) for a combined failure of both systems:

> *A baseball bat and ball cost together one dollar and 10 cents. The bat costs one dollar more than the ball. How much does the ball cost?*
> Almost everyone reports an initial tendency to answer '10 cents' because the sum $1.10 separates naturally into $1 and 10 cents, and 10 cents is about the right magnitude. [. . .] many intelligent people yield to this immediate impulse: 50% (47/93) of Princeton students, and 56% (164/293) of students at the University of Michigan gave the wrong answer. (p. 451)

According to dual process theory, the fast-reacting S1 "hijacks" the subject's attention and jumps automatically and immediately with the answer of 10 cents, since the numbers one dollar and 10 cents are salient, and since the orders of magnitude are roughly appropriate. For many people, S1's conclusions are accepted uncritically; thus, in a sense, they "behave irrationally." For others, S1 had also immediately jumped with this answer, but in the next stage, their S2 interfered critically and made the necessary adjustments to give the correct answer (5 cents). Significantly, the way S1 worked here, namely, coming up with a very quick decision based on salient features of the problem and of rough sense of what is appropriate in the given situation, usually gives good results under natural conditions, such as searching for food or avoiding predators.

Researchers in cognitive psychology do not usually consider educational implications of their research, so we offer instead our own thoughts on the implications of dual-process theory. Because many of the misconceptions come from combined failure of both S1 and S2, we propose that the most important educational implication is the need to train people *to be aware of the way S1 and S2 operate, and to include this awareness in their problem-solving toolbox.* This suggestion has an interesting (almost paradoxical) recursive nature: It in effect implies that S2 should monitor not only the operation of S1 (its standard role), but the S1-S2 interaction as well; thus, S2 has to monitor its own functioning in monitoring S1. In a way, we might say that an operation of a "System 3" is needed here (to monitor S2), but in practice, this function is recursively assumed by S2 itself.[12] While monitoring and critiquing S1 is one of the reasons S2 has evolved in the first place, monitoring the S1-S2 interaction seems to be what Geary (2002; see next section) has called *biologically secondary skills,* one which will not normally develop without explicit instruction.

When cognitive psychologists do mention education, it may sound a bit naïve to practitioners in the educational field:

> The characteristics that determine analytic system intervention, other than cognitive ability, are *dispositional.* People may *choose* to engage in effortful analytic thinking because they are inclined to do so by strong deductive reasoning instructions [. . .] or, perhaps, because they have personal motivation.
> [. . .]
> This dispositional aspect of the analytic system also provides encouragement to those who believe that our educational systems can and should encourage people to think in a more "rational," abstract, and decontextualized manner [. . .]. (Evans, 2006, p. 383)

While this advice is backed up by empirical data from the psychologist's lab, experienced mathematics educators facing the realities of the mathematics classroom usually prefer to deal with misconceptions through appropriate activities and discussions rather than instructions.

AN EVOLUTIONARY PSYCHOLOGY PERSPECTIVE: ERRORS AS A CLASH BETWEEN HUMAN NATURE AND MODERN CIVILIZATION

We take from the young discipline of evolutionary psychology (EP) the scientific view of *human nature* as a collection of universal, reliably developing, cognitive and behavioral abilities—such as walking on two feet, face recognition, and the use of language—that are spontaneously acquired and effortlessly used by all people under normal development (Cosmides & Tooby, 1992, 1997; Pinker, 1997, 2002; Ridley, 2003; Tooby & Cosmides, 2005). We also take from EP the evolutionary origins of human nature; hence, the frequent mismatch between the ancient ecology to which it is adapted and the demands of modern civilization. We all do manage, however, to learn many modern skills (such as writing or driving, or some mathematics) because of our mind's plasticity and its ability to "co-opt" ancient cognitive mechanisms for new purposes (Bjorklund & Pellegrini, 2002; Geary, 2002). However, this is easier for some skills than for others, and nowhere are these differences more manifested than in the learning of mathematics. The ease of learning in such cases is determined by the *accessibility* of the co-opted cognitive mechanisms rather than the complexity of the task.

We emphasize that what is part of human nature need not be innate: we are not born walking or talking. What seem to be innate are the motivation and the ability to engage the species-typical physical and social environment in such a way that the required skill will develop (Geary, 2002). This is the ubiquitous mechanism that Ridley (2003) has called "nature *via* nurture." We also emphasize that what is *not* part of human nature, or even what goes *against* human nature, need not be unlearnable. Individuals in all cultures have always accomplished prodigious feats such as juggling 10 balls while riding a bicycle, playing a Beethoven piano sonata, or proving an abstract mathematical theorem in a formal language. However, research on people's reasoning, and on mathematical thinking in particular, usually deals with what most people are able to accomplish under normal conditions. Under such conditions, many people will produce nonnormative answers if the task requires reasoning that goes against human nature. In terms of mathematical education, this means that learning successfully such skills will require a particularly high level of motivation and perseverance—conditions that are hard to achieve for a long time and for many people in the standard classroom. Finally, it is in order to note here that EP is a hotly debated discipline. Much of the criticism leveled at EP is ideologically or emotionally motivated, but see, e.g., Stanovich and West (2003) and Evans (2003) for a sample of scientifically respectable alternative views.

We take Geary (2002) as an example of the EP community's effort to create a discipline of "evolutionary educational psychology." Geary (2002) distinguishes between *biologically primary skills,* which emerge spontaneously and naturally in all people under normal development, i.e., they are part of human nature as conceived by evolutionary psychologists, and *biologically secondary skills,* which require more effort and more motivation to learn, usually within teaching-oriented situations.

> The principles of evolutionary educational psychology will provide a much needed anchor for guiding instructional research and practice. An evolutionarily informed science of academic development is in fact the only perspective that readily accommodates basic observations that elude explanation by other theoretical perspectives, such as constructivism [. . .]
>
> For instance, it follows logically from the principles of evolutionary educational psychology that children will learn language easily and without formal instruction, and years later many of these children will have difficulty learning to read and write even with formal instruction. (p. 340)

Recently, the EP framework has been applied to explain a curious phenomenon concerning functions (Paz & Leron, 2009), which may offer some insight into the relationship between mathematical thinking and human nature. Paz & Leron study this relationship in the case of what they call the action-on-objects scheme (AOS: the Piagetian notion that *when you act on an object it changes its properties, but still remains the same object*) and the function concept. Based on their empirical findings, they propose that the same parts of human nature (the AOS in this case) that may initially support the learning of a mathematical concept (functions in this case) can later clash with more advanced versions of that same concept. This offers an intriguing view of the (cultural) evolution of mathematical concepts, as being initially anchored in human nature, but then developing in ways that clash with the very intuitive foundations that gave rise to them in the first place.

We conclude this concise review of EP with a word about the relation of human nature to dual-process theory (Section 3.3). Human nature consists, by definition, of a more-or-less fixed collection of traits and behaviors that all human beings in all cultures acquire spontaneously and automatically under normal developmental conditions. System 1, in our view, contains all the traits and behaviors that comprise human nature but, on top of that, also all the traits and behaviors that have become S1 for a particular culture or a particular person because of specific (nonuniversal) developmental conditions. For example, learning language is part of human nature and, thus, part of S1 for all human beings under normal developmental conditions; in contrast, speaking English is not part of human nature but is an S1 skill for people whose mother tongue is English. Similarly, driving a car is not part of human nature but has become an S1 skill for experienced drivers, as evidenced by their ability to hold an intellectual conversation (an S2 task, fully engaging the working memory resources) while driving.

An interesting corollary of an EP perspective is that the persistent errors of the kind discussed here stem not from weakness of the human cognitive system but from its *strength*. Indeed, human nature is a collection of all the skills people are naturally good at, and the errors stem from the clash between the requirements of modern society with these mechanisms that are adapted to the ecology of our Stone-Age ancestors.

Putting the Four Perspectives to Work: Statistical Thinking

We now apply the four perspectives for the analysis of the medical diagnosis problem and the data presented in Section 2.1. As noted before, the two familiar perspectives (mathematics and mathematics education) will be only mentioned briefly, as a basis for comparison with the two relatively novel perspectives (cognitive and EP).

A mathematical point of view would point to an error in statistical thinking: students neglect to take into account the base rate, or, alternatively, fail to correctly apply Bayes' theorem.

Since we did not find in the *mathematics education* literature an analysis of the medical diagnosis problem, we will make up a possible interpretation with the hope of capturing some of the spirit of that community (or one of its subcommunities). One interpretation (in the spirit of Nesher & Teubal, 1975, or Leron & Hazzan, 1997) is that people do not look deeply into the problem but, instead, do some routine calculations based on verbal cues. For example, because of the meaning of "5% false positive," they may classify this problem as a "subtraction problem," and just do the subtraction 100%–5%, which leads to the observed base-rate neglect. Another interpretation is that people actually do take into account *some* base rate, but not necessarily the tiny one (0.001) postulated in the problem. In our experience, some people tend to personalize the problem ("if *I* tested positive . . ."), bringing in a whole baggage of realistic conditions that are abstracted away in the original formulation. For example, they would not normally take the test if they did not have an a priori *serious* worry that they might have the disease, that is, if they did not assume a very high base rate. In this case, a positive test would indeed mean a high probability of having the disease. Of course one could devise empirical tests to decide between these interpretations (Barby & Sloman, 2007).

We now proceed to analyze the medical diagnosis problem from a *cognitive psychology* perspective, more specifically, from a dual-process theory perspective. An initial analysis may look rather similar to the bat-and-ball analysis in Section 3.3. S1 quickly and effortlessly generates the 95% response because of its accessibility (subtracting the 5% error rate from 100%), and, because of the automatic, intuitive interpretation of "false positive," the base rate of 1/1,000 is completely ignored. As in the bat-and-ball analysis, too, "the dormant S2" failed to catch the error in its role as critic and monitor of S1's output. The difference between the two problems lies in the complexity of what S2 is required to notice and correct: adjusting the *difference* between the costs of the bat and ball in the former vs attending to the multiple *nested-set relationships* in the latter (the network of subsets among the whole population, the people who have the disease, the ones who were tested and the ones who tested positive). Thus, for S2 to notice and correct S1's response needs only a simple alert in the first case but a much greater effort and skill in the second, which may account for the difference in the percentage of correct responses.

More specifically, according to Evans (2006), in order to solve the medical diagnosis problem correctly, subjects must integrate all the information in a *single* mental model, and this is facilitated in formulations that make the nested-set structure salient (including, e.g., the frequentist formulation explained next).[13]

> It appears that heuristic processes cannot lead to correct integration of diagnostic and base rate information, and so Bayesian problems can only be solved analytically. This being the case, problem formats that cue construction of a single mental model that integrates the information in the form of nested sets appears to be critical, (p. 391)

> Researchers with *evolutionary and ecological orientation* (Cosmides & Tooby, 1996; Gigerenzer, & Todd, the ABC research group, 1999) claim that people are "good statisticians after all" if only the input and output are given in "natural frequencies" (integers instead of fractions or percentages).

> In this article, we will explore what we will call the "frequentist hypothesis"—the hypothesis that some of our inductive reasoning mechanisms do embody aspects of a calculus of probability, but they are designed to take frequency information as input and produce frequencies as output. (Cosmides & Tooby, 1996, p. 3)

The EP explanation is that the brains of our hunter-gatherers ancestors developed such a module because it was vital for survival and reproduction and because this is the format that people would naturally use under those conditions. The statistical formats of today, however, are the result of the huge amount of information that is collected, processed, and shared by modern societies with modern technologies. To demonstrate a typical EP theorizing, it is worth quoting at some length from Cosmides and Tooby (1996):

> In our natural environment, the only database available from which one could inductively reason was one's own observations, and possibly

those communicated by the handful of other individuals one lived with. More critically, the "probability" of a single event is intrinsically unobservable.

No sense organ can discern that if we go to the north canyon, there is a .25 probability that today's hunt will be successful. Either it will or it won't; that is all one can observe. As useful as a sense organ for detecting single-event probabilities might be, it is theoretically impossible to build one. No organism can evolve cognitive mechanisms designed to reason about, or receive as input, information in a format that did not regularly exist.

What *was* available in the environment in which we evolved was the encountered frequencies of actual events—for example, that we were successful 5 out of the last 20 times we hunted in the north canyon. Our hominid ancestors were immersed in a rich flow of observable frequencies that could be used to improve decision-making, given procedures that could take advantage of them. So if we have adaptations for inductive reasoning, they should take frequency information as input.

Once frequency information has been picked up, why not convert it into a single-event probability? Why not store the encountered frequency—"5 out of the last 20 hunts in the north canyon were successful"—as a single-event probability—"there is a .25 chance that a hunt in the north canyon will be successful"? There are advantages to storing and operating on frequentist representations because they preserve important information that would be lost by conversion to a single-event probability. (pp. 15–17)

Because of such considerations, EP researchers have sometimes been accused of telling "just-so stories." However, this accusation is misconceived. As usual in EP methodology, such evolutionary theorizing is not taken as evidence but only as a theoretical framework for generating and explaining hypotheses. The test of the hypotheses is done under the standard psychological methodologies. Indeed, Cosmides and Tooby (1996) have replicated the experiment of Casscells et al. (1978), but with natural frequencies replacing the original fractional formats, and the base-rate neglect has all but disappeared:

Although the original, non-frequentist version of Casscells et al.'s medical diagnosis problem elicited the correct bayesian answer of "2%" from only 12% of subjects tested, pure frequentist versions of the same problem elicited very high levels of bayesian performance: an average of 76% correct for purely verbal frequentist problems and 92% correct for a problem that requires subjects to construct a concrete, visual frequentist representation. (ibid, p. 58)

These data, and the evolutionary claims accompanying them, have been consequently challenged by other researchers (see Samuels et al., 1999; Evans, 2008). In particular, Evans (2008) claims that what makes the subjects in these experiments achieve such a high success rate is not the frequency format per se, but that "there is now much evidence that what facilitates Bayesian reasoning is a problem structure that cues explicit mental models of nested-set relationships" (p. 6.13). However, the fresh perspective offered by EP has been seminal in reinvigorating the discussion of statistical thinking in particular, and of cognitive biases in general. The very idea of the frequentist hypothesis, and the exciting and fertile experiments that it has engendered by supporters and opponents alike, would not have been possible without the novel evolutionary framework. Here is how Samuels et al. (1999) summarize the debate:

> But despite the polemical fireworks, there is actually a fair amount of agreement between the evolutionary psychologists and their critics. Both sides agree that people do have mental mechanisms which can do a good job at bayesian reasoning, and that presenting problems in a way that makes frequency information salient can play an important role in activating these mechanisms. (p. 101)

Putting the Four Perspectives to Work: Mathematical Reasoning

We now apply the four perspectives for the analysis of students' use and misuse of Lagrange's theorem and the data presented in Section 2.2.

A *mathematical analysis* of the students' errors in the above application of Lagrange's theorem would include, for example: Students do not understand how to use the theorem, fail to check the initial conditions, or confuse between the theorem and its converse. From *the mathematical education perspective,* Hazzan and Leron (1996) and Leron and Hazzan (1997) give detailed analyses of the Lagrange's theorem data, both from a "misconceptions" perspective, and—using the novel tool of virtual monologue—from a "coping" perspective.

From the *cognitive psychology perspective,* Leron and Hazzan (2006), applying dual-process theory, proposed that some cases of students' misuse of Lagrange's theorem reflect a combined S1-S2 failure. The analysis closely resembles Kahneman's analysis of the bat-and-ball data, except for the somewhat surprising demonstration that S1 can "hijack" cognitive behavior even in advanced mathematical settings, where the name of the game is explicitly reasoning and analytical thinking (i.e., S2 mode).

As usual, the S1 response is invoked by what is most immediately accessible to the students in the situation, which also looks roughly appropriate to the task at hand. Specifically, the students know that using a theorem in such situations is expected; they also know more-or-less immediately and effortlessly that Lagrange's theorem says something about groups and subgroups and the divisibility of their orders (it is the *details and logic* of what the theorem says that requires the effortful and pedantic intervention of S2); finally, the appearance of the two numbers 3 and 6 as orders of the groups Z_3 and Z_6, and the fact that 3 divides 6, immediately and automatically cues Lagrange's theorem, yielding the answer, "Z_3 is a subgroup of Z_6 by Lagrange's theorem, because 3 divides 6." This is a striking example for an answer that looks entirely appropriate by the "logic" of S1, but is extremely inappropriate by the logic of S2.

In addition to S1's inappropriate reaction, S2 too fails in its role as critic of S1, since there is nothing in the task situation to alert the monitoring function of S2. The missing judgment—mainly that Lagrange's theorem cannot be used to establish the existence of a subgroup but only its absence—clearly requires S2 processes. It is important to note that some of the students may well have the knowledge required to produce the right answer, had they only stopped to think more (that is, invoke S2). The problem is, rather, that they have no reason to suspect that the answer is wrong; thus, the "permissive System 2" (Kahneman, 2002) remains dormant:

> [An] evaluation of the heuristic attribute comes immediately to mind, and [. . .] its associative relationship with the target attribute is sufficiently close to pass the monitoring of a permissive System 2. (p. 469)

Just as in the bat-and-ball situation, the final (erroneous) response in this case is a combination of S1's quick and effortless reaction, together with S2's failure to take a corrective action in its role as critic and monitor. Since the operation of S1 is effortless and that of S2 so effortful, students will not make the extra effort unless something in the situation alerts them to such need. It is a feasible (and eminently researchable) hypothesis that, at least for some of the students, a small nonmathematical cue would be enough to set them on the path for a correct answer. They may already have all the necessary (S2) knowledge to solve this problem correctly, but a nudge by the interviewer (even just raising an eyebrow or looking doubtful) is needed to mobilize this knowledge. This shows, incidentally, that the dual-system framework leads not only to new explanations but also (like all good theories) to interesting new research *questions*.

We now turn to look at students' misuse of Lagrange's theorem from an *evolutionary psychology perspective* by referring to the *logic of social exchange*. Cosmides and Tooby (1992, 1997) have used the *Wason card selection task*

(Wason & Johnson-Laird, 1972) to uncover what they refer to as people's evolved reasoning "algorithms." In a typical example of the Wason task, subjects are shown four cards, say $\boxed{A}$, $\boxed{6}$, $\boxed{T}$, and $\boxed{3}$, and are told that each card has a letter on one side and a number on the other. The subjects are then presented with the rule, "if a card has a vowel on one side, then it has an even number on the other side," and are asked the following question: *What card(s) do you need to turn over to see if any of them violate this rule?* The notorious result is that roughly 90% of the subjects, including science majors in college, give an incorrect answer. Many similar experiments have been carried out, using rules of the same logical form "if P then Q," but varying the content of P and Q. The error rate has varied depending on the particular context, but mostly remained high (over 50%).

The motivation behind the original Wason experiment was partly to see if people naturally behave in accordance with the Popperian paradigm that science advances through *refutation* of held beliefs (rather than their confirmation). The normative response to the Wason task depends on the question: What will *refute* the given rule? The answer is that the rule is violated if and only if a card has a vowel on one side but an *odd* number on the other. Thus, according to mathematical logic, the cards you need to turn are $\boxed{A}$ (to see if it has an odd number on the other side) and $\boxed{3}$ (to see if it has a vowel on the other side). Most subjects, however, choose the $\boxed{A}$ card and sometimes also $\boxed{6}$, but rarely $\boxed{3}$.

Cosmides and Tooby (1992, 1997) have also presented their subjects with many versions of the task, all having the usual logical form "if P then Q," but varying widely in the contents of P and Q and in the background story. While the classical results of the Wason task show that most people perform poorly on it, Cosmides & Tooby demonstrated that their subjects performed significantly better on tasks involving conditions of *social exchange*. In social exchange situations, the individual receives some benefit and is expected to pay some cost. On theoretical grounds, and from what is known about the evolution of cooperation, certain kinds of social skills are expected to have conferred evolutionary advantages on those who excelled in them and, thus, would be naturally selected during evolutionary history. In the Wason task, social exchange situations are represented by statements of the form "if you get the benefit, then you pay the cost" (e.g., if you give me your watch, then I give you $20). A *cheater* is someone who takes the benefit but do not pay the cost. Cosmides & Tooby argue that when the Wason task concerns social exchange, a correct answer amounts to detecting a cheater. Since subjects performed correctly and effortlessly in such situations, and since evolutionary theory clearly shows that cooperation cannot evolve in a community if cheaters are not detected and punished, Cosmides & Tooby have concluded that our mind contains evolved "cheater detection algorithms."

Significantly for the Lagrange's theorem task discussed here, Cosmides & Tooby also tested their subjects on the "switched social contract" (mathematically, the converse statement "if Q then P"), in which the correct answer by the logic of social exchange is different from that of mathematical logic (Cosmides & Tooby, 1992, pp. 187–193). As predicted, their subjects overwhelmingly chose the former over the latter: *When conflict arises, the logic of social exchange overrides mathematical logic.* In other words, in a social exchange situation, people will mostly interpret the Wason task statement in a symmetrical way, as if it were an "if and only if" statement, rather than a directional way as required by mathematical logic.

This EP theoretical and empirical framework adds a new level of support, prediction and explanation to the many findings that students are prone to confusing between mathematical propositions and their converse and, in particular, to our Lagrange's theorem data. Importantly, in the EP view, people fail not because of a weakness in their cognitive apparatus, but because of its *strength*: our impressive skill in negotiating social relationships. Unfortunately for mathematics education, this otherwise adaptive skill sometimes happens to clash with the requirements of modern mathematical thinking.

Conclusion

In this article, we have proposed to view mathematical thinking in the wider context of human cognition in general, specifically, focusing on situations of conflict between intuitive and analytical thinking. We have surveyed research from cognitive psychology and EP that bears on two questions of interest to mathematics education theory and practice: How is mathematical thinking enabled by our general cognitive system, and, on the other hand, why is mathematical thinking so difficult for so many people? More specifically, we have discussed four perspectives on the sources of recurring mathematical errors and on how to deal with them in the classroom. We believe that looking at these questions from several perspectives— complementary rather than competing—may enrich our understanding and offer new insights and new research directions. As for teaching and curriculum planning, these perspectives give additional weight to the importance of teaching as much as possible in ways that are consonant with intuition (more specifically, with "human nature" in its evolutionary psychological sense), and the challenge of finding ways to bridge the gap where formal mathematics clashes with this intuition. Finally, the cognitive and evolutionary perspectives may shed a new light (Leron, 2003) on the question "Is mathematical thinking a natural extension of common sense?"

Acknowledgements

The first author gratefully acknowledges a Bellagio Residency grant from the Rockefeller Foundation, where some of the ideas presented in this paper were developed.

Notes

1. Gigerenzer (2005), p. 1.
2. See also Nisan and Schocken (2005) for such a multilevel view of computer science.
3. This is meant to describe a typical view in the community. Some individuals or sub-communities may, of course, hold different views.
4. We refer here mainly to the reasoning and decision-making subcommunities.
5. The quote is from Cosmides and Tooby (1996), p. 2.
6. This means that 5% of the healthy population will test positive.
7. Here is Cosmides and Tooby's (1996, p. 4) explanation:

> In science, what everyone really wants to know is the probability of a hypothesis given data—$p(H \mid D)$. That is, *given these observations, how likely is this theory to be true?* This is known as an *inverse probability* or a *posterior probability*. The strong appeal of Bayes' theorem arises from the fact that it allows one to calculate this probability:
>
> $p(H \mid D) = p(H)p(D \mid H)/p(D)$, where $p(D) = p(H)p(D \mid H) + p(\sim H)p(D \mid \sim H)$.

Bayes' theorem also has another advantage: it lets one calculate the probability of a single event, for example, the probability that a particular person, Mrs. X, has breast cancer, given that she has tested positive for it.

8. Adapted from Hazzan and Leron (1996).
9. In the present context, these conditions are equivalent to the standard definition of a group.
10. The elements of the groups Z_n are often taken to be equivalence classes, not numbers as in our definition, which would lead to a different mathematical analysis of the task. The present analysis, however, is the one relevant for the version that our students have learned.
11. Hazzan and Leron (1996) discuss data on two more tasks, which show that this misuse of Lagrange's theorem is deeper and more prevalent than might appear merely from the data presented here.
12. See Stanovich (2008) for a recent attempt to formulate a tri-process theory.
13. In the following quotation, Evans uses *heuristic processes* instead of Stanovich & Kahneman's System 1 and *analytic processes* instead of System 2.

References

Barbey, A. K., & Sloman, S. A. (2007). Base-rate respect: From ecological rationality to dual processes. *The Behavioral and Brain Sciences, 30*(3), 241–254.

Barkow, J., Cosmides, L., & Tooby, J. (Eds.) (1992). In *The adapted mind: Evolutionary psychology and the generation of culture*. Oxford University Press: Oxford.

Bjorklund, D. F., & Pellegrini, A. D. (2002). *The origins of human nature: Evolutionary develop mental psychology.* Washington, D.C.: American Psychological Association Press.

Buss, D. M. (2005). *The handbook of evolutionary psychology.* Hoboken: Wiley.

Casscells, W., Schoenberger, A., & Grayboys, T. (1978). Interpretation by physicians of clinical laboratory results. *New England Journal of Medicine, 299,* 999–1000.

Cosmides, L., & Tooby, J. (1992). Cognitive adaptations for social exchange. In J. Barkow, L. Cosmides, & J. Tooby (Eds.), *The adapted mind: Evolutionary psychology and the generation of culture* (pp. 163–228). Oxford: Oxford University Press.

Cosmides, L., & Tooby, J. (1996). Are humans good intuitive statisticians after all? Rethinking some conclusions from the literature on judgment under uncertainty. *Cognition, 58,* 1–73. doi:10.1016/0010-0277(95)00664-8.

Cosmides, L., & Tooby, J. (1997). *Evolutionary psychology: A primer.* Center for Evolutionary Psychology, University of California, Santa Barbara. http://www.psych.ucsb.edu/research/cep/primer.html. Accessed 5 May 2010.

Davis, R. B., Maher, C. A. & Noddings, N. (Eds.) (1990). Constructivist views on the teaching and learning of mathematics. *Journal for Research in Mathematics Education,* Monograph no. 4.

Evans, J. S. B. T. (2003). In two minds: dual-process accounts of reasoning. *Trends in Cognitive Sciences, 7* (10), 454–459. doi:10.1016/j.tics.2003.08.012.

Evans, J. S. B. T. (2006). The heuristic–analytic theory of reasoning: extension and evaluation. *Psychonomic Bulletin & Review, 13,* 378–395.

Evans, J. S. B. T. (2008). Dual-processing accounts of reasoning, judgment, and social cognition. *Annual Review of Psychology, 59,* 61–624.

Gallian, J. A. (1990). *Contemporary abstract algebra* (2nd ed.). Lexington: D. C. Heath.

Geary, D. (2002). Principles of evolutionary educational psychology. *Learning and Individual Differences, 12,* 317–345. doi:10.1016/S1041-6080(02)00046-8.

Gigerenzer, G. (2005). I think, therefore I err. *Social Research, 72*(1), 1–24. doi:10.1007/s11205-004-4512-5.

Gigerenzer, G., Todd, P. M., & ABC research group (1999). *Simple heuristics that make us smart.* New York: Oxford University Press.

Gilovich, T., Griffin, D., & Kahneman, D. (Eds.) (2002). In *Heuristics and biases: The psychology of intuitive judgment.* Cambridge: Cambridge University Press.

Hazzan, O., & Leron, U. (1996). Students' use and misuse of mathematical theorems: the case of Lagrange's theorem. *For the Learning of Mathematics, 16,* 23–26.

Kahneman, D. (Nobel Prize Lecture, December 8, 2002). Maps of bounded rationality: A perspective on intuitive judgment and choice. In T. Frangsmyr (Ed.), *Les Prix Nobel* (pp. 416–499). http://www.nobel.se/economics/laureates/2002/kahnemann-lecture.pdf. Accessed 5 May 2010.

Kahneman, D., & Frederick, S. (2005). A model of heuristic judgment. In K. Holyoak & R. G. Morrison (Eds.), *The Cambridge handbook of thinking and reasoning,* pp. 267–294. Cambridge: Cambridge University Press.

Leron, U. (2003). *Origins of mathematical thinking: A synthesis, Proceedings CERME3,* Bellaria, Italy, March, 2003. http://www.dm.unipi.it/~didattica/CERME3/proceedings/Groups/TG1/TG1_leron_cerme3. pdf. Accessed 5 May 2010.

Leron, U., & Hazzan, O. (1997). The world according to Johnny: a coping perspective in mathematics education. *Educational Studies in Mathematics, 32,* 265–292. doi:10.1023/A:1002908608251.

Leron, U., & Hazzan, O. (2006). The rationality debate: application of cognitive psychology to mathematics education. *Educational Studies in Mathematics, 62*(2), 105–126. doi:10.1007/s10649-006-4833-1.

Looren de Jong, H. (2002). Levels of explanation in biological psychology. *Philosophical Psychology*, *15*(4), 441–462. doi:10.1080/0951508021000042003.

Nesher, P., & Teubal, E. (1975). Verbal cues as an interfering factor in verbal problem solving. *Educational Studies in Mathematics*, *6*, 41–51. doi:10.1007/BF00590023.

Nisan, N., & Schocken, S. (2005). *The elements of computing systems: Building a modern computer from first principles.* Cambridge: MIT.

Paz, T., & Leron, U. (2008). The slippery road from actions on objects to functions and variables. *Journal for Research in Mathematics Education*, *40* (1) 18–39.

Pinker, S. (1997). *How the mind works.* New York: Norton.

Pinker, S. (2002). *The blank slate: The modern denial of human nature.* Nyborg: Viking.

Ridley, M. (2003). *Nature via nurture: Genes, experience, and what makes us human.* London: Harper Collins.

Samuels, R., Stitch, S., & Tremoulet, P. (1999). Rethinking rationality: From bleak implications to Darwinian modules. In E. LePore & Z. Pylyshyn (Eds.), *What is cognitive science?* (pp. 74–120). Oxford: Blackwell.

Stanovich, K. (2008). Distinguishing the reflective, algorithmic, and autonomous minds: Is it time for a tri-process theory? In J. Evans & K. Frankish (Eds.), *In two minds: Dual processes and beyond.* Oxford: Oxford University Press.

Stanovich, K. E., & West, R. F. (2000). Individual differences in reasoning: Implications for the rationality debate. *The Behavioral and Brain Sciences*, *23*, 645–726. doi:10.1017/S0140525X00003435.

Stanovich, K. E., & West, R. F. (2003). Evolutionary versus instrumental goals: How evolutionary psychology misconceives human rationality. In D. E. Over (Ed.), *Evolution and the psychology of thinking: The debate* (pp. 171–230). Hove: Psychology Press.

Stein, R. (1996). *Without good reason: The rationality debate in philosophy and cognitive science.* Oxford: Oxford University Press.

Tooby, J., & Cosmides, L. (2005). Conceptual foundations of evolutionary psychology. In D. M. Buss (Ed.), *The handbook of evolutionary psychology* (pp. 5–67). New York: Wiley.

Wason, P., & Johnson-Laird, P. (1972). *The psychology of reasoning: Structure and content.* Cambridge: Harvard University Press.

History and Philosophy
of Mathematics

Why Did Lagrange "Prove" the Parallel Postulate?

JUDITH V. GRABINER

Introduction

We begin with an often-told story from the *Budget of Paradoxes* by Augustus de Morgan: "Lagrange, in one of the later years of his life, imagined" that he had solved the problem of proving Euclid's parallel postulate. "He went so far as to write a paper, which he took with him to the [Institut de France], and began to read it."

But, De Morgan continues, "something struck him which he had not observed: he muttered 'Il faut que j'y songe encore.' [I've got to think about this some more] and put the paper in his pocket" [8, p. 288].

Is De Morgan's story true? Not quite in that form. But, as Bernard Cohen used to say, "Truth is more interesting than fiction." First, according to the published minutes of the Institut for 3 February 1806, "M. Delagrange *read* an analysis of the theory of parallels" [25, p. 314; italics added]. Those present are listed in the minutes: Lacroix, Cuvier, Bossut, Delambre, Legendre, Jussieu, Lamarck, Charles, Monge, Laplace, Haüy, Berthollet, Fourcroy—a most distinguished audience!

Furthermore: Lagrange did not throw his manuscript away. It survives in the library of the Institut de France [32]. There is a title page that says, in what looks to me like Lagrange's handwriting, "On the theory of parallels: memoir read in 1806," together with the signatures of yet more distinguished people: Prony and Poisson, along with Legendre and Lacroix. The first page of text says, again in Lagrange's handwriting, that it was "read at the Institut in the meeting of 3 February 1806."

It is true that Lagrange never did publish it, so he must have realized there was something wrong. In another version of the story, told by Jean-Baptiste Biot, who claims to have been there (though the minutes do not list his name), everybody there could see that something was wrong, so Lagrange's talk was followed by a moment of complete silence [2, p. 84]. Still, Lagrange kept the manuscript with his papers for posterity to read.

This episode raises the three questions I will address in this article. First, what did Lagrange actually say in this paper? Second, once we have seen how he "proved" the parallel postulate, why did he do it the way he did? And last, above all, why did Joseph-Louis Lagrange, the consummate analyst, creator of the *Analytical Mechanics*, of Lagrange's theorem in group theory and the Lagrange remainder of the Taylor series, pioneer of the calculus of variations, champion of pure analysis and foe of geometric intuition, why did Lagrange risk trying to prove Euclid's parallel postulate from the others, a problem that people had been unsuccessfully trying to solve for 2000 years? Why was this particular problem in geometry so important to him?

I think that the manuscript is interesting in its own right, but I intend also to use it to show how Lagrange and his contemporaries thought about mathematics, physics, and the universe. As we will see, this was not the way we view these topics today.

The Contents of Lagrange's 1806 Paper

First, we look at the contents of the paper Lagrange read in 1806. The manuscript begins by asserting that the theory of parallels is fundamental to all of geometry. Notably, that includes the facts that the sum of the angles of a triangle is two right angles, and that the sides of similar triangles are proportional. But Lagrange agreed with both the ancients and moderns who thought that the parallel postulate should not be assumed, but needed to be proved.

To see why people wanted to prove the parallel postulate, let us recall Euclid's five geometric postulates [9, pp. 154–155]. The first is that a straight line can be drawn from any point to any other point; the second, that a finite straight line can be produced to any length; the third, that a circle can be drawn with any point as center and any given radius; the fourth, that all right angles are equal; and the fifth, the so-called parallel postulate, which is the one in question. Euclid's parallel postulate is not, as a number of writers wrongly say (e.g., [5, p. 126]), the statement that only one line can be drawn parallel to a given line through an outside point. Euclid's postulate states that, if a straight line falls on two straight lines making the sum of the interior angles on the same side of that line less than two right angles, then the two straight lines eventually meet on that side. Euclid used Postulate 5 explicitly only once: to prove that if two lines are parallel, the alternate interior angles are equal. Of course, many later propositions rest on this theorem, and thus presuppose the parallel postulate.

Already in antiquity, people were trying to prove Postulate 5 from the others. Why? Of course one wants to assume as little as possible in a demonstrative science, but few questions were raised about Postulates 1–4. The historical focus on the fifth postulate came because it felt more like the kind of thing that gets proved. It is not self-evident, it requires a diagram even to explain, so it might have seemed more as though it should be a theorem. In any case, there is a tradition of attempted proofs throughout the Greek and then Islamic and then eighteenth-century mathematical worlds. Lagrange followed many eighteenth-century mathematicians in seeing the lack of a proof of the fifth postulate as a serious defect in Euclid's *Elements*. But Lagrange's criticism of the postulate in his manuscript is unusual. He said that the assumptions of geometry should be demonstrable "just by the principle of contradiction"—the same way, he said, that we know the axiom that the whole is greater than the part [32, p. 30R]. The theory of parallels rests on something that is not self-evident, he believed, and he wanted to do something about this.

Now it had long been known—at least since Proclus in the fifth century—that the "only one parallel" property is an easy consequence of Postulate 5. In the eighteenth century, "only one parallel" was adopted as a postulate by John Playfair in his 1795 textbook *Elements of Geometry* and by A.-M. Legendre in his highly influential *Elements of Geometry* [34]. So this equivalent to Postulate 5 had long been around; in the 1790s people focused on it, and so did Lagrange. But Lagrange, unlike Playfair and Legendre, didn't assume the uniqueness of parallels; he "proved" it. Perhaps now the reader may be eager to know, how did Lagrange prove it?

Recall that Lagrange said in this manuscript that axioms should follow from the principle of contradiction. But, he added, besides the principle of contradiction, "There is another principle equally self-evident," and that is Leibniz's principle of sufficient reason. That is: nothing is true "unless there is a sufficient reason why it should be so *and not otherwise*" [42, p. 31; italics added]. This, said Lagrange, gives as solid a basis for mathematical proof as does the principle of contradiction [32, p. 30V].

But is it legitimate to use the principle of sufficient reason in mathematics? Lagrange said that we are justified in doing this, because it has already been done. For example, Archimedes used it to establish that equal weights at equal distances from the fulcrum of a lever balance. Lagrange added that we also use it to show that three equal forces acting on the same point along lines separated by a third of the circumference of a circle are in equilibrium [32, pp. 31R–31V].

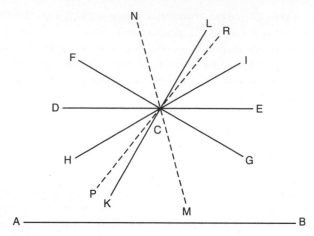

FIGURE 1. Lagrange's proof.

Now we are ready to see how Lagrange deduced the uniqueness of parallels from the principle of sufficient reason.

Suppose DE is drawn parallel to the given line AB through the given point C. Now suppose the parallel DE isn't unique. Then we can also draw FG parallel to AB.

But everything ought to be equal on each side, Lagrange said, so there is no reason that FG should make, with DE, the angle ECG on the right side; why not also on the left side? So the line HCI, making the angle DCH equal to angle ECG, ought also to be parallel to AB. One can see why the argument so far seemed consistent with Lagrange's views on sufficient reason.

By the same procedure, he continued, we can now make another line KL that makes angle HCK equal to angle ICG, but placed on the other side of the new parallel line HI (see Figure 1); and we can keep on in this way to make arbitrarily many in this fashion, which, as he said, "is evidently absurd" [32, pp. 32V–33R].

The modern reader may object that Lagrange's symmetry arguments are, like the uniqueness of parallels, equivalent to Euclid's postulate. But the logical correctness, or lack thereof, of Lagrange's proof is not the point. (In this manuscript, by the way, Lagrange went on to give an analogous proof—also by the principle of sufficient reason—that between two points there is just one straight line, because if there were a second straight line on one side of the first, we could then draw a third straight line on the other side, and so on [32, pp. 34R—34V]. Lagrange, then, clearly liked this sort of argument.)

Why Did He Attack the Problem This Way?

It is now time to address the second, and more important question: Why did he do it in the way he did?

I want to argue this: Lagrange's arguments from sufficient reason were shaped by properties of space, space as it was believed to be in the seventeenth and eighteenth centuries. These properties are profoundly Euclidean. To eighteenth-century thinkers, space was infinite, it was exactly the same in all directions, no direction was privileged, it was like the plane in having no curvature, and symmetrical situations were equivalent. Lagrange himself explicitly linked his symmetry arguments to Leibniz's principle of sufficient reason, but—as we will see—these ideas are also historically linked to Giordano Bruno's arguments for the infinite universe, Descartes' view of space as indefinite material extension, the projective geometry used to describe perspective in Renaissance art, various optimization arguments like "light travels in straight lines because that is the shortest path," and, above all, the Newtonian doctrine of absolute space. As we will soon see, these properties were essential to physical science in the seventeenth and eighteenth centuries: both physics and philosophy promoted the identification of space with its Euclidean structure.

This goes along with a shift in emphasis concerning what Euclidean geometry is about. Geometry, in ancient times, was the study of geometric figures: triangles, circles, parallelograms, and the like, but by the eighteenth century it had become the study of space [41, chapter 5]. The space eighteenth-century geometry was about was, in Henri Poincaré's words, "continuous, infinite, three-dimensional, homogeneous and isotropic" [44, p. 25]. Bodies moved through it preserving their sizes and shapes. The possible curvature of three-dimensional space did not even occur to eighteenth-century geometers. Their space was Euclidean through and through.

Why did philosophers conclude that space had to be infinite, homogeneous, and the same in all directions? Effectively, because of the principle of sufficient reason. For instance, Giordano Bruno in 1600 argued that the universe must be infinite because there is no reason to stop at any point; the existence of an infinity of worlds is no less reasonable than the existence of a finite number of them. Descartes used similar reasoning in his *Principles of Philosophy*. "We recognize that this world . . . has no limits in its extension. . . . Wherever we imagine such limits, we . . . imagine beyond them some indefinitely extended space" [28, p. 104]. Similar arguments were used by other seventeenth-century authors, including Newton. Descartes identified space and the extension of matter, so geometry was, for him, about real physical space. But geometric space, for Descartes, had to be Euclidean. This is because the theory of parallel lines is crucial for

Descartes' analytic geometry—not for Cartesian coordinates, which Descartes did not have, but because he needed the theory of similar figures in order to give meaning to expressions of arbitrary powers of x [23, p. 197]. Descartes was the first person to justify using such powers. An expression like x^4 for Descartes is not the volume of a 4-dimensional figure, but a line, which can be defined as the fourth proportional to the unit line, x, and x^3. That is, $1/x = x^3 / x^4$. They are all lines, and since all powers of x are lines, they can all be constructed geometrically—but only if we have the theory of similar triangles, for which we need the theory of parallels.

Now let us turn from seventeenth-century philosophy to seventeenth-century physics. Descartes, some 50 years before Newton published his first law of motion, was a co-discoverer of what we call linear inertia: that in the absence of external influences a moving body goes in a straight line at a constant speed. Descartes called this the first law of nature, and for him, this law follows from what we now recognize as the principle of sufficient reason. Descartes said, "Nor is there any reason to think that, if [a part of matter] moves . . . and is not impeded by anything, it should ever by itself cease to move with the same force" [30, p. 75]. And the straight-line motion of physical moving bodies obviously requires the indefinite extendability of straight lines and thus indefinitely large, if not infinite, space [23, p. 97].

Descartes' contemporary Pierre Gassendi, another co-discoverer of linear inertia, used "sufficient reason" to argue for both inertia and the isotropy of space. Gassendi said, "In principle, all directions are of equal worth," so that in empty spaces, "motion, in whatever direction it occurs . . . will neither accelerate nor retard; and hence will never cease" [29, p. 127].

Artists, too, helped people learn to see space as Euclidean. We see the space created in the paintings and buildings of the Renaissance and later as Euclidean. Renaissance artists liked to portray floors with rectangular tiles and similar symmetric architectural objects—to show how good they were at perspective. These works of art highlight the observations that parallel lines are everywhere equidistant, that two lines perpendicular to a third line are parallel to each other. And our experience of perspective in art and architecture helps us shape the space we believe we live in. We have seen pictures like these many times, but consider them now as conditioning people to think in a particular way about the space we live in: as Euclidean, symmetric, and indefinitely extendible—going on to infinity [11].

Artist-mathematicians like Piero della Francesca began the development of the subject of projective geometry, but the first definitive mathematical treatise on it is that of Girard Desargues in the 1630s. Seventeenth-century projective geometry used the cone (like the artist's rays of sight or light) to prove properties of all the conic sections as projections of the circle. For instance, geometers treated the ellipse as the circle projected to a plane not

FIGURE 2. Pierro della Francesca (1410/1420–1492), "The Ideal City."

FIGURE 3. Leonardo da Vinci (1452–1519), "The Last Supper."

perpendicular to the cone. And the parabola, as Kepler pointed out, behaves projectively like an ellipse with one focus at infinity. So projective geometry explicitly brought infinity into Euclidean geometry: planes and lines go to infinity; parallel lines meet at the point at infinity.

And the geometry of perspective and projective geometry reinforced Euclideanness in a wide variety of other ways, from the role of Euclid's *Optics* in the humanistic classical tradition to the use of the theory of parallels to draw military fortifications from 2-dimensional battlefield sketches [12, p. 24].

Although these Euclidean views prevailed, perhaps they didn't have to. There were alternatives suggested even in the eighteenth century [22]. Is visual space Euclidean? Not necessarily. Bishop Berkeley, for instance, said that we don't "see" distance at all; we merely infer it from the angles we do see. And Thomas Reid pointed out that a straight line right in front of you looks exactly like a circle curved with you at the center—or even a circle curved away from you in the other direction. Reid gave a set of rules for

FIGURE 4. Raphael (1483–1520), "The School of Athens."

visual space—he called this the "geometry of visibles"—which clearly are not Euclid's rules; a modern philosopher has called Reid's geometry of visibles "the geometry of the single point of view" [46, p. 396].

And there are other alternatives to Euclideanness. Cultures other than the western often speak about space differently and order their perceptions differently: particular directions have special connotations, and "closeness" can be cultural as well as metrical. Many cultures do not use the idea of an outside abstract space at all; instead—as Leibniz did—they recognize only the relations between bodies [3], [35], [36]. So, as a matter of empirical fact, abstract Euclidean space is not something that all human thinkers do use, let alone that all humans must use.

In the twentieth century, experimental psychologists showed that when people in a dark room are asked to put luminous points into two equidistant lines, or two parallel lines, the people are satisfied when the lines in fact curve away from the observer. As a result, Rudolf Luneburg in the 1940s claimed that visual space is a hyperbolic space of constant curvature; later psychological experiments suggest that visual space is not represented by any consistent geometry [48, pp. 30–31].

FIGURE 5. Parmigianino (1503–1540), "Self Portrait in a Convex Mirror."

Even in the Renaissance, some painters portrayed what we now recognize as 3-dimensional non-Euclidean spaces, using reflections in a convex mirror, notably Parmigianino's (1524) "Self Portrait in a Convex Mirror," and, most famously, the "Arnolfini Wedding" by Jan van Eyck (1434). In the spaces in these mirrors, parallel lines are not everywhere equidistant.

A modern physicist, John Barrow, has said that if people had paid more attention to these mirrors, non-Euclidean geometry might have been discovered much sooner [1, p. 176]. But I am not so sure. I think that these artists viewed convex mirrors as presenting an especially difficult problem in portraying 3-dimensional Euclidean space on a 2-dimensional Euclidean canvas; for instance, J. M. W. Turner included such drawings in his strongly Euclidean lectures [45] on perspective.

The winning view, I think, is that expressed by the Oxford art historian Martin Kemp, who says that from the Renaissance to the nineteenth century, the goal of constructing "a model of the world as it appears to a rational, objective observer" [27, p. 314] was shared by scientists and artists alike. Virtually unanimously, artists, armed with Euclid's *Optics*, have long helped teach us to "see" a Euclidean world.

The Crucial Argument: Newtonian Physics

Let us now return to physics and to the most important seventeenth-century argument of all for the reality of infinite Euclidean space: Newtonian

FIGURE 6a. Jan van Eyck (c. 1390–1441), "The Arnolfini Wedding."

mechanics. Newton needed absolute space as a reference frame, so he could argue that there is a difference between real and apparent accelerations. He wanted this so he could establish that the forces involved with absolute (as opposed to relative) accelerations are real, and thus that gravity is real. Newton's absolute space is infinite and uniform, "always similar and immovable" [38, p. 6], and he described its properties in Euclidean terms. And it is real; it has a Platonic kind of reality.

Leibniz, by contrast, did not believe in absolute space. He not only said that spatial relations were just the relations between bodies, he used the principle of sufficient reason to show this. If there were absolute space, there would have to be a reason to explain why two objects would be related in one way if East is in one direction and West in the opposite direction, and

FIGURE 6b. Detail from "The Arnolfini Wedding."

related in another way if East and West were reversed [24, p. 147]. Surely, said Leibniz, the relation between two objects is just one thing! But Leibniz did use arguments about symmetry and sufficient reason in philosophy and science—sufficient reason was his principle, after all. Thus, although Descartes and Leibniz did not believe in empty absolute space and Newton did, they all agreed that what I am calling the Euclidean properties of space are essential to physics.

In the eighteenth century, the Leibniz-Newton debate on space was adjudicated by one of Lagrange's major intellectual influences, Leonhard Euler. In his 1748 essay "Reflections on Space and Time," Euler argued that space must be real; it cannot be just the relations between bodies as the Leibnizians claim [10]. This is because of the principles of mechanics—that is, Newton's first and second laws. These laws are beyond doubt, because of the "marvelous" agreement they have with the observed motions of bodies. The inertia of a single body, Euler said, cannot possibly depend on the behavior of other bodies. The conservation of uniform motion in the same direction makes sense, he said, only if measured with respect to immovable space, not to various other bodies. And space is not in our minds, said Euler; how can physics—real physics—depend on something in our minds? So space for Euler is real.

The philosopher Immanuel Kant was influenced by Euler's analysis [14, pp. 29, 207]. Kant agreed that we need space to do Newtonian physics. But in his *Critique of Pure Reason* of 1781, Kant placed space in the mind nonetheless. We order our perceptions in space, but space itself is in the mind, an intuition of the intellect. Nevertheless, Kant's space turned out to be Euclidean too. Kant argued that we need the intuition of space to prove theorems in geometry. This is because it is in space that we make the constructions necessary to prove theorems. And what theorem did Kant use as an example? The sum of the angles of a triangle is equal to two right angles, a result whose proof requires the truth of the parallel postulate [26, "Of space," p. 423].

Even outside of mathematics and physics, explicit appeals to sufficient reason, symmetry, parallels, and infinity pervade eighteenth-century thought, from balancing chemical equations to symmetry in architecture to the balance of powers in the U.S. Constitution.

Let me call one last witness from philosophy: Voltaire. Like many thinkers in the eighteenth century, Voltaire said that universal agreement was a marker for truth. Religious sects disagree about many things, he said, so on these topics they are all wrong. But by contrast, they all agree that one should worship God and be just; therefore that must be true. Voltaire pointed out also that "There are no sects in geometry" [47, p. 195]. One does not say, "I'm a Euclidean, I'm an Archimedean." What everyone agrees on: that is what is true. "There is but one morality," said Voltaire, "as there is but one geometry" [47, p. 225].

The Argument from Eighteenth-Century Mathematics and Science

Now let us turn to eighteenth-century mathematics and science. Eighteenth-century geometers tended to go beyond Euclid himself in assuming Euclideanness. As a first example, look at the 1745 *Elémens de Géométrie* by Alexis-Claude Clairaut. Clairaut grounded geometry not on Euclid's postulates but on the capacity of the mind to understand clear and distinct ideas. For instance, Euclid had defined parallel lines as lines in the same plane that never meet. Clairaut, less interested in proof than in Euclidean plausibility, defined parallel lines as lines that are everywhere equally distant from one another [6, p. 10]. The great French *Encyclopedia* [7, vol. 11, pp. 905–906] defined parallel lines as "lines that prolonged to infinity never get closer or further from one another, or *that meet at an infinite distance*" [italics added] assuming, then, a uniform, flat Euclidean space infinitely extended. The "equidistant" definition of parallels is reinforced by ordinary language, as we speak of parallel developments, or, more geometrically, ships on parallel courses, and even of parallels of latitude.

And speaking of latitude raises the question of why the fact that the geometry on the surface of a sphere, with great circles serving as "lines," is not Euclidean—there are no parallels, for example—did not shake mathematicians' conviction that all of Euclid's postulates are true and mutually consistent. Lagrange himself is supposed to have said that spherical trigonometry does not need Euclid's parallel postulate [4, pp. 52–53]. But the surface of a sphere, in the eighteenth-century view, is not non-Euclidean; it exists in 3-dimensional Euclidean space [20, p. 71]. The example of the sphere helps us see that the eighteenth-century discussion of the parallel postulate's relationship to the other postulates is not really about what is logically possible, but about what is true of real space.

Now, let us turn to eighteenth-century physics. As we will see, Euclideanness, especially the theory of parallels and the principle of sufficient reason, was essential to the science of mechanics in the eighteenth century, not only to its exposition, but to its progress.

Johann Heinrich Lambert was one of the mathematicians who worked on the problem of Postulate 5. Lambert explicitly recognized that he had not been able to prove it, and considered that it might always have to remain a postulate. He even briefly suggested a possible geometry on a sphere with an imaginary radius. But Lambert also observed that the parallel postulate is related to the law of the lever [20, p. 75]. He said that a lever with weightless arms and with equal weights at equal distances is balanced by a force in the opposite direction at the center equal to the sum of the weights, and that all these forces are parallel. So either we are using the parallel postulate, or perhaps, Lambert thought, some day we could use this physical result to prove the parallel postulate.

Lagrange himself in his *Analytical Mechanics* [31, pp. 4–5] gave an argument about balancing an isosceles triangle similar to, but much more complex than, Lambert's discussion. Lagrange himself did not explicitly link the law of the lever to the parallel postulate, but the geometry of the equilibrium situation that Lagrange was describing nonetheless requires it [4, pp. 182–183]. In a similar move, d'Alembert had tried to deduce the general law of conservation of momentum purely from symmetry principles [13, pp. 821–823]. And in the 1820s, J.-B. Fourier, from a very different philosophical point of view, also said that the parallel postulate could be derived from the law of the lever. From this Fourier concluded that geometry follows from statics and so geometry is a physical science [20, pp. 78–79]. But note that it is still Euclidean geometry.

Let us now concentrate further on Lagrange's mechanics. His deepest conviction was that a subject must be seen in its full generality. Like many Enlightenment thinkers only more so, Lagrange wanted to reduce the vast number of laws and principles to a single fundamental general principle, preferably one that is independent of experience. "Sufficient reason" was such a principle.

Although he did not explicitly cite Leibniz's principle in his *Analytical Mechanics* [43, p. 146], Lagrange used it frequently. For instance, he wrote, "The equilibrium of a straight and horizontal lever with equal weights and with the fulcrum at its midpoint is a self-evident truth because there is no reason that either of the weights should move." [31, Section 1, Paragraph 2].

Another key example of Euclideanness as physical argument is the use of parallelograms to find resultant forces. Lagrange, in his *Analytical Mechanics,* used the principle of sufficient reason, Euclid's theory of parallels, and the infinity of space and its Euclidean nature to discuss the composition of forces [31, p. 17]. He said that a body which is moved uniformly in two different directions simultaneously must necessarily traverse the diagonal of the parallelogram whose sides it would have followed separately. So parallels are needed. Lagrange continued, "with regard to the direction in the case of two equal forces, it is obvious that there is no reason that the resultant force should be nearer to one than the other of these two equal forces; therefore it must bisect the angle formed by these two forces" [31, p. 21]. Lagrange also used the principle of composition of forces to get the conditions of equilibrium when two parallel forces are applied to the extremities of a straight lever. He suggested that we imagine "that the directions of the forces extend to infinity" and then, on this basis, we can prove that "the resultant force must pass through the point of support." In effect this is the parallelogram argument with the corner of the parallelogram at infinity. Lagrange tried to reduce even his own fundamental physical principle—the principle of virtual velocities—to levers and parallelograms of forces, and after him Ampère, Carnot, Laplace, and Poisson tried to do the same [39, p. 218].

Pierre-Simon Laplace, too, related *a priori* arguments, including sufficient reason and Euclid's theory of parallels, to argue that physical laws had to be the way they were. For instance he said that a particle on a sphere moves in a great circle because "there is no reason why it should deviate to the right rather than the left of that great circle"—notice, "not a word about the forces acting on this particle on this sphere" [15, p. 104]. Laplace also asked why gravitation had to be inverse-square, and gave a geometric answer [4, p. 53], [16, p. 310]. He said that inverse-square gravitation implies that if the size of all bodies and all distances in the whole universe were to decrease proportionally, the bodies would describe the same curves that they do now, so that universe would still look exactly the same. The observer, then, needs to recognize only the ratios. So, Laplace said, even though we haven't proved Euclid's fifth postulate, we know it must be true, and so the theorems deduced from it must also be true. For Laplace, then, the idea of space includes the following self-evident property: *similar figures have proportional sides* [4, p. 54]. The Newtonian physical universe requires similar figures to have proportional sides, and this of course requires the theory of parallels, and thus geometry must be Euclidean [33, p. 472].

These men did not want to do mechanics, as, say, Newton had done. They wanted to show not only that the world was this way, but that it necessarily had to be. A modern philosophical critic, Helmut Pulte, has said that Lagrange's attempt to "reduce" mechanics to analysis strikes us today as "a misplaced endeavour to mathematize . . . an empirical science, and thus to endow it with infallibility" [39, p. 220]. Lagrange would have responded, "Right! That's just exactly what we are all doing." Lagrange thought these two things: Geometry is necessarily true; mechanics is mathematics. He needed them both.

Why Did It Matter So Much?

And now, we are ready for the last question. Why did actually proving Postulate 5 matter so much to Lagrange and to his contemporaries? I trust I have convinced the reader of the central role of Euclideanness in the eighteenth century. But still, if it were just a matter of simple logic, surely after 2000 years people should have concluded: we have been trying as hard as possible, we cannot imagine how to prove this, so let us just concede defeat. It can't be done. Euclid was right in deciding that it had to be assumed as a postulate. Why did eighteenth-century geometers not settle for this, and, in particular, why didn't Lagrange?

Because there was so much at stake. Because space, for Newtonian physics, has to be uniform, infinite, and Euclidean, and because metaphysical principles like that of sufficient reason and optimality were seen both as Euclidean and as essential to eighteenth-century thought. How could all of this rest on a mere assumption? So, many eighteenth-century thinkers believed that it was crucial to shore up the foundations of Euclid's geometry, and we can place Lagrange's manuscript in the historical context of the many attempts in the eighteenth century to cure this "blemish" in Euclid by proving Postulate 5.

Also, Lagrange was not just any eighteenth-century mathematician. Lagrange was, mathematically speaking, a Cartesian and a Leibnizian. His overall philosophy of mathematics was to reduce each subject to the most general possible principle. In calculus, as I have argued at length in two books [17], [18], Lagrange wanted to reduce all the ideas of limits and infinites and infinitesimals and rates of change or fluxions to "the algebraic analysis of finite quantities." In algebra, Lagrange said that even Newton's idea of algebra as "universal arithmetic" wasn't general enough; algebra was the study of systems of operations. In mechanics, his goal was to reduce everything to the principle of virtual velocities—and then to use "only algebraic operations subject to a regular and uniform procedure" [31, preface]. Lagrange even composed his *Analytical Mechanics* without a single diagram, precisely so he could show he had reduced physics to pure analysis. Geometry, then, ought also to be reducible to self-evident principles, to clear, distinct, and general ideas.

Finally, there are social causes to be considered. First, the social background will help answer this question: Why was Lagrange doing this in 1806, as opposed, say, to the 1750s when he taught mathematics at the military school in Turin or in the 1770s and 1780s when he was the leading light of the Berlin Academy of Sciences? One reason is that in about 1800 there was a revival of interest in synthetic geometry in France. There was a Parisian school in synthetic geometry including Monge, Servois, Biot, Lacroix, Argand, Lazare Carnot and his students, and Legendre. Important reasons for this were partly practical, partly ideological [40, p. 450]. The practical needs are related to Monge's championing of descriptive geometry, so clearly useful in architecture and in military planning. Monge also helped directly to pique Lagrange's interest, writing two letters to him in the early 1790s soliciting his assistance on problems involving the geometry of perspective [37].

As for the ideology promoting geometry in France after the Revolution, as Joan Richards has written, "the quintessentially reasonable study of universally known space had a central role to play in educating a rational populace" [40, p. 454]. Lagrange himself articulated such views throughout his lifetime, writing as early as 1775 that synthetic geometry was sometimes better than analytic because of "the luminous clarity that accompanies it" [19, p. 135], and, near the end of his life, telling his friend Frédéric Maurice that "geometric considerations give force and clarity to judgement" [19, p. 1295].

So the French geometers would not have been favorably disposed to inventing a non-Euclidean geometry. It is no wonder that only comparative outsiders like the Hungarian Janos Bolyai and the Russian Nikolai Ivanovich Lobachevsky were the first to publish on this topic. Even Gauss, who out of fear of criticism did not publish his own invention of the subject that he christened "non-Euclidean geometry," was somewhat outside the French mathematical mainstream.

The British would not have found inventing non-Euclidean geometry enticing either. Even William Rowan Hamilton, who in the 1840s was to devise the first noncommutative algebra, wrote in 1837, "No candid and intelligent person can doubt the truth of the chief properties of Parallel Lines, as set forth by Euclid in his Elements, two thousand years ago. . . . The doctrine involves no obscurity nor confusion of thought and leaves in the mind no reasonable ground for doubt" [21, p. 354]. In fact, even after Hermann von Helmholtz and W. K. Clifford had introduced non-Euclidean geometry into Victorian Britain, some British thinkers continued to maintain that real space had to be Euclidean. There was a great deal at stake in

Britain: the doctrine of the unity of truth, the established educational program based on the Euclidean model of reason, and the attitudes toward authority that this entailed.

The authority and rigor of Euclid, both in Britain and in France, were part and parcel of the established intellectual order. And—one last social point—non-Euclidean geometry even in the twentieth century was culturally seen as anti-establishment, partly through its association with relativity theory. For instance, surrealist artists used it that way: misunderstood, perhaps, but still explicitly part of their assault on traditional artistic canons. Two examples are Yves Tanguy's "Le Rendez-vous des parallèles" (1935) and Max Ernst's "Young Man Intrigued by the Flight of a Non-Euclidean Fly" (1942–1947).

Conclusion

I cannot explain why Lagrange initially thought that his proof was a good one, but I hope it is clear why he thought he needed to prove the parallel postulate, and why he tried to prove it using the techniques that he used.

The story I have told reminds us that, although the great eighteenth-century mathematicians are our illustrious forbears, our world is not theirs. We no longer live in a world of certainty, symmetry, and universal agreement. But it was only in such a world that the work of Lagrange and Laplace, Fourier and Kant, Euler and d'Alembert could flourish. That space must be Euclidean was part of the Cartesian, Leibnizian, Newtonian, symmetric, economical, and totally rationalistic world view that underlies all of Lagrange's mathematics and classical mechanics—ideas that, from Newton and Leibniz through Kant and Laplace, buttressed the whole eighteenth-century view of the universe and the laws that govern it. And the certainty of Euclidean geometry was the model for the whole Enlightenment program of finding universally-agreed-upon truth through reason.

Thus, though Lagrange's illustrious audience in Paris may have realized that his proof was wrong, their world-view made them unable to imagine that the parallel postulate couldn't be proved, much less to imagine that the world itself might be otherwise.

Acknowledgments

I thank the Department of the History and Philosophy of Science, University of Leeds, England, for its hospitality and for vigorous discussions of this research. I also thank the Bibliothèque de l'Institut de France for permission to study Lagrange's manuscripts, the donors of the Flora Sanborn Pitzer Professorship at Pitzer College for their generous support, the Mathematical Association of America for inviting me to talk about this topic at Math-Fest 2007, and Miss Kranz, my trigonometry teacher at Fairfax High School in Los Angeles, who once on a slow day in class revealed to us all that there was such a thing as non-Euclidean geometry.

References

1. J. D. Barrow, Outer space, in *Space: In Science, Art and Society*, F. Penz, G. Radick, and R. Howell, eds., Cambridge University Press, Cambridge, 2004, 172–200.

2. J.-B. Biot, Note historique sur M. Lagrange, in *Mélanges Scientifiques et Littéraires*, vol. III, Michel Lévy Frères, Paris, 1858, 117–124.

3. P. Bloom et al., *Language and Space*, MIT Press, Cambridge, MA, 1996.

4. R. Bonola, *Non-Euclidean Geometry*, Dover, New York, 1955.

5. R. Carnap, *An Introduction to the Philosophy of Science*, Dover, New York, 1995; reprint of Basic Books, New York, 1966.

6. A.-C. Clairaut, *Élémens de Géométrie*, Par la Compagnie des Libraires, Paris, 1765.

7. J. D'Alembert and D. Diderot, eds., *Encyclopédie, ou Dictionnaire Raisonné des Sciences, des Arts et des Métiers*, Briasson, Paris, 1762–1772.

8. A. De Morgan, *A Budget of Paradoxes*, Longmans Green, London, 1872.

9. Euclid, *The Thirteen Books of Euclid's Elements*, T. L. Heath, ed., Cambridge University Press, Cambridge, 1956.

10. L. Euler, Reflexions sur l'espace et le tems, *Mémoires de l'académie de Berlin* **4** (1750) 324–333.

11. J. V. Field, *The Invention of Infinity: Mathematics and Art in the Renaissance*, Oxford University Press, Oxford, 1997.

12. J. V. Field and J. J. Gray, *The Geometrical Work of Girard Desargues*, Springer, New York, 1987.

13. J. Franklin, Artifice and the natural world: Mathematics, logic, technology, in *Cambridge History of Eighteenth Century Philosophy*, K. Haakonssen, ed., Cambridge University Press, Cambridge, 2006, 815–853.

14. M. Friedman, *Kant and the Exact Sciences*, Harvard University Press, Cambridge, MA 1992.

15. E. Garber, *The Language of Physics: The Calculus and the Development of Theoretical Physics in Europe, 1750–1914*, Birkhäuser, Boston, 1999.

16. C. C. Gillispie, *Pierre-Simon Laplace, 1749–1827: A Life in Exact Science*, in collaboration with R. Fox and I. Grattan-Guinness, Princeton University Press, Princeton, 1997.

17. J. V. Grabiner, *The Calculus as Algebra: J.-L. Lagrange, 1736–1813*, Garland, New York, 1990.

18. ———, *The Origins of Cauchy's Rigorous Calculus*, MIT Press, Cambridge, MA, 1981.

19. I. Grattan-Guinness, *Convolutions in French Mathematics, 1800–1840*, 3 vols., Birkhäuser, Basel, 1990.

20. J. Gray, *Ideas of Space: Euclidean, Non-Euclidean and Relativistic*, 2nd ed., Clarendon Press, Oxford, 1989.

21. T. L. Hankins, Algebra as pure Time: William Rowan Hamilton and the foundations of algebra, in *Motion and Time, Space and Matter*, P. Machamer and R. Turnbull, eds., Ohio State University Press, Columbus, OH, 1976, 327–359.

22. R. J. Herrnstein and E. G. Boring, *A Source Book in the History of Psychology*, Harvard University Press, Cambridge, MA, 1966.

23. L. Hodgkin, *A History of Mathematics*, Oxford University Press, Oxford, 2005.

24. N. Huggett, ed., *Space from Zeno to Einstein: Classic Readings with a Contemporary Commentary*, MIT Press, Cambridge, MA, 1999.

25. Institut de France, Académie des Sciences, *Procès-Verbaux des Séances de l'Académie, 1804–1807*, vol. III, Académie des Sciences, Hendaye, 1913.

26. I. Kant, *Critique of Pure Reason* (trans. F. M. Müller), Macmillan, New York, 1961.

27. M. Kemp, *The Science of Art: Optical Themes in Western Art from Brunelleschi to Seurat*, Yale University Press, New Haven, CT, 1990.

28. A. Koyré, *From the Closed World to the Infinite Universe*, Johns Hopkins University Press, Baltimore, MD, 1957.

29. ———, *Metaphysics and Measurement*, Harvard University Press, Cambridge, MA, 1968.

30. ———, *Newtonian Studies*, University of Chicago Press, Chicago, IL, 1965.

31. J.-L. Lagrange, *Analytical Mechanics*, A. Boissonnade and V. N. Vagliente, trans. and eds., Kluwer, Dordrecht, 1997; from J.-L. Lagrange, *Mécanique analytique*, 2nd ed., Courcier, Paris, 1811–1815. In *Oeuvres de Lagrange*, Gauthier-Villars, Paris, 1867–1892, vol. XI.

32. ———, Sur la Théorie des Parallèles, Mémoire lu en 1806. Unpublished manuscript in the Bibliothèque de l'Institut de France, Inst MS 909, ff 18–35.

33. P.-S. Laplace, *Exposition du système du monde*, Cercle-Social l'An IV, Paris, 1796, in *Ouevres complètes de Laplace*, Gauthier-Villars, Paris, 1878–1912, vol. VI.

34. A.-M. Legendre, *Eléments de Géométrie*, Didot, Paris, 1794.

35. S. C. Levinson, Language and mind: Let's get the issues straight, in *Language in Mind: Advances in the Study of Language and Thought*, D. Gertner and S. Goldin-Meadow, eds., MIT Press, Cambridge, MA, 2003, 25–46.

36. ———, Frames of reference and Molyneux's question: Crosslinguistic evidence, in *Language and Space*, MIT Press, Cambridge, MA, 1996, 109–170.

37. G. Monge, Two letters to Lagrange, n. d., *Oeuvres de Lagrange*, Gauthier-Villars, Paris, 1867–1892, vol. XIV, 308–310, 311–314.

38. I. Newton, *Sir Isaac Newton's Mathematical Principles of Natural Philosophy and His System of the World*, trans. A. Motte, rev. F. Cajori, University of California Press, Berkeley, 1960.

39. H. Pulte, 1788: Joseph Louis Lagrange, *Mechanique analitique*, in *Landmark Writings in Western Mathematics, 1640–1940*, I. Grattan-Guinness, ed., Elsevier, Amsterdam, 2005, 208–224.

40. J. L. Richards, The Geometrical Tradition: Mathematics, Space, and Reason in the Nineteenth Century, in *The Modern Physical and Mathematical Sciences*, M. J. Nye, ed., Vol. 5 of the *Cambridge History of Science*, Cambridge University Press, Cambridge, 2003, 449–467.

41. B. A. Rosenfeld, *A History of Non-Euclidean Geometry: Evolution of the Concept of a Geometric Space*, Springer, Berlin and Heidelberg, 1988.

42. B. Russell, *A Critical Exposition of the Philosophy of Leibniz*, Allen and Unwin, London, 1937.

43. R. Taton, Lagrange et Leibniz: De la théorie des functions au principe de raison suffisante, in *Beiträge zur Wirkungs- und Rezeptionsgeschichte von Gottfried Wilhelm Leibniz*, A. Heinekamp, ed., Franz Steiner Verlag, Stuttgart, 1986, 139–147.

44. R. Torretti, *Philosophy of Geometry from Riemann to Poincaré*, D. Reidel, Boston, 1978.

45. J. M. W. Turner, drawings, in A. Fredericksen, *Vanishing Point: The Perspective Drawings of J. M. W. Turner*, Tate, London, 2004.

46. J. Van Cleve, Thomas Reid's geometry of visibles, *Philosophical Review* **111** (2002) 373–416.

47. F. M. Arouet de Voltaire, Philosophical Dictionary, articles "Sect" and "Morality," in *The Portable Voltaire*, B. R. Redman, ed., Viking, New York, 1949.

48. M. Wagner, *The Geometries of Visual Space*, Erlbaum, Mahwah, NJ, 2006.

Kronecker's Algorithmic Mathematics

Harold M. Edwards

I wonder if it is as widely believed by the younger generation of mathematicians, as it is believed by my generation, that Leopold Kronecker was the wicked persecutor of Georg Cantor in the late nineteenth century and that, to the benefit of mathematics, by the end of the century the views of Cantor had prevailed and the narrow prejudices of Kronecker had been soundly and permanently repudiated.[*]

I suspect this myth persists wherever the history of mathematics is studied, but even if it does not, an accurate understanding of Kronecker's ideas about the foundations of mathematics is indispensable to understanding constructive mathematics, and the contrast between his conception of mathematics and Cantor's is at the heart of the matter.

It is true that he opposed the rise of set theory, which was occurring in the years of his maturity, roughly from 1870 until his death in 1891. Set theory grew out of the work of many of Kronecker's contemporaries—not just Cantor, but also Dedekind, Weierstrass, Heine, Méray, and many others. However, as Kronecker told Cantor in a friendly letter written in 1884, when it came to the philosophy of mathematics he had always recognized the unreliability of philosophical speculations and had taken, as he said, "refuge in the safe haven of actual mathematics." He went on to say that he had taken great care in his mathematical work "to express its phenomena and truths in a form that was as free as possible from philosophical concepts." Further on in the same letter, he restates this goal of his work and its relation to philosophical speculations saying, "I recognize a true scientific value—in the field of mathematics—only in concrete mathematical truths, or, to put it more pointedly, only in mathematical formulas."

Certainly, this conception of the nature and substance of mathematics restricts it to what is called "algorithmic mathematics" today, and it is

---

[*] This essay is a lecture presented at "Computability in Europe 2008," Athens, June 19, 2008.

what I had in mind when I chose my title "Kronecker's Algorithmic Mathematics." Indeed, these quotations from Kronecker show that my title is a
redundancy—for Kronecker, that which was not algorithmic was not mathematics, or, at any rate, it was mathematics tinged with philosophical concepts that he wished to avoid.

At the time, I don't think that this attitude was in the least unorthodox.
The great mathematicians of the first half of the nineteenth century had,
I believe, similar views, but they had few occasions to express them, because such views were an understood part of the common culture. There is
the famous quote from a letter of Gauss in which he firmly declares that
infinity is a *façon de parler* and that completed infinites are excluded from
mathematics. According to Dedekind, Dirichlet repeatedly said that even
the most recondite theorems of algebra and analysis could be formulated as
statements about natural numbers. One needs only to open the collected
works of Abel to see that for him mathematics was expressed, as Kronecker
said, in mathematical formulas. The fundamental idea of Galois theory, in
my opinion, is the theorem of the primitive element, which allowed Galois
to deal concretely with computations that involve the roots of a given polynomial. And Kronecker's mentor Kummer—whom Kronecker credits in
his letter to Cantor with shaping his view of the philosophy of mathematics—developed his famous theory of ideal complex numbers in an altogether algorithmic way.

It is an oddity of history that Kronecker enunciated his algorithms at a
time when there was no possibility of implementing them in any nontrivial
cases. The explanation is that the algorithms were of theoretical, not practical, importance to him. He goes so far as to say in his major treatise *Grundzüge
einer arithmetischen Theorie der algebraischen Grössen* that, by his lights, the notion of *irreducibility* of polynomials lacks a firm foundation (*entbehrt einer
sicheren Grundlage*) unless a *method* is given that either factors a given polynomial or proves that no factorization is possible.

When I first encountered this opinion of Kronecker's, I had to read it
several times to be sure I was not misunderstanding him. The opinion was
so different from my mid-twentieth century indoctrination in mathematics
that I could scarcely believe he meant what he said. Imagine Bourbaki saying that the notion of a nonmeasurable set lacked a firm foundation until a
method was given for measuring a given set or proving that it could not be
measured!

But he did mean what he said and, as I have since learned, there are other
indications that the understanding of mathematical thought in that time was
very different from ours. Another example of this is provided by Abel's
statement in his unfinished treatise on the algebraic solution of equations
that "at bottom" (*dans le fond*) the problem of finding all solvable equations

was the same as the problem of determining whether a given equation was solvable. It would be explicable if he had said that the proof that an equation is solvable is "at bottom" the problem of solving it, but he goes much further: If you know how to decide whether any given equation is solvable, you know how to find all equations that are solvable.

To be honest, I don't feel I fully understand these extremely constructive views of mathematics—I am a product of my education—but knowing that a mathematician of Abel's caliber and experience saw mathematics in this way is an important phenomenon that a viable philosophy of mathematics needs to take into account.

So Kronecker did mean it when he said that a method of factoring polynomials with integer coefficients is essential if one is to make use of irreducible polynomials, and he took care to outline such a method. I won't go into any explanation of his method—I doubt that it was original with him, but his treatise is the standard reference—except to say that it is pretty impractical even with modern computers and to say that in his day it was utterly out of the question even for quite small examples.

This observation makes it indisputable that the objective of Kronecker's algorithm had to do with the *meaning* of irreducibility, *not* with practical factorization. It is a distinction that at first seems paradoxical but that arises in many contexts. If you are trying to find a specific root of a specific polynomial, Newton's method is almost certainly the best approach, but if you want to prove that every polynomial has a complex root, Newton's method is useless. In practice, it converges very rapidly, but the error estimates are so unwieldy that you can't prove that it will converge at all until you are able to prove that there is a root for it to converge *to*, and for this you need a more plodding and less effective method.

More generally, we all know that in practical calculations clever guesswork and shortcuts can play important roles, and Monte Carlo methods are everywhere. These are important topics in algorithmic mathematics, but not in *Kronecker's* algorithmic mathematics. I am not aware of any part of his work where he shows an interest in practical calculation. Again, his interest was in mathematical *meaning*, which for him was *algorithmic* meaning.

I have always fantasized that Euler would be ecstatic to have access to modern computers and would have a wonderful time figuring out what he could do with them, factoring Fermat numbers and computing Bernoulli numbers. Kronecker, on the other hand, I think would be much cooler toward them. In my fantasy, he would feel that he had *conceived of* the calculations that interested him and had no need to carry them out in any specific case. His attitude might be the one Galois expressed in the "preliminary discourse" to his treatise on the algebraic solution of equations: ". . . I need only to indicate to you the method needed to answer your question, without wanting to make myself or anyone else carry it out. In a word, the calculations are impractical." (. . . *je n'aurai rien à y faire que de vous indiquer le moyen de répondre à votre question, sans vouloir charger ni moi ni personne de le faire. En un mot les calculs sont impraticables.*) This somewhat provocative statement was omitted from the early publications of Galois's works. See page 39 of the critical edition (1962) of Galois's works. Galois's mathematics, like Kronecker's, was algorithmic but not practical. That's why it is not so surprising that all of this algorithmic mathematics—we could call it impractical algorithmic mathematics—was developed at a time when computers didn't exist.

This, in my opinion, was Kronecker's conception of mathematics—that which his predecessors had accomplished and that which he wanted to advance. What generated the oncoming tide of set theory that was about to engulf this conception?

Kronecker wrote about the rising tendency in very few places, but when he did write about it, he identified the motive for its development: Set theory was developed in an attempt to encompass the notion of *the most general real number*.

In 1904, after Kronecker had been dead for more than a dozen years, Ferdinand Lindemann published a reminiscence about Kronecker that has become a part of the Kronecker legend and that is surely wrong. According to Lindemann, Kronecker asked him, apparently in a jocular way, "What is the use of your beautiful researches about the number π? Why think about such problems when irrational numbers do not exist?"

We can only guess what Kronecker said to Lindemann that Lindemann remembered in this way, but I am confident that he would not have said that irrational numbers did not exist. To be persuaded of this, one only needs to know that Kronecker refers in his lectures on number theory (the ones edited and published by Kurt Hensel) to "the transcendental number π from geometry," which he describes by the formula $\frac{\pi}{4} = 1 - \frac{1}{3} + \frac{1}{5} - \frac{1}{7} + \cdots$. Note that Kronecker introduces π in his *first* lecture on *number theory*. Note also that he accepts π not only as an irrational number but as a *transcendental* number; the proof of the transcendence of π was of course the achievement for which Lindemann was, and remains, famous. (His later belief that he had proved Fermat's Last Theorem is benignly neglected.)

Kronecker, as one of the great masters of analytic number theory, made frequent use of transcendental methods and would have had no qualm about real numbers. His qualm—and he stated it explicitly—had to do with the conception of the *most general real number*.

My colleague Norbert Schappacher of the University of Strasbourg has discovered a document that states Kronecker's qualm about the most general real number in a different way and confirms Kronecker's statement to

Cantor that his notions about the philosophy of mathematics were taught him by Kummer. The document is a letter of Kummer in which he states that he and Kronecker are in agreement in their belief that the effort to create enough individual points to fill out a continuum—that is, enough real numbers to fill out a line—is as vain as the ancient efforts to prove Euclid's parallel postulate. (The quotation occurs in a letter from Kummer to his son-in-law H. A. Schwarz, dated March 15, 1872, in the Nachlass Schwarz of the archives of the Berlin-Brandenburg Academy of Sciences, folder 977.)

In our time, when young students are routinely told that "the real line" consists of uncountably many real numbers and that it is "complete" as a topological set, this opinion of Kummer and Kronecker is heresy in the most literal sense—it denies the truth of what young people are told has the agreement of all authorities.

So Kronecker, along with Kummer, saw what was going on—saw the push to describe the most general real number, saw, as it were, the wish on the part of his colleagues to talk about "the set of all real numbers." Moreover, he responded to it. His response was: *It is unnecessary*.

I have said that Kronecker says very little about the foundations of mathematics in his writings. But in the few words he does say, this message is clear: It is unnecessary. One of the main goals of his mathematical work was to *demonstrate* that it was unnecessary by, as he told Cantor, expressing the truths and phenomena of mathematics in ways that were as free as possible from philosophical concepts. That would most certainly exclude any general theory of real numbers. He wished to show such a theory was unnecessary by doing without it.

In view of the Kummer passage found by Schappacher, we see that he also believed there was a special importance to his belief that the construction of the set of all real numbers was not necessary, because he believed it was doomed to fail.

In all likelihood you are now hearing for the first time the opinion that "the real line" may not be a well-founded concept, so I probably have no realistic hope of convincing you that this view may be justified. I won't make a serious effort to do so. I will let it pass with just a brief reference to complications like Russell's paradox, Gödel's incompleteness theorem, the independence of the continuum hypothesis and the axiom of choice, nonstandard models of the real numbers, and, coming at it from a different direction, Brouwer's free choice sequences. There is a long history of unsuccessful efforts to wrestle with infinity in a rigorous way, efforts which, so far as I have ever been able to see, have been consistently frustrated. As Kummer and Kronecker foresaw.

But even if one accepts that one day it will succeed—or that it long ago did succeed, except for uninteresting nitpicking—it seems to me that

Kronecker's main message is still worth hearing and considering: It is unnecessary. Mathematics should proceed without it to the maximum extent possible. Kronecker was confident that in the end its exclusion would prove to be no impediment at all.

Well, of course modern mathematics has painted itself into a corner in which dealing with infinity in a rigorous manner *is* necessary. If mathematics is defined to be that which mathematicians do, then dealing with the real line is essential to mathematics. If mathematics insists on talking about "properties of the real line" as though the real line were a given, there is no room for the belief that it is unnecessary.

Inevitably, then, Kronecker's assertion is an assertion about the nature and domain of mathematics itself. It asserts that that which lies outside the Kroneckerian conception of mathematics is unnecessary. (Instead of the Kroneckerian conception, I would prefer to call it the classical conception of mathematics in deference to Euler and Gauss and Dirichlet and Abel and Galois, but somehow "classical mathematics" has come to mean the Cantorian opposite of this; therefore I am forced to call it the Kroneckerian conception.)

With this meaning of "Kronecker's algorithmic mathematics" in mind, we can perhaps agree that it is unnecessary to attempt to embrace the most general real number—to embrace "the real line." What is lost by adopting this view of mathematics?

I often hear mention of what must be "thrown out" if one insists that mathematics needs to be algorithmic. What if one is throwing out error? Wouldn't that be a good thing rather than the bad thing the verb "to throw out" insinuates? I personally am not prepared to argue that what is being thrown out is *error,* but I think one can make a very good case that a good deal of confusion and lack of clarity are being thrown out.

The new ways of dealing with infinity that set theory brought into mathematics can be seen in the method used to construct an integral basis in algebraic number theory. Kronecker gave an algorithm for this construction. You could write a computer program following his plan, and the program would work, although it might be very slow. Hilbert in his *Zahlbericht* approaches the same problem in a different, and outrageously nonconstructive, way. He imagines all numbers in the field written as polynomials with rational coefficients in a particular generating element α. The polynomials are then of degree less than m, where m is the degree of α. Moreover, there is a common denominator for all the *integers* in the field when they are written in this way. Hilbert has the *chutzpah* to say: For each $s = 1, 2, \ldots, m$, choose an integer in the field which is represented as a polynomial of degree less than s, and in which the numerator of the leading coefficient is the greatest common divisor of all numerators that occur in such integers.

Such a choice is to be carried out for each s; the m integers in the field "found" in this way are an integral basis.

Let me try to state in as simple a way as possible the process he is indicating: The integers in the field are a countable set, so it is legitimate to regard them as listed in an infinite sequence. The entries in the sequence are polynomials in α of degree less than m whose coefficients are rational numbers with a fixed denominator D. For each s, Hilbert wants us to first strike from the list all polynomials of degree s or greater, and, from among those that remain, choose one in which the numerator of the coefficient of α^{s-1} is nonzero, but otherwise is as small as possible in absolute value. (Hilbert looks at the greatest common divisor of the numerators rather than the absolute value, but the effect is the same.) So, not once but m times, we are to survey an infinite list of integers and pick out a nonzero one that has the smallest possible absolute value.

To put this in perspective, let me describe an analogous situation. Imagine an infinite sequence of zeros and ones is given by some unknown rule. Would it be reasonable for me to ask you to record a 1 if the sequence contains infinitely many ones and otherwise to record a 0? In twentieth-century mathematics, one was asked to do such things all the time. Therefore, it is perhaps difficult to deny, as I would like to do, that it is a reasonable thing to ask. But surely *no one* would contend that it is an *algorithm*.

No doubt Hilbert regarded his as a simplification of Kronecker's construction. But only someone indoctrinated in the nonconstructive Hilbertian orthodoxy, as I was, and as many of you surely were, could hear it called a "construction" without leaping from his or her chair in protest.

To "throw out" from mathematics arguments of this type should be regarded as ridding it of ideas that are at best sloppy thinking and at worst delusions. And in this particular case, the argument for throwing out Hilbert's argument is all the stronger because Kronecker had already shown many years earlier that it was, in truth, unnecessary.

This contrast, between Kronecker's algorithm for constructing an integral basis and Hilbert's nonconstructive proof (can it be called a proof?) of the existence of an integral basis, illustrates the fork in the road that mathematics encountered at the end of the nineteenth century: To follow Kronecker's algorithmic path, or to choose instead the daring new set-theoretic path proposed by Dedekind, Cantor, Weierstrass, and Hilbert.

You all understand very well which path was taken and you all understand as well how I feel about the choice that was made.

But now, in the twenty-first century, I hope mathematicians will begin to reconsider that fateful choice. Now that there are conferences devoted to "Computability in Europe" and mathematicians in their daily practice are dealing more and more with algorithms, approaching problems more and more by asking themselves how they can use their powerful computers to gain insight and find solutions, the climate of opinion surely will change. How can anyone who is experienced in serious computation consider it important to conceive of the set of all real numbers as a mathematical "object" that can in some way be "constructed" using pure logic? For computers, there are no *irrational* numbers, so what reason is there to worry about the most general *real* number? Let us agree with Kronecker that it is best to express our mathematics in a way that is as free as possible from philosophical concepts. We might in the end find ourselves agreeing with him about set theory. It is unnecessary.

Indiscrete Variations on Gian-Carlo Rota's Themes

Carlo Cellucci

Introduction

I never met Gian-Carlo Rota but I have often made references to his writings on the philosophy of mathematics, sometimes agreeing, sometimes disagreeing.

In this paper I will discuss his views concerning four questions: the existence of mathematical objects, definition in mathematics, the notion of proof, the relation of philosophy of mathematics to mathematics.

The Existence of Mathematical Objects

Although in the twentieth century the main question in the philosophy of mathematics has been the justification of mathematics, next to it there has been the question of the existence of mathematical objects (see, for example, Cellucci, 2006).

There are four possible answers to this question: (i) Mathematical objects exist; (ii) No they don't; (iii) We don't know; (iv) The question is irrelevant to mathematical practice or meaningless.

IRRELEVANCE OF THE EXISTENCE OF MATHEMATICAL OBJECTS

Rota's answer to the question of the existence of mathematical objects is of kind (iv).

For he states that "it does not matter whether mathematical items exist, and probably it makes little sense to ask the question" (Rota, 1997, p. 161). If "someone proved beyond any reasonable doubt that mathematical items do not exist", that would not "affect the truth of any mathematical statement" (*ibid.*). Discussions concerning the existence of mathematical objects "are motivated by deep-seated emotional cravings for permanence which are of psychiatric rather than philosophical interest" (*ibid.*).

Rota's answer seems a very sensible one. A similar answer was given in the Seventeenth and the Eighteenth century by such disparate thinkers as Descartes, Locke, Hume. For example, Descartes stated that "arithmetic, geometry, and other such disciplines" are "indifferent as to whether these things do or do not in fact exist" (Descartes, 2006, p. 11).

True, in the Twentieth century Brouwer attempted to develop an alternative mathematics, banning certain objects on the ground that they did not exist and replacing them by certain other objects on the ground that they did exist. His attempt, however, ended in failure, for his alternative mathematics was so awkward that nobody managed to use it for any essential purpose. Moreover, it excluded certain objects, such as discontinuous real functions, which are essential to physics. Thus Brouwer's attempt turned out to be a purely ideological one. Indeed, Brouwer explicitly stated that he did not care a bit for the applicability of mathematics to physics, because he rejected "expansion of human domination over nature" (Brouwer, 1975, p. 483). For him mathematics was a search for beauty, and in applicable mathematics "beauty will hardly be found" (*ibid.*; for more on this, see Cellucci, 2007, pp. 76–81).

While Rota's answer seems a very sensible one, what Rota positively says about the nature of mathematical objects seems less convincing.

He states that such things as "prices, poems, values, emotions, Riemann surfaces, subatomic particles, and so forth" are "ideal objects" (Rota, 1986a, p. 169). The method of logical analysis inaugurated by Husserl sets itself "the purpose of construction (rather than dissection) of ideal objects to be subjected to yet-to-be-discovered ideal laws and relations" (Rota, 1986a, p. 172).

Thus for Rota mathematical objects are ideal objects, and mathematical laws and relations concern such objects.

Now, an idealization simplifies certain real items by ignoring some features which make only a small difference in practice, while retaining other features which are basic. Therefore, ideal objects should retain the basic features of the real items they are said to idealize. Then Rota's statement that mathematical objects are ideal objects contrasts with the fact that several mathematical objects have nothing in common with real objects. For example, infinite sets have no basic feature in common with concrete physical aggregates, so they cannot be said to be an idealization of them.

Moreover, stating that mathematical objects are ideal objects trivializes the question of the applicability of mathematics to the physical world. For it makes all mathematical statements vacuously true of it.

For example, consider the statement: in any triangle, the interior angles are equal to two right angles. Such statement is of the form $P(x) \rightarrow Q(x)$, where $P(x)$ expresses 'x is a triangle' and $Q(x)$ expresses 'The interior angles of x are equal to two right angles'. Now, $P(x)$ is an ideal statement which is false of the physical world for triangles do not exist in it. Therefore $P(x) \rightarrow Q(x)$ is vacuously true of the physical world. For the same reason, even the false statement 'In any triangle the interior angles are not equal to two right angles', being of the form $P(x) \rightarrow Q(x)$, is vacuously true of the physical world.

Furthermore, Rota pushes his view that mathematical objects are ideal objects to the extreme. For he states that "the ideal of all science, not only of mathematics, is to do away with any kind of synthetic a posteriori statement and to leave only analytic trivialities in its wake" (Rota, 1997, p. 119). Science is "the transformation of synthetic facts of nature into analytic statements of reason" (*ibid.*).

Thus, according to Rota, not only mathematical objects are ideal objects, but mathematical statements, and all scientific statements generally, are ultimately analytic.

Holding that scientific statements are ultimately analytic, Rota seems to be attracted by the view that the laws of the world can be derived a priori. He even goes so far as saying that the mathematician "forces the world to obey the laws his imagination has freely created" (Rota, 1997, p. 70). Thus, for Rota, not only the laws of the world are ultimately analytic, but the mathematician enforces them on the world.

However, the laws of the world are not ultimately analytic, for the facts of nature are not truths of reason. Nor the mathematician enforces such laws on the world, for the laws in question are just a means by which humans make the world understandable to themselves. Science is what humans grasp of the world in their own terms, and this essentially depends on synthetic facts of nature. Moreover, the creations of the mathematician are not completely free, for they essentially depend on the mathematician's biological constitution.

MATHEMATICAL OBJECTS AS HYPOTHESES

A more satisfactory account of the nature of mathematical objects can be given stating that mathematical objects are hypotheses tentatively introduced to solve mathematical problems.

A mathematical object is the hypothesis that a certain condition is satisfiable. For example, an even number x is the hypothesis that the condition $x = 2y$ is satisfiable for some integer y.

If, in the course of reasoning, the condition turns out to be satisfiable, we say that the object 'exists', if it turns out to be unsatisfiable, we say that it 'does not exist'. Thus speaking of 'existence' is just a metaphor.

That the condition turns out to be unsatisfiable typically occurs in proofs by *reductio ad absurdum*.

For example, suppose that, to solve the problem whether, in Euclidean geometry, in any triangle the interior angles are equal to two right angles, we tentatively introduce a new kind of objects: triangles whose interior angles are not equal to two right angles. We say: let *ABC* be any such triangle. Then we draw a line through one of its vertices parallel to the opposite side and we see that the interior angles are actually equal to two right angles. We thus have a contradiction. Therefore we conclude that a triangle such as *ABC* cannot exist, and hence that, in Euclidean geometry, in any triangle the interior angles are equal to two right angles.

There is no more to mathematical existence than the fact that mathematical objects are hypotheses tentatively introduced to solve mathematical problems. Such hypotheses are in turn a problem to be solved, it will be solved by introducing other hypotheses, and so on. Thus solving a mathematical problem is a potentially infinite task (see Cellucci, 2002, Ch. 22; Cellucci, 2008b).

The view that mathematical objects are hypotheses is related to Plato's view that "those who practice geometry, arithmetic and similar sciences hypothesize the odd, and the even, the geometrical figures, the three kinds of angle, and any other thing of that sort which are relevant to each subject" (Plato, *Republic*, VI 510 c 2–5). Thus they hypothesize mathematical objects. But, in addition to them, they also hypothesize properties of such objects. Therefore mathematical hypotheses concern either mathematical objects or their properties.

Those who practice geometry, arithmetic and similar sciences, however, do not confine themselves to making hypotheses, but also give an account of them by introducing other hypotheses. For "when you had to give an account of the hypothesis itself, you would give it in the same way, once again positing another hypothesis" (Plato, *Phaedo*, 101 d 5–7). And so on. This is Plato's 'dialectical method' or 'dialectic'.

It is often claimed that "Plato thinks of the method of mathematics as one that starts by assuming some hypotheses and then goes 'downwards' from them (i.e. by deduction), whereas the method of philosophy (i.e. dialectic) is to go 'upwards' from the initial hypotheses, finding reasons for them (when they are true), until eventually they are shown to follow from an 'unhypothetical first principle' " (Bostok, 2009, pp. 13–14).

Actually, quite the opposite is true. According to Plato, the method that starts by assuming some hypotheses and then goes 'downwards' from them, that is, the axiomatic method, is a degeneration of the genuine method of mathematics, which is the same as the method of philosophy, that is, the dialectical method.

Mathematicians who practice the axiomatic method assume their hypotheses as starting-points (principles) without giving any account of them, and go downwards from them. But, giving no account of them, they do not really know their principles. And, "when a man does not know the principle, and when the conclusion and intermediate steps are also constructed out of what he does not know, how can he imagine that such a fabric of convention can ever become science?" (Plato, *Republic*, VII 533 c 3–6).

Therefore, criticizing the axiomatic method, Plato opposes the method of philosophy not to the method of mathematics but rather to a degeneration of that method.

Admittedly, according to Plato, mathematical objects in their true nature are independently existing entities, which can be known only by means of intellectual intuition. For only the latter is capable of grasping the unhypothetical first principle. But intellectual intuition is possible only if we "leave our body and contemplate the things themselves with the soul by itself" (Plato, *Phaedo*, 66 e 1–2). Until then, we may only try to state more and more general hypotheses. In this way we will get better and better approximations to mathematical objects, but will never arrive at grasping them fully. On the other hand, however, this is the only way available to us.

HYPOTHESES VS. FICTIONS

The view that mathematical objects are hypotheses must not be confused with fictionalism—the view that mathematical objects are like characters in fiction.

According to fictionalism, all there is to mathematics is that "we have a good story about natural numbers, another good story about sets, and so forth" (Field, 1989, p. 22). The sense in which "'2 + 2 = 4' is true is pretty much the same as the sense in which 'Oliver Twist lived in London' is true: the latter is true only in the sense that it is true according to a certain well-known story, and the former is true only in that it is true according to standard mathematics" (Field, 1989, p. 3).

Fictionalism is inadequate in several respects, which I cannot discuss here (see Cellucci, 2007, pp. 109–14). I will only point out that hypotheses are essentially different from fictions for the following two reasons.

1) While fictions are stated with the awareness that they are not real, hypotheses are aimed at reality. They are meant to provide an adequate approach to a still unknown or not perfectly known reality, although, as in the case of proofs by *reductio ad absurdum*, they eventually may fail to provide an adequate approach. Admittedly, the determination of reality given by a hypothesis is provisional. But one will try to make it relatively stable showing that the hypothesis is plausible, that is, compatible with the facts of experience.

2) While fictions are merely thinkable, hypotheses are supposed to be possible, so they must agree with the facts of experience. Only then hypotheses can be said to provide an adequate approach to reality. On the contrary, fictions are not supposed to be possible, and hence are not rejected if they do not agree with the facts of experience.

EXISTENCE AND IDENTITY

Hypotheses provide an only partial and provisional characterization of mathematical objects. The latter may receive new determinations through interactions between hypotheses and experience.

Rota proposes an apparently related view when he states that "a full description of the logical structure of a mathematical item lies beyond the reach of the axiomatic method" (Rota, 1997, p. 156). For example, "the real line, or any mathematical item, is not fully given by any one specific axiom system" and "allows an open-ended sequence of presentations by new axiomatic systems. Each such system is meant to reveal new features of the mathematical item" (Rota, 1997, p. 157).

On the other hand, however, Rota also states that "a mathematical item is 'independent' of any particular axiom system" in much the same way as "an idea is 'independent'" of "the words that are used to express the idea" (Rota, 1997, p. 160).

Rota acknowledges that from this one might be tempted to conclude that he assumes that "mathematical items 'exist' independently of axiom systems" (ibid.). But he rejects such conclusion stating that, "in discussing the properties of 'mathematical items,'" he is "in no way required to take a position as to the 'existence' of mathematical items" (Rota, 1997, p. 161). He is only concerned with the question of the identity of mathematical items, and "identity does not presuppose existence" (ibid.).

Rota's distinction between identity and existence is an important one. What he says about the identity of mathematical objects, however, is somewhat indefinite.

On the one hand, he states that "there is no way to 'reduce' identity to any mental process" (Rota, 1997, p. 187). Thus identity is not mental. On the other hand, he states that, if I kick a stone, "it is a mistake to believe that what I kick is a material stone", rather "'the stone' is an item that has no existence, but has an identity" (Rota, 1997, p. 186). Thus identity is not physical either.

But, if identity is neither mental nor physical, what is it? Rota does not tell us. And he could not tell us, for he claims that "identity is the 'undefined term'" (ibid.).

Moreover, Rota claims that "the properties of identity are the axioms from which we 'derive' the world" (ibid.). But this is impossible, for the properties of identity are purely logical, and from purely logical properties one cannot derive any feature of the world, even less 'the world'.

Thus on Rota's view mathematical objects have a somewhat unsettled status.

THE INEXHAUSTIBILITY OF MATHEMATICAL OBJECTS

This problem is avoided if mathematical objects are hypotheses. Admittedly, fixing properties of mathematical objects, hypotheses characterize their identity but say nothing about their existence. The latter requires a further investigation. But their identity is by no means, as Rota states, the undefined term. On the contrary, it is characterized by the hypothesis, and characterized differently by different hypotheses.

For example, according to Euclid, the identity of the real line is characterized by the hypothesis that it is a breadthless length which lies evenly with the points on itself. According to Cantor and Dedekind, it is characterized by the hypothesis that it is the set of singleton real numbers.

Again, hypotheses do not characterize the identity of mathematical objects completely and conclusively but only partially and provisionally, therefore their identity is always open to new determinations. It is so because mathematical objects are open to interactions with other objects, from which new properties may arise.

Gödel states that "the creator necessarily knows all properties of his creatures, because they can't have any others except those he has given to them" (Gödel, 1995, p. 311).

But this is contradicted, for example, by the fact that an artist is not necessarily a good art critic, or a mathematician is not necessarily a good philosopher of mathematics. For the creator's creatures, when put in relation with other things, may be seen to have properties not included in those the creator has given to them.

Moreover, from the interactions of the given mathematical objects with other objects, new facts may emerge which may suggest to modify or completely replace the hypothesis by which the identity of the given mathematical objects had been characterized. This is a potentially endless process, for mathematical objects are always open to interactions with other objects.

Definition in Mathematics

In the last century the received view on mathematical definition—that is, the view that has been taken for granted without further criticism—has been that mathematical definitions are arbitrary stipulations which serve as abbreviations and hence engender no knowledge.

For example, Frege states that "no definition extends our knowledge. It is only a means for collecting a manifold content into a brief word or sign, therefore making it easier to handle. This and this alone is the use of definition in mathematics" (Frege, 1984, p. 274). Then "it would be inappropriate to count definitions among principles. For to begin with, they are arbitrary stipulations" (*ibid.*). Even when "what a definition has stipulated is subsequently expressed as an assertion, still its epistemic value is no greater than that of an example of the law of identity $a = a$," and "one would hardly wish to accord the status of an axiom to every single instance, to every example, of the law" (*ibid.*). Therefore, "by defining, no knowledge is engendered" (*ibid.*).

DEFINITION, DESCRIPTION AND ANALYSIS OF CONCEPTS

Rota criticizes the received view arguing that, while "mathematicians take mischievous pleasure in faking the arbitrariness of definition", actually "no mathematical definition is arbitrary" (Rota, 1997, p. 97).

In fact "a lot of mathematical research time is spent in finding suitable definitions to justify statements that we already know to be true" (Rota, 1997, p. 50). For example, the Euler's formula for polyedra was known to be true "long before a suitable general notion of polyedron could be defined" (*ibid.*).

Definition must not be confused with description, for "description and definition are two quite different enterprises" (Rota, 1997, p. 48). In ages past, "mathematical objects were described before they could be properly defined," and "the mathematics of the past centuries confirms the fact that mathematics can get by without definitions but not without descriptions" (Rota, 1997, pp. 48–9). For example, "the lack of definition of a tensor did not stop Einstein, Levi-Civita and Cartan from doing some of the best mathematics in this century" (Rota, 1997, p. 97).

Rota is quite right in criticizing the received view, for the fact that mathematicians often spend so much research time in finding suitable definitions to justify statements that they already know to be true would be incomprehensible if definitions were arbitrary stipulations. As a matter of fact, mathematicians often use concepts for a long time before a suitable definition is found.

When what is defined is a concept which had previously been used without a precise definition, the definition contains an analysis of the concept, and may therefore express an advance in knowledge. In such cases, a definition gives definiteness to a concept which had previously been more or less vague. Presumably, what Rota calls 'description' is the more or less vague statement of the concept on the basis of which that concept was previously used.

That definitions may contain an analysis of concepts which had previously been more or less vague, explains why mathematicians often spend so much time in looking for definitions to justify theorems which are already known. Finding an adequate definition is not the result of an arbitrary stipulation but rather of an investigation.

DEFINITION IN MATHEMATICS AND IN PHILOSOPHY

Rota, however, makes some statements about definitions which seem to be in conflict with the view that mathematical definitions are not arbitrary stipulations but may contain an analysis of concepts.

For example, he asks: "Doesn't mathematics begin with definitions and then develop the properties of the objects that have been defined by an admirable and infallible logic?" (Rota, 1997, p. 97).

It would be tempting to consider this question ironic, but it is not. For Rota opposes definition in mathematics to definition in philosophy, stating that, salutary as the injunction 'Define your terms!' "may be in mathematics, it has had disastrous consequences when carried over to philosophy. Whereas mathematics starts with a definition, philosophy ends with a definition" (*ibid.*).

Rota would have been surprised to learn that the same view is put forward by Dummett, one of the most eminent representatives of that analytic philosophy which he criticizes.

For Dummett states that the aims of mathematicians "differ from those of philosophers. They care little whether the definitions they arrive at capture the concept as we implicitly understand it in ordinary life: they are concerned only to formulate a precise concept under which it may be reasonably claimed that every case determinately either falls or does not fall" (Dummett, 2010, p. 11). On the contrary, "the philosopher's reasoning takes place on the basis of our existing implicit understanding; it appeals to that understanding and hence is not carried out, as the mathematician's is, within a framework of concepts already made precise" (*ibid.*). The philosopher's aim is "the analysis of concepts we already possess, but about which we are confused; he seeks to remove that confusion" (*ibid.*).

Thus, like Rota, Dummett claims that, while mathematics starts with a definition, philosophy ends with a definition.

There is, however, little credibility in the view that, unlike philosophy, mathematics starts with a definition. If it were so, then it would be incomprehensible why mathematicians often try to find new definitions for concepts for which definitions are already available.

In mathematics, as in philosophy, definitions are not a starting point but rather an arrival point of the investigation. (In the sixteenth century

Zabarella put forward a similar view of mathematical definitions; see Cellucci, 1989.) Definitions are not arbitrary stipulations but rather hypotheses and, as all hypotheses, are means of discovery. Two definitions of the same concept may be not equally adequate as means of discovery (see Cellucci, 2002, Ch. 36).

THE ALLEGED CIRCULARITY OF DEFINITIONS AND THEOREMS

Rota also claims that "the theorems of mathematics motivate the definitions as much as the definitions motivate the theorems. A good definition is 'justified' by the theorem that can be proved with it, just as the proof of the theorem is 'justified' by appealing to a previously given definition. There is, thus, a hidden circularity in formal mathematical exposition. The theorems are proved starting with definitions; but the definitions themselves are motivated by the theorems that we have previously decided ought to be correct" (Rota, 1997, p. 97).

This circularity, however, only affects definitions in the axiomatic method. If definitions are hypotheses, then there is no circularity between definitions and theorems. For a definition is not justified, as Rota claims, by the theorem that can be proved with it. As any other hypothesis, it is justified by its plausibility, that is, compatibility with the existing data. Therefore, it is justified by things distinct from the theorems which are established by means of the definition (see also Cellucci, 2008a, p. 209).

The Notion of Proof

In the last century the received view on proof has been the axiomatic one: a proof is a deductive derivation of a proposition from primitive premises that are true, in some sense of 'true'.

Strictly related is the notion of formal proof. In fact, according to the Hilbert-Gentzen Thesis, every axiomatic proof can be represented by a formal proof (see Cellucci, 2008b).

PROOF AS THE OPENING UP OF POSSIBILITIES

Rota criticizes the received view arguing that, while "some mathematicians will go as far as to pretend that mathematics is the axiomatic method", the "mistaken identification of mathematics with the axiomatic method has led to a widespread prejudice among scientists that mathematics is nothing but a pedantic grammar, suitable only for belaboring the obvious" (Rota, 1997, p. 142).

The error of the received view "lies in assuming that a mathematical proof, say the proof of Fermat's last theorem, has been devised for the explicit purpose of proving what it purports to prove" (Rota, 1997, p. 143). On the contrary, the point of proof "is to open up new possibilities for mathematics," for example, "this opening up of possibilities is the real value of the proof of Fermat's conjecture" (Rota, 1997, p. 144).

Indeed, the value "of Wiles' proof lies not in what it proves, but in what it has opened up, in what it will make possible," specifically, in "a host of new techniques that will lead to further connections between number theory and algebraic geometry. Future mathematicians will discover new applications, they will solve other problems, even problems of great practical interest, by exploiting Wiles' proof and techniques" (*ibid.*).

Rota is quite right in criticizing the received view, arguing that the point of proof is not so much to prove what it purports to prove as rather to open up new possibilities for mathematics. In fact, the main point of proof is to discover plausible hypotheses which not only prove what they purport to prove, but also establish connections between different areas of mathematics—connections that may lead to new discoveries. Briefly, the main point of proof is its heuristic value (see Cellucci, 2002, Ch. 24; Cellucci, 2008b).

On the other hand—surprisingly enough in view of his criticism of the received view—Rota also states that "a mathematical theory begins with definitions and derives its results from clearly agreed-upon rules of inference" (Rota, 1997, p. 90). For "there is at present no viable alternative to axiomatic presentation if the truth of mathematical statements is to be established beyond reasonable doubt" (Rota, 1997, p. 142).

Rota's statement is in conflict with his view that the point of proof is to open up new possibilities. For, in an axiomatic theory all possibilities are implicitly contained in the axioms, which are given from the very beginning. Therefore, an axiomatic proof cannot open up new possibilities establishing connections between different areas of mathematics. Axiomatic theories are closed systems (see Cellucci, 1998, pp. 192–203; Cellucci, 2002, Ch. 7).

AXIOMATIC PRESENTATION AND GÖDEL'S INCOMPLETENESS THEOREMS

Rota's statement that there is at present no viable alternative to axiomatic presentation if the truth of mathematical statements is to be established beyond reasonable doubt, is also in conflict with Gödel's incompleteness results. (For a philosophically motivated treatment of these results, see Cellucci, 2007).

By Gödel's first incompleteness theorem, for each axiomatic theory satisfying certain minimal conditions, there are elementary sentences which are true but cannot be deduced from the axioms of the theory. Thus, contrary to Rota's statement, for each axiomatic presentation of a given area of mathematics, there are mathematical truths of that area that cannot be proved by means of the axioms of that axiomatic presentation.

Moreover, by Gödel's second incompleteness theorem, for each axiomatic theory satisfying certain minimal conditions, the truth of the axioms of the theory, and hence of the results which can be deduced from them, cannot be established by absolutely reliable means. Thus, contrary to Rota's statement, for each axiomatic presentation of a given area of mathematics, that presentation does not guarantee that the truth of the results proved by means of it can be established beyond reasonable doubt.

Rota's statement is surprising because it takes no account of the implications of Gödel's incompleteness results for the axiomatic method.

This, however, is not uncommon among mathematicians. Many of them believe that Gödel's results have "almost no relevance to the work of most mathematicians" (Davies, 2008, p. 88). These mathematicians, however, provide no evidence for their belief, which is then a purely emotional one—a matter of faith, not of reason. In particular, they do not explain how they reconcile the belief that the method of mathematics is the axiomatic method with the belief that Gödel's results have almost no relevance to their work.

This is due to the fact that, as Rota himself acknowledges, among scientists it is not uncommon to "believe in unrealistic philosophies of science" (Rota, 1997, p. 108).

Doubts may also be raised about the effectiveness of axiomatic presentation for teaching and learning. Such doubts were already raised by Descartes and Newton (see Cellucci, 2008b, pp. 23–24).

ARE THERE DEFINITIVE PROOFS?

Another difficulty concerning proof is raised by Rota's claim that there are definitive proofs.

According to Rota, "after a new theorem is discovered, other simpler proofs of it will be given until a definitive one is found" (Rota, 1997, p. 146). The "first proof of a great many theorems is needlessly complicated," and "it takes a long time, from a few decades to centuries, before the facts that are hidden in the first proof are understood" (*ibid.*). This "gradual bringing out of the significance of a new discovery takes the appearance of a succession of proofs, each one simpler than the preceding. New and simpler versions of a theorem will stop appearing when the facts are finally understood" (*ibid.*). Then a definitive proof will have been reached.

Rota is quite right in saying that a salient feature of mathematical practice, from ancient times to the present, has been the looking for new proofs of results for which proofs were already known. This shows that the concern of mathematicians is not truth. Otherwise a single proof would be enough for each result, and there would be no point in looking for new proofs.

Rota, however, assesses this feature of mathematical practice in terms of understanding, and assumes that new proofs of a given result will stop appearing when the facts are finally understood. This is in conflict with the fact that often finding a new proof does not lead to any improvement in the understanding of the old proofs, for the new proof is based on different ideas.

Looking for new proofs makes sense only if the main point of proof consists in its heuristic value. Specifically, it consists in the discovery of hypotheses which not only allow one to prove the result, but also establish connections between different areas of mathematics, where such connections may lead to new discoveries. The result is looked at from several different perspectives, each of which may suggest a different hypothesis and hence a different proof.

Since, at least potentially, new perspectives are always possible, this means that new proofs are always possible. Therefore there are no definitive proofs.

Rota's claim that there are definitive proofs is also in conflict with his statement that any mathematical item allows an open-ended sequence of presentations by new axiomatic systems, each of which is meant to reveal new features of that mathematical item. For example, "the real line has been axiomatized in at least six different ways, each one appealing to different areas of mathematics, to algebra, number theory, or topology. Mathematicians are still discovering new axiomatizations of the real line" (Rota, 1997, p. 156). And "the theory of groups may be axiomatized in innumerable ways" (Rota, 1997, p. 155).

Indeed, each new axiomatization is based on new hypotheses, which may allow to give a new proof of a given result. Therefore, again, there are no definitive proofs.

PROOF AND EVIDENCE

Instead of saying that the main point of proof is its heuristic value, Rota states that "the purpose of proof is to make it possible for one mathematician to bring back to life the same evidence that had been previously reached by another mathematician" (Rota, 1997, p. 169). This ensures truth, for truth "is the recognition of the identity of my evidence with someone else's evidence" (*ibid.*).

Thus, according to Rota, the main point of proof is to provide evidence. In his view, "evidence is the primary logical concept. Evidence is the condition of possibility of truth. Truth is a derived notion" (Rota, 1997, p. 168). Therefore, "whereas mathematicians may claim to be after truth, their work belies this claim. Their concern is not truth, but evidence" (Rota, 1997, p. 171). Looking for new proofs of results for which proofs are already known is motivated by the search for evidence, because "a proof that shines with the light of evidence comes long after discovery" (*ibid.*).

Rota, however, provides no convincing account of evidence. He claims that evidence is aimed at understanding, for "all understanding is the fulfillment of some evidence" (*ibid.*).

But evidence cannot be aimed at the understanding of the result, for in most cases, including Fermat's Last Theorem, the result is clear enough.

As Kant points out, "nothing can be comprehended more than that what the mathematician demonstrates, e.g., that all lines in the circle are proportional. And yet he does not comprehend how it happens that such a simple figure has these properties" (Kant, 1992, p. 570).

On the other hand, evidence cannot be aimed at the understanding of proof either, because, as it has been already mentioned, often finding a new proof does not lead to any improvement in the understanding of the old proofs, for the new proof is based on different ideas.

For these reasons, saying that evidence is aimed at understanding does not seem convincing.

Moreover, Rota's view of evidence risks paving the way to irrationalism. For Rota claims that "the experience that most mathematicians will corroborate, that a statement 'must' be true, is not psychological. In point of fact, it is not an experience. It is the condition of possibility of our experiencing the truth of mathematics" (Rota, 1997, p. 168). Or rather, it is a sort of "primordial experience" which is "constitutive of mathematical reasoning" (*ibid.*).

Now, Rota's claim depends on the principle "I feel it is so; therefore it must be true", and this principle is one of the "faulty beliefs that are common among people with anxiety disorders" (Hyman-Pedrick, 2006, p. 49). In fact, to say: 'I feel it is so; therefore it must be true', amounts to saying: 'I hereby declare that this is true, and I am unwilling to discuss it further'.

Rota also claims that "after reaching evidence we go back to our train of thought and try to pin down the moment when evidence was arrived at. We pretend to locate the moment we arrived at evidence in our conscious life. But such a search for a temporal location is wishful thinking" (Rota, 1997, pp. 168–9).

In this way Rota makes mathematics ultimately depend on a ur-experience which is ineffable and cannot even be located in time.

The Relation of Philosophy of Mathematics to Mathematics

In the last century, the received view on the relation of philosophy of mathematics to mathematics has been that the philosophy of mathematics must be normative: it must provides a canon for mathematical practice.

This depends on the assumption that the philosophy of mathematics must provide a justification of mathematics. To assume this amounts to assuming that in mathematics there are justifiable and unjustifiable methods, proofs, practices, etc., and acceptable results are those which are obtainable by justifiable methods. The task of the philosophy of mathematics is then to determine which methods are legitimate and hence justifiably used, thus what an acceptable mathematics should be like. For that reason, the philosophy of mathematics must be normative.

THE DESCRIPTIVE CHARACTER OF THE PHILOSOPHY OF MATHEMATICS

Rota opposes the received view stating that in the philosophy of mathematics "a strictly descriptive attitude is imperative", therefore "all normative assumptions shall be weeded out" (Rota, 1997, pp. 135–6).

Rota is quite right in opposing the received view. The philosophy of mathematics must be descriptive since it must account for mathematical practice. This seems to be a necessary requirement for any philosophy of mathematics deserving the name.

From the fact that the philosophy of mathematics must be descriptive, however, Rota draws the conclusion that the question 'What is mathematics?' "requires no answer," it is "simply an inadequate expression of that wonder one feels contemplating the majestic edifice of mathematics" (Rota, 1999, pp. 104–5).

This conclusion seems unwarranted. An answer to the question 'What is mathematics?' is important because it may have an impact on mathematical practice, for example, on methods used in mathematics.

THE NEED FOR A DRASTIC OVERHAUL OF LOGIC

Rota draws a further conclusion from the fact that the philosophy of mathematics must be descriptive: to account for mathematical practice requires an approach to logic different from mathematical logic.

Indeed, Rota states that "that magnificent clockwork mechanism that is mathematical logic is slowly grinding out the internal weaknesses of the system" (Rota, 1986b, p. 180). The notions on which present-day logic is based "were invented one day for the purpose of dealing with a certain model of the world", and that model of the world is "inadequate to the needs of the new sciences" (*ibid.*). Today, "in all circumstances imaginable, including mathematical reflection (the true one, not the one of a posteriori reconstructions), logic shines for its absence" (Rota, 1999, p. 94).

In fact, mathematical logic has proved inadequate not only to the needs of the new sciences developed in the last century—from computer science and artificial intelligence to life sciences and social sciences—but also to the primary use for which Frege had designed it: to be "a tool in the philosophy of mathematics; just as other mathematics, for example the theory of partial differential equations, is a tool in what used to be called natural philosophy" (Kreisel, 1967, p. 201).

From this Rota concludes that, "if we are to set the new sciences on firm, autonomous, formal foundations, then a drastic overhaul" of "logic is in order. This task is far more complex than the Galilean revolution" (Rota, 1986b, p. 180). It "will have to give the new sciences their theoretical autonomy" (Rota, 1999, p. 116).

Rota's plea for a drastic overhaul of logic seems justified. This task, however, need not be, as Rota claims, far more complex than the Galilean revolution. As Rota says, the latter "was inaugurated by a philosophical revolution" (Rota, 1999, p. 115). Like the Galilean revolution, then, the drastic overhaul of logic might 'only' require a philosophical revolution: a different outlook of the world.

Rota's Place in the Philosophy of Mathematics

Present-day philosophers of mathematics are currently distinguished into 'main-stream' and 'maverick' (see Kitcher-Aspray, 1988; Hersh, 1997; Mancosu, 2008).

Mainstream philosophers of mathematics view mathematics as a static body of knowledge, are mainly concerned with the question of justification of mathematical knowledge and set themselves within the analytic philosophy tradition.

Maverick philosophers of mathematics view mathematics as a dynamic body of knowledge, are mainly concerned with the question of the growth of mathematical knowledge, in particular with the dynamics of mathematical discovery, and are generally ill at ease with the analytic philosophy tradition.

The use of terms 'mainstream' and 'maverick', however, does not seem felicitous, for it suggests that the mainstream philosophers of mathematics are the orthodoxy whereas the maverick ones are the heresy—a heresy which could never become the orthodoxy. Using the terms 'static' and 'dynamic' seems preferable. For this reason, in what follows I will refer to the mainstream philosophers of mathematics as 'static' and to the maverick philosophers of mathematics as 'dynamic', and will speak of the 'static approach' and the 'dynamic approach'.

In addition to these two approaches, a 'third way' approach has also been proposed, characterized by attention to mathematical practice while remaining within the analytic philosophy tradition (see Mancosu, 2008).

However, on the one hand, mathematical practice is a main concern of the dynamic approach. Now, as soon as one begins to consider mathematical practice, it becomes obvious that mathematics is a dynamic body of knowledge, and questions about mathematical discovery immediately arise. Thus attention to mathematical practice necessarily leads to the dynamic approach.

On the other hand, the third way approach pays no or little attention to the dynamics of discovery. The latter is the real discriminant between the static and the dynamic approach. Therefore, the third way approach appears to be simply an extension of the static approach to other topics, retaining the basic features of that approach.

Rota does not fit in the static approach because he does not consider mathematics as a static body of knowledge, and criticizes the analytic philosophy tradition, for which he foresees an "increasing irrelevance followed by eventual extinction" (Rota 1997, p. 103). By contrast, he shares the view of the dynamic approach that discovery should be a main concern of the philosophy of mathematics. For he points out that "heuristic arguments are a common occurrence in the practice of mathematics. However, heuristic arguments do not belong to formal logic. The role of heuristic arguments has not been acknowledged in the philosophy of mathematics despite the crucial role they play in mathematical discovery" (Rota, 1997, p. 134).

In this connection, it is worth noting that Rota states that "it is a frequent experience among mathematicians" that "they cannot solve a problem unless they like it" (Rota, 1991, p. 262). Indeed, "motivation and desire are essential components of mathematical reasoning" (Rota, 1997, p. 160).

This is a valuable remark because it stresses the importance of emotion in mathematical research, and particularly in mathematical discovery. Hadamard made a similar point, stating that "an affective element is an essential part in every discovery or invention", indeed "no significant discovery or invention can take place without the will of finding" (Hadamard, 1996, p. 31). Hadamard's statement, however, has been generally neglected, so it is all the more important that Rota reminds us of the relevance of emotion in mathematical practice. (For more on this issue, see Cellucci, 2008c, Ch. 20).

Admittedly, Mancosu qualifies the dynamic approach in a way that is somewhat different from the one suggested above. For he characterizes it not in terms of attention to the dynamics of mathematical discovery but rather in terms of the following three features: "a) anti-foundationalism, i.e. there is no certain foundation for mathematics; mathematics is a fallible activity; b) anti-logicism, i.e. mathematical logic cannot provide the tool for an adequate analysis of mathematics and its development; c) attention to mathematical practice" (Mancosu, 2008, p. 5).

Nevertheless, Rota also fits in this alternative characterization of the dynamic approach for he states: a) "The lack of certainty, the insecurity of evidence are described by Husserl as features of all evidence whatsoever. Husserl's description is confirmed by observing mathematicians at work" (Rota, 1997, p. 171); b) Mathematical logic does not provide the tools for an adequate analysis of mathematics and has even "given up all claims of providing a foundation to mathematics" (Rota, 1997, pp. 92–3); c) "The philosophy of mathematics is beset with insistent repetitions of a few crude examples taken from arithmetic and from elementary geometry, in total disregard of the philosophical issues that are faced by mathematicians at work" (Rota, 1997, p. 163).

True, Rota's views on mathematics are not always homogeneous. As I have pointed out in this paper, sometimes they are even inconsistent with each other, probably because they were stated at different times. Moreover, his claim that evidence is the primary logical concept is unconvincing. Nevertheless, his criticism of the static approach and his penetrating remarks on some aspects of mathematical practice are an important contribution to the dynamic approach.

Acknowledgements

I am grateful to Miriam Franchella, Donald Gillies and Andrea Reichenberger for useful comments on an earlier draft of this paper, to Fabrizio Palombi for helping me to locate a quotation from Rota's MIT lectures, and to the audience of the Rota memorial conference for useful remarks.

References

Aspray, W., Kitcher, P. (eds.) (1988), *History and Philosophy of Modern Mathematics*, Minneapolis, University of Minnesota Press.

Bostock, D. (2009), *Philosophies of Mathematics. An Introduction*, Malden, MA, Wiley-Blackwell.

Brouwer, L. E. J. (1975), *Collected Works*, vol. I, A. Heyting (ed.), Amsterdam, North-Holland.

Cellucci, C. (1989), *De conversione demonstrationis in definitionem*, in G. Corsi, C. Mangione, and M. Mugnai (eds.), *Atti del convegno internazionale di storia della logica. Le teorie della modalità*, Bologna, CLUEB, pp. 301–6.

Cellucci, C. (1998), *Le ragioni della logica*, Rome-Bari, Laterza.

Cellucci, C. (2002), *Filosofia e matematica*, Rome-Bari, Laterza.

Cellucci, C. (2006), *Introduction* to *Filosofia e matematica,* in R. Hersh (ed.), *18 Unconventional Essays on the Nature of Mathematics*, Berlin, Springer, pp. 17–36.

Cellucci, C. (2007), *La filosofia della matematica del Novecento*, Rome-Bari, Laterza.

Cellucci, C. (2008a), *The Nature of Mathematical Explanation*, "Studies in History and Philosophy of Science", vol. 39, pp. 202–10.

Cellucci, C. (2008b), *Why Proof? What is a Proof?*, in R. Lupacchini and G. Corsi (eds.), *Deduction, Computation, Experiment. Exploring the Effectiveness of Proof*, Berlin, Springer, pp. 1–27.

Cellucci, C. (2008c), *Perché ancora la filosofia*, Rome-Bari, Laterza.

Davies, E. B. (2008), *Interview,* in V.F. Hendricks and H. Leitgeb (eds.), *Philosophy of Mathematics: 5 Questions*, New York, Automatic Press/VIP, pp. 87–99.

Descartes, R. (2006), *Meditations, Objections, and Replies*, R. Ariew and D.A. Cress (eds.), Indianapolis, IN., Hackett.

Dummett, M. (2010), *The Nature and Future of Philosophy,* New York, Columbia University Press.

Field, H. (1989), *Realism, Mathematics and Modality*, Oxford, Blackwell.

Frege, G. (1964), *The Basic Laws of Arithmetic. Exposition of the System*, M. Furth (ed.), Berkeley, University of California Press.

Frege, G. (1984), *Collected Papers on Mathematics, Logic, and Philosophy*, B. McGuinness (ed.), Oxford, Blackwell,

Gödel, K. (1995), *Collected Works, Volume III*, S. Feferman et al. (eds.), Oxford, Oxford University Press.

Hadamard, J. (1996), *The Mathematician's Mind: the Psychology of Invention in the Mathematical Field,* Princeton, Princeton University Press.

Hersh, R. (1997), *What is Mathematics, Really?*, Oxford, Oxford University Press.

Hyman, B. M., Pedrick, C. (2006), *Anxiety Disorders*, Minneapolis, MN., Lerner.

Kac, M., Rota, G.-C., Schwartz, J.T. (1986), *Discrete Thoughts. Essays on Mathematics, Science and Philosophy*, Boston, Basel, Berlin, Birkhäuser.

Kant, I. (1992), *Lectures on Logic*, J.M.Young (ed.), Cambridge, Cambridge University Press.

Kreisel, G. (1967), *Mathematical Logic: What has it Done for the Philosophy of Mathematics?*, in R. Schoenman (ed.), *Bertrand Russell, Philosopher of the Century*, London, Allen and Unwin, pp. 201–72.

Mancosu, P. (ed.) (2008), *The Philosophy of Mathematical Practice*, Oxford, Oxford University Press.

Rota, G.-C. (1986a), *Husserl and the Reform of Logic*, in (Kac, Rota, Schwartz, 1986), pp. 167–73.

Rota, G.-C. (1986b), *Husserl*, in (Kac, Rota, Schwartz, 1986), pp. 175–81.

Rota, G.-C. (1991), *The End of Objectivity. The Legacy of Phenomenology. Lectures at MIT*, Cambridge, MA., MIT Mathematics Department.

Rota, G.-C. (1997), *Indiscrete Thoughts*, F. Palombi (ed.), Boston, Basel, Berlin, Birkhäuser.

Rota, G.-C. (1999), *Lezioni napoletane,* F. Palombi (ed.), Neaples, La Città del Sole.

Circle Packing: A Personal Reminiscence

Philip L. Bowers

The Prehistory of Circle Packing

Two important stories in the recent history of mathematics are those of the geometrization of topology and the discretization of geometry. The first of these begins at a time when the mathematical world is entrapped by abstraction. Bourbaki reigns and generalization is the cry of the day. Coxeter is a curious doddering uncle, at best tolerated, at worst vilified as a practitioner of the unsophisticated mathematics of the nineteenth century.

THE GEOMETRIZATION OF TOPOLOGY

It is 1978 and I have just begun my graduate studies in mathematics in geometric topology. By my second year of study, there is some excitement in the air over ideas of Bill Thurston that purport to offer a way to resolve the Poincaré conjecture by using nineteenth century mathematics—specifically, the noneuclidean geometry of Lobachevski and Bolyai—to classify all 3-manifolds. These ideas appear in a set of notes from Princeton, and the notes are both fascinating and infuriating—theorems are left unstated and often unproved, chapters are missing never to be seen, the particular dominates—but the notes are bulging with beautiful and exciting ideas, often with but sketches of intricate arguments to support the landscape that Thurston sees as he surveys the topology of 3-manifolds. Thurston's vision is a throwback to the previous century, having much in common with the highly geometric, highly particular landscape that inspired Felix Klein and Max Dehn. These geometers walked around and within Riemann surfaces, one of the hot topics of the day, knew them intimately, and understood them in their particularity, not from the rarified heights that captured the mathematical world in general, and topology in particular, in the period from the 1930's until the 1970's. The influence of Thurston's Princeton notes on the development of topology over the next 30 years would be pervasive, not only in its mathematical content, but even more so in its vision of how to do mathematics. It gave a generation of topologists permission to get their collective hands dirty with the particular and to delve deeply into the study of specific structures on specific examples.

What has geometry to do with topology? Thurston reminded us what Klein had known, that the topology of manifolds is closely related to the geometric structures they support. Just as surfaces may be classified and categorized using the mundane geometry of triangles and lines, Thurston suggested that the infinitely richer, more intricate world of 3-manifolds could, just possibly, be classified using the natural 3-dimensional geometries, which he classified and of which there are eight. And if he was right, the resolution of the most celebrated puzzle of topology—the Poincaré Conjecture—would be but a corollary to this geometric classification.

The Thurston Geometrization Conjecture dominated the discipline of geometric topology over the next three decades. Even after its recent resolution by Hamilton and Perelman, its imprint remains embedded in the working methodology of topologists, who have geometrized not only the topology of manifolds, but the fundamental groups attached to these manifolds. Thus we have as legacy the young and very active field of geometric group theory that avers that the algebraic and combinatorial properties of groups are closely related to the geometries on which they act. This seems to be a candidate for the next organizing principle in topology.

The decade of the eighties was an especially exciting and fertile time for topology as the geometric influence seemed to permeate everything. In the early part of the decade, Jim Cannon, inspired by Thurston, took up a careful study of the combinatorial structure of fundamental groups of surfaces and 3-manifolds, principally cocompact Fuchsian and Kleinian groups, constructing by hand on huge pieces of paper the Cayley graphs of example after example. He has relayed to me that the graphs of the groups associated to hyperbolic manifolds began to construct themselves, in the sense that he gained an immediate understanding of the rest of the graph, after he had constructed a large enough neighborhood of the identity. There was something automatic that took over in the construction and, after a visit with Thurston at Princeton, automatic group theory emerged as a new idea that has found currency among topologists studying fundamental groups. In this work, Cannon anticipated the thin triangle condition as the sine qua non of negative curvature, itself the principal organizing feature of Thurston's classification scheme. He studied negatively curved groups, rather than negatively curved manifolds, and showed that the resulting geometric structure on the Cayley graphs of such groups provides combinatorial tools that make the structure of the group amenable to computer computations. This was a marriage of group theory with both geometry and computer science, and had immediate ramifications in the topology of manifolds.

Ultimately these ideas led to one of the most beautiful conjectures left still unresolved by Perelman's final resolution of the Geometrization Conjecture, namely, that a negatively curved group with 2-sphere boundary is, essentially, a cocompact Kleinian group, the fundamental group of a compact hyperbolic 3-manifold.[1,2] Cannon's attempt to resolve this conjecture in collaboration with his colleagues, Bill Floyd and Walter Parry, has produced some of the most elegant geometric results in recent memory, especially in their elucidation of finite subdivision rules and their use as combinatorial constructs in producing conformal structures on surfaces; see Fig. 1. Activity surrounding Cannon's program led in the nineties to two discretizations of the classical Riemann Mapping Theorem, the Discrete Riemann Mapping Theorem of Cannon-Floyd-Parry [10] that offers a discrete analogue, proved independently by Schramm [19], and the Combinatorial Riemann Mapping Theorem of Cannon [9] that constructs a classical conformal structure on a space that carries an appropriate combinatorial superstructure. We will return to this development later, but we first mention another strand of the story.

In 1987, Mikhail Gromov increased the excitement and accelerated this project in geometrizing topology and group theory with the publication of his essay *Hyperbolic Groups* [12] in which he presents his grand, but rather sparsely argued, view of negative curvature in group theory. He followed this in 1993 with the influential *Asymptotic Invariants of Infinite Groups* [13] accomplishing a similar task for nonpositive curvature. Almost immediately after the respective publications of these two works, groups of mathematicians offered detailed proofs elucidating the big ideas of Gromov, and graduate students in succeeding years wrote theses whose geneses lay in these essays, sometimes explaining, with detailed proofs, a remark of a single sentence from one of the essays. I recall sometime in the mid-nineties at an AMS conference a conversation between two graduate students about their soon to be submitted dissertations. Both acknowledged their debt to Gromov, one stating something along the lines that "yea, mine is based on a secondary comment to a question Gromov posed on page 89 of *Asymptotic* . . . ," this in response to the other's explanation that his was based on an unproved claim in Gromov's *Hyperbolic*. . . ."

Finally, it is interesting that the resolution of the Geometrization Conjecture in the past five years relies less on the spirit of nineteenth century hyperbolic geometry, which lies at the heart of Thurston's original insights, and more on the modern theory of differential equations. Thurston had proved the conjecture for large classes of manifolds, including Haken manifolds, for which he received the Fields Medal in 1982. That same year,

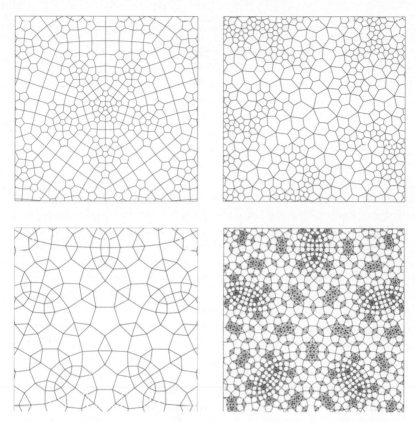

FIGURE 1. Examples of circle packing-generated embeddings of expansion complexes of Cannon-Floyd-Parry that tile the complex plane and are generated by finite subdivision rules with pentagonal combinatorics. The upper left tiling is particularly interesting as it is regular in the sense that each of the pentagons the eye sees at *every* level of scaling is conformally regular—conformally equivalent to a regular pentagon preserving vertices. Moreover, the whole infinite tiling can be generated from any single elementary tile, just by repeated anti-conformal reflections across edges. See [8].

Richard Hamilton proposed a method for solving part of the conjecture using Riemannian geometry and the flow of Ricci curvature [14], and later generalized this to a proposal for proving the general conjecture. Hamilton pushed his program forward over the next two decades, but it took the insights of Grigori Perelman to overcome difficulties that arose from singularities of the Ricci flow, in which the principal ingredient is the modern theory of differential equations and global analysis. Perelman famously turned down the offer of a Fields Medal for this work in 2006.

CIRCLE PACKING: THE BEGINNING

It was into and from this setting that circle packing was born. Its antecedents lay in Koebe's Theorem [17] of 1936 and Andre'ev's Theorems [1, 2] of 1970. The analyst Paul Koebe had proved that every triangulation of a disk produces a circle packing of the unit disk in $\mathbb{C}$, a pattern of circles within the unit disk, each corresponding to a vertex of the triangulation, with circles tangent when the corresponding vertices are adjacent, and boundary circles internally tangent to the boundary of the unit disk. Moreover, this pattern is unique up to Möbius transformations of the disk or, what is the same, up to isometries of the disk realized as a model of the hyperbolic geometry of Lobachevski and Bolyai. This seeming novelty was promptly forgotten and it did not emerge among the circle packing community that Koebe had proved this until sometime in the early nineties. From a completely different direction, the geometer E.M. Andre'ev in a first paper of 1970 gave a description of the cusped 3-dimensional hyperbolic polyhedra, and in a second described more general hyperbolic polyhedra, ultimately giving a description of certain 3-dimensional finite-volume hyperbolic polyhedra.

In the late seventies, Thurston, unaware of both the Koebe and Andre'ev results, proved a generalization of Koebe's Theorem that allowed for specified overlaps among the circles rather than tangency, gave a proof subject to computations, and generalized to surfaces of arbitrary genus. He then observed that the results on 3-dimensional hyperbolic polyhedra discovered by Andre'ev follow from the generalized Koebe results. The connection with hyperbolic 3-space is made by realizing the complex plane as the boundary of the upper half-space of $\mathbb{C} \times \mathbb{R}$ a model of 3-dimensional hyperbolic geometry, augmenting the circle packing with a dual pattern of circles mutually perpendicular to the given ones, and taking the intersection of the appropriate hyperbolic half-spaces bounded by these circles to yield a convex hyperbolic polyhedron with cusps. The generalization to prescribed overlap angles among circles allowed Thurston to include the finite-volume results of Andre'ev in his scheme. Thurston has reported that he proved this theorem one morning, and later that afternoon found the Andre'ev results. It took another fifteen years for Koebe's contribution to be recognized. These results were not peripheral to his program for classifying 3-manifolds, but were a central ingredient to a part of the program. The now-called Koebe-Andre'ev-Thurston Theorem has been generalized further to include cases where the objects being packed are not circles, but other convex figures, to cases where circles wrap around themselves multiple times, to infinite triangulations of open disks, and even to cases where adjacent vertices represent circles that 'overlap' with imaginary angles.

Though Thurston included his circle packing results in the infamous Chapter 13 of the Princeton notes—infamous because many copies sent out omitted Chapter 13—the topic did not generate any heat until 1985 when Thurston suggested in a lecture at Purdue University that the original Koebe Theorem could be used to give an effective computational algorithm for approximating the Riemann mapping of a proper simply-connected domain of $\mathbb{C}$ to the unit disk. In the audience sat the analyst Ken Stephenson, now author of the definitive text on circle paking[3] and one of the current leaders in the field, then a circle packing novice astonished and intrigued by Thurston's conjectural association of Koebe's Theorem with one of the most important results of complex analysis. Stephenson had been trained as a complex analyst, but always had sympathies for the more geometric aspects of the theory. Thurston's talk, and the subsequent verification of Thurston's suggestion by Burt Rodin and Dennis Sullivan [18], began the transformation of Stephenson from classical complex analyst to discrete conformal geometer. It led to a flurry of research by several groups and individuals in the geometry of circle packings, their existence and uniqueness, connection to conformal and analytic mappings, applications to solving open mathematical problems, amenability to computations, and relationships to random walks and other discrete phenomena.

Circle packing as a discipline quickly attracted the interest of researchers in analysis, combinatorics, geometry, and topology, and initially developed along "two not disparate branches. The one branch may be characterized broadly as analytic and combinatorial in style with particular attention focused on the topic of Thurston's Purdue talk; namely, the relationship of circle packing to the approximation of conformal mappings. The emphasis is on the connections of circle packing to classical complex analysis, which motivates and legitimizes its study, and in particular on the construction of both discrete analogs and proofs of some classical theorems of complex analysis. . . . The other branch may be characterized broadly as geometric and topological in style with particular interest in the pure geometry of circle packing. It needs no other motivation or justification than the pleasing interplay between geometry, topology, and combinatorics that exposes a certain rigidity of circle packing that is reminiscent of the rigidity so characteristic of complex analytic functions. The results are often beautiful and sometimes surprising, reminding the authors of their first encounters as graduate students with the surprising rigidity and striking inevitability that permeates this world of complex analytic functions."[4] This quote of the writer and his colleague Ken Stephenson was written in 1991 and gives a broad sense of the early directions of the subject.

THE DISCRETIZATION OF GEOMETRY

The initial work in circle packing, from 1985 through the mid-nineties, took place in the atmosphere of the geometrizing influence of the topologists. It was not central to their theme of classifying manifolds and understanding the geometric structure of groups, but it did offer an unexplored tributary open to the interested researcher. It was, in addition, amenable to computer exploration and, in fact, not a few mathematicians have been drawn into the subject by the beautifully intricate computer-generated pictures of circle packings and patterns that adorn many publications. Circle-Pack is the most extensive and sophisticated software package available for generating these pictures and is continuously updated by its author and the chief proselytizer of computer experimentation in circle packing—the analyst turned geometer, Ken Stephenson. This software has grown up hand-in-hand with theory, sometimes suggesting routes of theoretical inquiry, other times building intuition and displaying ideas graphically.

The computational tractability of circle packing has placed it squarely in the midst of our second important story in the recent history of mathematics, that of the discretization of geometry that has occured with increasing sophistication over the past decade and a half. Whereas our first story, and circle packing as one of its chapters, has its roots in theoretical, pure mathematics, our second has its in applied computational science, particularly in the various sorts of 2- and 3-dimensional imaging problems that have become accessible only recently with the advent of powerful desktop computers. These imaging problems led to challenging mathematical problems in collecting, representing, manipulating, and deforming representations of embedded surfaces and solids in $\mathbb{R}^3$. This ultimately meant that tesselations of 2- and 3-dimensional objects, sometimes triangulations, sometimes quadragulations, needed to be represented and manipulated as data in computations. The worry was to preserve as much as possible in the combinatorial data that encoded the geometry of these objects the original metric data—distances, intrinsic geodesics, angles, and intrinsic and extrinsic curvature. Often this meant calculating geodesics and curvature on combinatorial objects rather than on the original smooth ones, and the usual tools of Riemannian geometry are rendered ineffective in this combinatorial setting.

This story is so unlike the first story with its few prominent protagonists (Thurston, Cannon, Gromov), guided by an overarching paradigm (the Geometrization Conjecture and negative curvature), with revered texts as blueprints (Thurston's Princeton Notes and Gromov's *Hyperbolic Groups*). Discretization of geometry was and is driven by the habitual need to see and manipulate images, and its impetus came from a varied assortment of sources, from the pure mathematics of minimal surfaces and their images, to computer vision and target recognition, from medical imaging of anatomical data to surface manipulation in manufacturing design. It became obvious to several groups of mathematical and computer scientists that the problems they faced were not going to be solved by mere approximation. What was needed was a discrete version of Riemannian geometry faithful in its domain to the spirit of the classical field, representing the continuous objects whose properties they mirrored in some faithful way, but whose computational tools stand on their own and not as approximation to the continuous. This led to the identification of natural analogues in the discrete setting of the usual tools of differential geometry—polyhedral surfaces and spaces, discrete normal vector fields, discrete curvature operators, discrete Laplace-Beltrami operators, discrete geodesics, discrete Gauss maps. One of the interesting features of this effort is that often there are several discrete analogues of the same continuous tool, each of which approximates the continuous appropriately in terms of convergence as mesh size shrinks, but that capture different properties of the discrete objects of study. The paradigm though is different from that of classical numerical approximation where the *raison d'être* of the discrete operations and calculations is the approximation of the continuous. Rather, the standard in this emerging field of discrete differential geometry is that the discrete theory is a self-contained whole, with natural, exact tools leading to exact calculations, not approximations. Though the classical theory emerges in the limit of small mesh size, this often is not the overriding interest.

Boris Springborn, Peter Schröder, and Ulrich Pinkall state the paradigm this way: "instead of viewing discretization as a means of making the smooth problem amenable to numerical methods, we seek to develop on the discrete level a geometric theory that is as rich as the analogous theory for the smooth problem. The aim is to discretize the whole theory, not just the equations. Instead of asking for an approximation of the smooth problem, we are thus guided by questions like: What corresponds to a Riemannian metric and Gaussian curvature in an analogous theory for triangle meshes?"[5]

This effort has progressed through the work of many research teams with a variety of expertise—expertise in the pure mathematics of Riemannian geometry, including geometry on infinite-dimensional manifolds, in combinatorial topology and metric geometry, in computer science, symbolic computations, databases and programming, in statistics and engineering, and in a variety of scientific fields. The effort is really vast and multifaceted, and its story does not lend itself to a linear telling, so I am content to name a sampling of notable groups with whom I have some personal familiarity and from which the interested reader may gain a sense of the scope and power of these new ideas in discrete differential geometry. Each of the following groups maintains extensive web archives that explain and detail their considerable work in this emerging field: the Multi-Res Modeling Group at Caltech under the direction of Peter Schröder, the Mathematical Geometry Processing Group at Freie Universität Berlin under the direction of Konrad Polthier, the German Matheon Research Center and particularly its Polyhedral Surfaces Unit under the coordination of Alexander Bobenko, the Discrete Geometry Group at Technische Universität Berlin under the direction of Günter Ziegler, and the Computer Vision Lab at Florida State University under the direction of Xiuwen Liu and Washington Mio.

Circle Packing Comes of Age

Though the origin of circle packing lies in topology, it began to emerge as a separate discipline in the mid-nineties and only recently has found its natural home in the setting of discrete differential geometry. That part of discrete differential geometry that concerns the discretization, in the sense already articulated, of classical complex analysis and conformal geometry is now subsumed under the heading of discrete conformal geometry. Circle packing is one approach to discretizing parts of these classical disciplines and has both advantages and disadvantages when compared to others. Its greatest success is in providing a faithful discrete analogue of classical complex analytic function theory in the complex plane, with discrete versions of classical analytic mappings that in their particularities share essential features and qualities of their classical counterparts. Many of the classical theorems of complex analysis have circle packing analogues that actually imply the truth of the classical theorems in the limit as circle radii approach zero. This is not a two way street as the truth of the classical rarely implies that of the discrete. It is as if the discrete is more fundamental, more primitive, with the classical theory derivative of the discrete. These comments apply as well to other subdisciplines within discrete differential geometry.

A fitting metaphor comes from physics. The discrete theory—circle packing and discrete analytic functions—is the quantum theory from which the classical theory of analytic functions emerges. Classical analytic functions are continuous deformations of the classical complex plane and can be very complicated, but when viewed at the atomic scale, i.e., from the tangent plane, they are local complex dilations that preserve infinitesimal circles. Discrete analytic functions preserve actual circles and locally model the behavior of their classical counterparts. The salient large scale features of the continuous classical functions arise from this atomic scale circle-preserving property of the discrete functions. When we look closely enough at what we thought was the continuous, we find lurking underneath the discrete.

CIRCLE PACKING AS DISCRETE GEOMETRY

To indicate how discrete combinatorial data can carry continuous geometric information, even without any metric or angular data decorating the combinatorics, consider an arbitrary triangulation K of a genus g compact oriented surface S. I stress that S carries no metric structure, it is merely a topological surface. Then there is a unique conformal structure on S and, in that conformal structure, a circle packing $C = \{C_v : v \in V(K)\}$, a collection of circles in S indexed by the vertex set $V(K)$ of K, unique up to Möbius transformations, with C_v tangent to C_w whenever the vertices v and w of K are adjacent. To unwrap this a bit, when the genus $g = 0$, S is a topological 2-sphere and up to equivalence there is only one conformal structure on S, the one identifying S as the Riemann sphere $\hat{C}$. The circles C_v are then usual circles in the Riemann sphere and this can be seen as a restatement of Koebe's Theorem. When the genus is $g = 1$, then S is a topological torus, and there is a continuous 2-dimensional family of pairwise nonequivalent conformal structures available for S. The triangulation K chooses exactly one such structure, and it carries with it an essentially unique flat metric, and the circles C_v are circles with respect to this metric. The generic case is when $g \geq 2$. Then there is a continuous $(6g - 6)$-dimensional family of pairwise nonequivalent conformal structures available for S and the combinatorics of K chooses exactly one such structure. This structure carries a unique metric of constant curvature -1, a hyperbolic metric, and the circles C_v are hyperbolic circles. The packing C is unique up to hyperbolic isometry. I emphasize the fact that none of the other uncountably many conformal structures with their hyperbolic metrics supports a circle packing whose tangencies are encoded in the combinatorics of K.

This is a striking example of discrete combinatorial data encoding continuous geometric information. The result may be proved by several different methods and was one of the key theorems that stimulated early research in circle packing. It is also one of the facts buried in Chapter 13 of Thurston's Princeton Notes and has been rediscovered and reworked and generalized by several mathematicians. The early years were very active in exploring and generalizing from this result and articulating a background theory of circle packing surfaces. The subject becomes more difficult and intricate when the surfaces or packings are more general—noncompact surfaces with nonempty boundary, or packings that overlap, or wrap around singularities, or infinite packings—but the interplay between combinatorics and geometry remains the organizing principle.

Circle packing has advanced to the point that it offers a discrete theory of general complex analytic mappings of planar domains, and a theory of polynomial mappings of the sphere. Discrete rational mappings of the sphere and discrete conformal mappings of Riemann surfaces are areas of current research. A new direction is in circle packing of surfaces with more general structures than conformal ones; for example, circle packing of projective surfaces has enjoyed recent attention.

Koebe's Theorem was derived by Koebe as a corollary of his uniformization theorem that says that every finitely connected planar domain is conformally equivalent to the complement in $\mathcal{C}$ of a finite number of disks and points, unique up to Möbius transformations. This Koebe Uniformization Theorem is a generalization of the Riemann Mapping Theorem and gives supporting evidence for the very seductive Koebe Conjecture, which avers that *every* domain in the Riemann sphere is conformally equivalent to a circle domain, defined as the complement in $\hat{\mathcal{C}}$ of a closed set, each component of which is either a point or a round disk, and this is unique up to Möbius transformations.

Despite concerted attempts to prove the Koebe Conjecture over the previous sixty years, not a great deal of success had been recorded until Zheng-Xu He and Oded Schramm in their 1993 *Annals of Mathematics* paper [15] offered an intricate proof based on circle packing that applied to all domains with countably many complementary components. The general conjecture remains unresolved a decade and a half later, and the He-Schramm result is still the state of the art on the conjecture.

Returning now to Cannon's Conjecture described earlier, that a negatively curved group with 2-sphere boundary is, essentially, a cocompact Kleinien group, circle packing has been instrumental as an experimental tool in Cannon, Floyd, and Parry's program. Their program uses combinatorial data on the 2-sphere boundary that arise as the 'shadows' of half-spaces to attempt to construct a conformal structure in which these 'shadows' are almost round, which they have shown is sufficient to guarantee the conjectured result. They have given conditions that would guarantee the existence of such conformal structures and have modeled the combinatorics-to-conformal-structure procedure by finite subdivision rules, a purely discrete, combinatorial construct. Circle packing techniques have been used as an experimental probe for suggesting this requirement of almost roundness in several important cases in their program. Fig. 2 shows the results of circle packing techniques applied to the dodecahedral subdivision rule that demonstrate how 'roundness' emerges from combinatorial data. The underlying combinatorial structure recognized in the example is purely a product of the iteration of the simple subdivision rules illustrated in the top of Fig. 2. Circle packing is then used to place a natural geometric structure on the

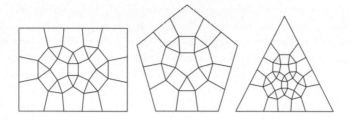

FIGURE 2. The figure is generated by applying the dodecahedral subdivision rule to a rectangle three times in succession [original in color]. At each application, the rectangles are subdivided according to the pattern shown in the upper left, the pentagons according to that of the upper middle, and the triangles according to that of the upper right. Circle packing generates the embedding, in which the almost roundness at multiple scales emerges. This is an approximate view of the shadows cast on the sphere at infinity by the first, second, and third generation hyperbolic half-spaces determined by the dodecahedral tiling of hyperbolic 3-space. (Cannon-Floyd-Parry)

combinatorics in which the roundness—the pattern of circles that the eye sees—miraculously appears at multiple scales of resolution. Mario Bonk and Bruce Kleiner [5, 6] have offered another approach to Cannon's Conjecture using analysis on metric spaces that, in at least one place, uses the Koebe-Andre'ev-Thurston Theorem.

I mention one last example where circle packing has made an impact in pure mathematics, and that is in offering new tools as well as new settings in which to study types of graphs. Infinite graphs offer a discrete setting in which to study random walks. Plane triangulation graphs are infinite graphs that offer not only the transient-recurrent dichotomy of random walks, but also the hyperbolic-parabolic dichotomy of circle packing. This hyperbolic-parabolic dichotomy arises since each plane triangulation graph encodes the tangency combinatorics of a circle packing in either the hyperbolic plane or the euclidean plane, but never both. How the circle packing type— hyperbolic or parabolic—compares with the random walk type—transient or recurrent—has led to fruitful interplay between circle packing and classical random walks, which has enriched both fields.

SOME APPLICATIONS

One of the interesting applications of circle packing and pattern theory is in discrete conformal flattening where a smooth surface in $\mathbb{R}^3$ is represented by a planar surface in a way that preserves, as much as possible, the conformal structure of the original. Though there are flattening algorithms based on classical numerical analysis that provide quick, accurate conformal maps, the advantage the circle packing algorithms bring is that, once flattened, the full range of the discrete theory of analytic and conformal mappings is available. Stephenson's software package has been used in anatomical flattenings in several studies; see, for example, [16].

Bobenko and Springborn [3] have initiated exploration recently of new algorithms based on circle patterns whose tangencies are encoded using square, rather than triangular, combinatorics. The inspiration for this comes from the physics of integrable systems, and has been successful, not only at conformal flattening, but in an elegant discrete representation of minimal embedded surfaces in $\mathbb{R}^3$ by Bobenko, Hoffmann, and Springborn [4], as in Fig. 3.

Finally, I mention the modest appearance of circle packing in the recently articulated mathematical theory of origami used to design and fold incredibly intricate origami figures, and their use in engineering design. I highly recommend Robert Lang's recent TED lecture of February 2008 found on the TED website at http://www.ted.com/.

Epilogue

This story began in an era of rarified abstraction in mathematics. The influence of mathematicians like Thurston and Cannon and Gromov brought a

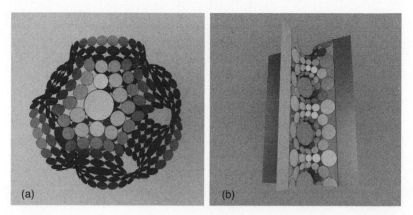

FIGURE 3. Discrete (a) Schwarz and (b) Scherk minimal surfaces generated by circle patterns based on square combinatorics; Bobenko-Hoffmann-Springborn (2006).

relaxation of that abstraction in geometry and topology and gave us permission to practice geometry again with our hands in the way of the great Felix Klein and his contemporaries. Circle packing represents one example of this more hands on approach to geometry, and its success is a testament to the power of the particular in mathematics. The elegant complexity that arises from the simplicity of elementary geometry—for what is more elementary than the circle—has fascinated a generation of circle packers, and promises to continue to fascinate new generations as the theory is extended and applied. All this from the humble circle!

Acknowledgments

This article is dedicated to the memory of Oded Schramm, who worked in circle packing before his discovery of stochastic Loewner evolution and its applications to critical phenomena. This extraordinary mathematician's untimely death on 01 September 2008 in a hiking accident was a great loss for our community.

Notes

1. Precisely, the group acts properly discontinuously, cocompactly, and isometrically on hyperbolic 3-space.
2. For a beautifully presented description of this work, see [11].

3. [21], *Introduction to Circle Packing: the Theory of Discrete Analytic Functions*, published in 2005 by Cambridge University Press.
4. [7], pp. 157–8.
5. [20], p. 77:1.

References

[1] E.M. Andre'ev, *Convex polyhedra in Lobacevskii space*, Math. USSR Sbornik **10** (1970), pp. 413–440.

[2] E.M. Andre'ev, *Convex polyhedra of finite volume in Lobacevskii space*, Math. USSR Sbornik **12** (1970), pp. 255–259.

[3] A.I. Bobenko and B.A. Springborn, *Variational principles for circle patterns and Koebe's theorem*, Trans. AMS **356** (2003), pp. 659–689.

[4] A.I. Bobenko, T. Hoffmann, and B.A. Springborn, *Minimal surfaces from circle patterns: geometry from combinatorics*, Annals of Math. **164:1** (2006), pp. 231–264.

[5] M. Bonk and B. Kleiner, *Quasisymmetric parametrizations of two-dimensional metric spheres*, Invent. Math. **150** (2002), no. 1, pp. 127–183.

[6] M. Bonk and B. Kleiner, *Conformal dimension and Gromov hyperbolic groups with 2-sphere boundary*, Geometry & Topology **9** (2005), pp. 219–246.

[7] P.L. Bowers and K. Stephenson, *Circle packings in surfaces of finite type: An in situ approach with applications to moduli*, Topology **32** (1993), pp. 157–183.

[8] P.L. Bowers and K. Stephenson, *A "regular" pentagonal tiling of the plane*, Con. Geom. and Dynamics **1** (1997), pp. 58–86.

[9] J.W. Cannon, *The combinatorial Riemann mapping theorem*, Acta. Math. **173** (1994), 155–234.

[10] J.W. Cannon, W.J. Floyd, and W.R. Parry, *Squaring rectangles: the finite Riemann mapping theorem*, in The *Mathematical Heritage of Wilhelm Magnus—Groups, Geometry, and Special Functions*, Contemporary Mathematics, vol. 169, Amer. Math. Soc., Providence, 1994, pp. 133–212.

[11] J.W. Cannon, W.J. Floyd, and W.R. Parry, *The length-area method and discrete Riemann mappings*, unpublished manuscript available from Bill Floyd that is based on a talk given by J. Cannon at the Ahlfors Celebration at Stanford University in September, 1997 (1998).

[12] M. Gromov, *Hyperbolic Groups*, in *Essays in Group Theory*, G.M. Gersten, ed., MSRI Publ. 8, 1987, pp. 75–263.

[13] M. Gromov, *Asymptotic invariants of infinite groups*, in *Geometric Group Theory*, Vol. 2 (Sussex, 1991), LMS Lecture Note Series, 182, Cambridge University Press, Cambridge, 1993, pp. 1–295.

[14] R. Hamilton, *Three-manifolds with positive Ricci curvature*, J. Diff. Geom. **17** (1982), pp. 255–306.

[15] Z-X. He and O. Schramm, *Fixed points, Koebe uniformization and circle packings*, Annals of Math. **137** (1993), pp. 369–406.

[16] M.K. Hurdal and K. Stephenson, *Cortical cartography using the discrete conformal approach of circle packings*, NeuroImage **23** (2004), Supplement 1, pp. S119–S128.

[17] P. Koebe, *Kontaktprobleme der Konformen Abbildung*, Ber. Sächs. Akad. Wiss. Leipzig, Math.-Phys. Kl. **88** (1936), pp. 141–164.

[18] B. Rodin and D. Sullivan, *The convergence of circle packings to the Riemann mapping*, J. Diff. Geom. **26** (1987), pp. 349–360.

[19] O. Schramm, *Square tilings with prescribed combinatorics*, Israel J. Math. **84** (1993), 97 118.

[20] B. Springborn, P. Schröder, and U. Pinkall, *Conformal equivalence of triangle meshes*, ACM Transactions on Graphics **27:3** [Proceedings of ACM SIGGRAPH 2008], Article 77, 2008.

[21] K. Stephenson, *Introduction to Circle Packing: the Theory of Discrete Analytic Functions*, Cambridge Univ. Press, 2005.

Applying Inconsistent Mathematics

MARK COLYVAN

Introduction

Inconsistent mathematics has a special place in the history of philosophy. The realisation, at the end of the nineteenth century, that a mathematical theory—naïve set theory—was inconsistent prompted radical changes to mathematics, pushing research in new directions and even resulted in changes to mathematical methodology. The resulting work in developing a consistent set theory was exciting and saw a departure from the existing practice of looking for self-evident axioms. Instead, following Russell (1907) and Gödel (1947), new axioms were assessed by their fruits.[1] Set theory shook off its foundationalist methodology. This episode is what philosophers live for. Philosophers played a central role in revealing the inconsistency of naïve set theory and played pivotal roles as new set theories took shape. This may have been philosophy's finest hour.[2]

Despite the importance of the crisis in set theory, inconsistent mathematics has received very little attention from either mathematicians or philosophers. Looking for inconsistency so that it might be avoided seems to be the extent of the interest. But inconsistent mathematics holds greater interest than merely providing an impetus for finding new, consistent theories to replace the old. Indeed, there are many reasons for taking inconsistent mathematical theories seriously and to be worthy of study in their own right. For example, there has been work on non-trivial, inconsistent mathematical theories such as finite models of arithmetic (Meyer, 1976; Meyer and Mortensen, 1984; Priest, 1997, 1998). While such mathematical theories might seem like mere curiosities, that's not the case. Chris Mortensen (1997, 2004) has argued that the best way to model inconsistent pictures (such as Penrose triangles and figures from Escher and Reutersvärd) is to invoke inconsistent geometry.[3] This work provides an interesting application for inconsistent mathematics. Although applications of inconsistent mathematics is the main theme of this chapter, I want to focus on other, more mundane applications of inconsistent mathematics—applications in modelling bits of the actual world (as opposed to Escher worlds). In particular, I will argue that there are a couple of puzzles arising from the applications of inconsistent mathematics, and in both cases, the puzzles have wider implications for philosophy of mathematics.

Indispensability of Inconsistent Mathematical Objects

The first puzzle concerns ontology. In particular, it seems that an indispensability argument can be mounted for inconsistent (mathematical) objects. To see this, it will be useful to recall a particular, inconsistent mathematical theory, namely the early calculus.

The early calculus was inconsistent in at least two ways. First, infinitesimals were taken to be zero and non-zero. Moreover, they were taken to be zero at one place and non-zero at another place within the same proof. When dividing by infinitesimals, they were taken to be non-zero (for otherwise the division was illegitimate) and at other times, they were taken to be equal to zero (for example, when an infinitesimal appeared as a term in a sum). Newton, at least, tried to address such concerns by giving an interpretation of infinitesimals (or fluxions) as changing quantities. But, alas, this interpretation was itself inconsistent. After all, if an infinitesimal, δ, is a changing quantity, it cannot appear in equations such as

$$a = a + \delta$$

where a is a constant. Why? Well, the term on the right is changing (since δ is changing) so it cannot equal anything fixed, such as a constant a. Yet early calculus required equations like the one above to hold.[4]

As it turns out, the calculus could be put on a firm basis, but that didn't come until the nineteenth century, when Bolzano, Cauchy, and Weierstrass developed a rigorous theory of limits and the ε-δ notation.[5] Now the puzzle is that in the interim—over 150 years—the calculus was widely used, both in mathematics and elsewhere in science. Indeed, it is hard to imagine a more widely used and applicable theory. This presents a problem for those (like me) who take indispensability to science to be a reason to believe in the entities in question.[6]

According to this line of thought, we should be committed to the existence of all and only the entities that are indispensable to our best scientific theories and, and, yet, for over 150 years, inconsistent mathematical entities—infinitesimals—were indispensable to these theories. This leads to the conclusion that we ought to have believed in the existence of inconsistent entities in the period between the late seventeenth century (by which time the calculus was finding widespread applications) and the middle of the nineteenth century (when the calculus was finally placed on a firm foundation). It seems that if one subscribes to the indispensability argument, there's a rather unpalatable conclusion beckoning: sometimes we ought to believe in the existence of inconsistent objects (Colyvan, 2008a, 2008b; Mortensen, 2008).[7] Indeed, it seems that the case for inconsistent mathematical objects (in the eighteenth century) was every bit as good as the case for believing in consistent mathematical objects.

A couple of comments on drawing ontological conclusions from inconsistent theories. Take any inconsistent theory along with classical logic and everything is derivable, including every other contradiction and the existence of all kinds of inconsistent objects. So what do we take to be the ontological commitments of an inconsistent theory? It is clear, and well-known, that in inconsistent settings like this, a paraconsistent logic is required.[8] With such a logic in place, triviality is avoided and we can make sense of specific inconsistent objects and conclusions being entailed by the theory in question.[9] Of course, Quine would have no truck with inconsistency and paraconsistent logics, but, nevertheless, what I'm arguing for here does seem to be a very natural extension of the Quinean approach to ontology. More importantly, it is at least plausible that scientists, when working with inconsistent theories, implicitly invoke a paraconsistent logic. Of course, most working scientists (even mathematicians) don't explicitly invoke a particular logic at all. The usual story that they all use classical logic is a rational (and heavily theory-laden) reconstruction of the practice. But it is interesting to note that when contradictions arise, working mathematicians do not derive results using the familiar C.I. Lewis proof,[10] even though such proofs are classically valid. This suggests, at least, that mathematical practice might be more appropriately modelled using a paraconsistent logic, in which such proofs are invalid (disjunctive syllogism is invalid in paraconsistent logics). Of course, there are other ways of explaining the practice. All I'm claiming here is that invoking paraconsistency is not as radical a move as it might first seem; it might be thought to be already implicit in mathematical practice.

On an historical note, it is interesting that both Newton and Leibniz believed that the methods of the calculus were in need of justification, and both sought geometric justifications. Newton took the justification task to be that of providing a geometric proof in place of each calculus proof—calculus for discovery, but geometry for justification. Leibniz, however, took the task to be that of providing a general justification of the methods of the calculus, then business as usual (Gaukroger, 2008). Although both Newton and Leibniz were thinking in terms of justification, they can also be seen to be offering two quite different anti-realist strategies in response to the indispensability argument I just presented. Newton was advocating a kind of eliminativist strategy, whereas Leibniz was seeking a non-revisionary account. Indeed, Leibniz's quest for a general justification of the methods of the calculus has a modern-day fellow traveller in Hartry Field (1980). Leibniz sought a general geometric limit account that would ensure that the calculus, despite being inconsistent, always gave the right answers on other matters. With a bit of massaging, we can see Leibniz as seeking something like a conservativeness proof: a demonstration that the calculus was a conservative extension of standard mathematics.[11]

I have argued elsewhere (Colyvan 2008b), that it is not clear what to make of this argument for the existence of inconsistent objects. Does it tell us that consistency should be an overriding constraint in such matters? If so, why?[12] I have also (tentatively) suggested that the apparently unpalatable conclusion should be accepted: There are times when we ought to believe in inconsistent objects. But before you dismiss such thoughts as madness or perhaps as a *reductio* of the original indispensability argument, it is important to make sure that other accounts of ontological commitments do not also fall foul of inconsistent objects. Both mathematical realists and anti-realists alike have always assumed the consistency of the mathematics in question. Considering inconsistent mathematical theories adds a new wrinkle to the debate over the indispensability argument, and the ontology of mathematics, more generally.

A Philosophical Account of Applied Mathematics

There is another, perhaps more disturbing, conclusion beckoning. If our only theories of space and time need to invoke inconsistent mathematical theories (as they did in the eighteenth century), this might be thought to give us reason to be realists about not just the inconsistent mathematical objects, but about the inconsistency of space and time themselves.[13] But putting such disturbing thoughts aside for the moment, let's assume that the world itself is consistent. Now there is a puzzle about how inconsistent mathematical models can be applied to the world. Again this is a new twist on an old problem.

The general problem is that of providing a philosophical account of the applicability of mathematics. This debate had its origins in the indispensability debate but has taken on a life of its own. The problem, in a nutshell, is as follows: how is it that mathematical structures can be so useful in modelling various aspects of the physical world.[14] The obvious answer is that when some piece of mathematics is applied to a physical system, the mathematics

is applicable because there are structural similarities between the mathematical structure and the structure of the physical system. So, for example,
there's no surprise that $\mathbb{R}^3$ is useful in modelling physical space, for the two
are isomorphic (putting aside relativistic curvatures). But in general, isomorphism is not the appropriate structural similarity—there is usually either more structure in the world or more in the mathematics. This is where
things get interesting. We need to explain how non-isomorphic structures
can be used to model one another and although there are several proposals
around (Batterman, 2010; Bueno and Colyvan, forthcoming; Leng, 2002,
2010; Pincock, 2004, 2007), none of these is complete. The realisation that
sometimes the mathematics in question is inconsistent, changes the way we
might approach the problem. Assuming that the world is consistent, the
problem is that of explaining how an inconsistent mathematical theory can
be used to model a consistent system. This seems much tougher than explaining cases where there's simply no isomorphism, and some of the proposals do not seem well-suited to dealing with this tougher problem.

Let me make a couple of suggestions about how this problem might be
solved. First, note that although the early calculus was inconsistent, it was
eventually put on a firm foundation. Indeed, even when calculus was first
developed, it might be argued that the consistent version existed, even
though the existence of the latter wasn't known at the time. It might be
further argued that this is all that's required; the inconsistent seventeenth
century calculus is useful in applications because of its similarity to a consistent latter-day calculus.[15] The idea here is that what matters in applying
mathematics is whether or not the mathematical model is capturing the
salient features of the empirical phenomena in question. The model can
achieve this irrespective of the knowledge of the modeller. An example
might help here.

Early electrical theory had it that when there was a potential difference
across a conductor, positively charged particles moved from the higher
potential to the lower. This, it turns out, is wrong in a couple of ways. First,
it's negatively-charged particles (electrons) that move, and in the opposite
direction to that of the proposed positive particles. Second, electrons do
not move very far in a conductor and they tend to oscillated—they certainly don't flow. The electrical current is the result of small movements of
the electrons compounding to a net drift. So why was the original theory,
which had all this wrong, so useful? It was useful because, for many purposes, these details are unimportant. The old, incorrect theory had a correct cousin—even though the latter was not known—and that's all that
matters. Electrons do not care whether electricians know about them or
not. There are electrons and there is (known or unknown) a correct theory
of them, so all that matters is that any useful theory of electricity resembles

the correct electron theory in certain respects. Clearly, positive particles flowing in one direction as opposed to negative particles flowing in the other, does not matter (unless one is specifically interested in the direction of particle movement), nor does it matter whether the particles in question flow or merely oscillate to ensure a net drift in one direction.

Returning to the case of the inconsistent calculus, we can see how the earlier suggestion might be fleshed out. It doesn't matter that the early calculus was inconsistent; it was, as a matter of fact, very similar to a consistent theory of calculus and it is this that explains the usefulness of the former. Indeed, on the account I'm proposing here, the usefulness of the calculus in itself suggests that there is a consistent theory in the offing. And this makes good sense of several key episodes in the history of mathematics where applications helped legitimise some questionable pieces of mathematics.[16] "It works" does seem like a very good response to suspicions about a new piece of mathematics.

This seems a promising start but there are some questions to be addressed. How can a consistent theory be similar to an inconsistent one? After all, it might appear that any given consistent theory is more like an arbitrary consistent theory than any inconsistent theory. And relatedly, it might seem that there is only one inconsistent theory since an inconsistent theory is trivial. The second worry is easily dealt with so let me tackle it first. In classical (and other explosive logics, such as intuitionistic logic), there is a sense in which there is only one inconsistent theory, namely, the trivial theory, where every proposition is true. But when dealing with inconsistency—or even potential inconsistency—we have already seen that we need to adopt a paraconsistent logic. Once this has been done, good sense can be made of *different* inconsistent theories. Seeing this also helps address the first question. Once we realise that we can discriminate between inconsistent theories, we can also determine which of these theories are similar to each other and to their consistent neighbours. It is not the case that all consistent theories are more like one another than they are to any inconsistent theory. Indeed, it is hard to see what would motivate such a thought, apart from the aforementioned mistake that there is only one inconsistent theory, namely, the trivial theory, and that this is radically unlike any consistent theory. There is still the difficult problem of how we compare theories, and nothing I've said here sheds any light on that more general problem. All I'm arguing for here is that inconsistent theories (in the context of a paraconsistent logic) can be compared in just the same way—whatever that is—to other theories, consistent and inconsistent.

The account just given seems right, in broad brush strokes, but further details will depend on the particular theory of applied mathematics adopted. So let me finish up by saying just a little about how some of the details might look in an account of the applications of mathematics I've recently developed with Otávio Bueno (forthcoming): *The Inferential Conception of Applied Mathematics*. The full details of this theory are not important for present purposes; the basic idea is that there are three separate stages of applying mathematics. First, there's the *immersion* step where an empirical set up is represented mathematically. The mathematics must be chosen in order to faithfully represent the parts of the empirical set up that are of interest. We do not require that the mathematics is isomorphic to the empirical set up. In general, the mathematical model and the empirical set up will not be isomorphic, but some structural features will be preserved in the mathematics. The second step is the inferential step where the mathematical model is investigated and various consequences of the model are revealed. The final step is the interpretation step where the results of the inferences conducted in the mathematical model in step two are interpreted back into the empirical set up. It is important to note that the interpretation step is constrained by the immersion step—mathematically representing some physical quantity in a specific way means that one must interpret the mathematics in question as a representation of the physical quantity—but the interpretation is not just the inverse of the immersion. For instance, at the interpretation step one is free to interpret more than what was initially represented in the immersion step. It is this feature of modelling that allows the mathematics to deliver novel phenomenon for investigation. It is one of the strengths of the inferential conception of applied mathematics that it is able to make sense of this important role mathematisation plays in science.[17]

With this framework in place, we can see how inconsistent mathematical theories, such as the early calculus, might be applied. First, we might explicitly treat the world as being inconsistent in the limit. That is, we treat the world as approximately inconsistent. (This is similar to when we make other idealisations, such as treating a fluid as being approximately incompressible and model it as such, despite holding that it really is compressible.) The inconsistent mathematics is then invoked to model this inconsistent picture of the world. Indeed, the inconsistent mathematics is essential here. No consistent mathematics could model the inconsistent limit being envisaged *as an inconsistent limit*. The inferential steps, of course, would need to be conducted using a paraconsistent logic, then the results of the inferences would be interpreted as being part of the nearest consistent story (if there is one).[18] There may be more than one consistent story in the neighbourhood. If there is, all such theories would need to be considered and the question of which theory to prefer would be settled by consideration of their theoretical virtues—simplicity, unificatory power, and so on.

Alternatively, we might only discover the (implicit) inconsistent assumptions about the world after the immersion and derivation steps. So,

for example, we might have an implicitly inconsistent theory of instantaneous change. But the inconsistencies in this theory might not be apparent
until the theory is represented using calculus, and some of the consequences
of the theory are revealed. Once it is realised that the theory is inconsistent,
we have no reason to force the interpretation to be consistent (as we did
in the case just considered). After all, in this case it was a discovery of the
mathematisation process that the underlying theory is inconsistent, so it
should come as no surprise that some inconsistent results are delivered.
The inconsistency here is more serious than in the previous case, and may
prompt further work on developing a consistent theory. In the meantime,
however, we can continue using the inconsistent theory (after employing a
paraconsistent logic). Again, the inconsistent mathematics is essential here.
The original theory of the empirical set up was (implicitly) inconsistent—
indeed, inconsistent in ways of interest (or so we are assuming here). In
order to faithfully represent the theory of the empirical set up, inconsistent
mathematics is needed. Any consistent mathematics used for the immersion
would not only hide the inconsistencies, but would render the mathemati-
cised theory consistent. This would make it more difficult to discover the
inconsistency of the original theory. But it might seem that that's preferable. At the end of the day, we are seeking a consistent theory, so using consistent mathematics might seem like a good way to facilitate this. But this is
to misunderstand the role of discovering the inconsistency. We are seeking
a consistent theory, but we are not seeking *any* consistent theory. The problem with this suggestion is that using consistent mathematics to model an
inconsistent theory simply renders the inconsistent theory consistent; it
does not reveal the inconsistency and it does not allow for careful reflection
on how best to resolve the inconsistency. It just papers over the problem.
Recognising an inconsistent theory as having a specific inconsistency is an
important step in securing the appropriate consistent theory.[19]

There is much more work to be done before we have an adequate philosophical account of the applications of mathematics. Although it might be
tempting to ignore cases of inconsistency—both inconsistent mathematics
and inconsistent empirical theories—when considering the applications of
mathematics, this would be a serious mistake. As I have just shown, considering applications of inconsistent mathematics forces attention onto the nature
of the structures in question and the relevant notion of similarity in a way
that is enlightening—we must be able to make sense of similarity between
consistent and inconsistent structures, for example. Considering the applications of inconsistent mathematics, it seems, will help shed light on the
general problem. And it is worth stressing that the inconsistent cases are not
mere test cases either. A great deal of one of the most important periods in
the history of science—the late seventeenth century to the mid-nineteenth

century—relied heavily on inconsistent mathematics. During this period, most scientists were working with an inconsistent mathematical theory and the theory was used almost everywhere. Ignoring inconsistent mathematics in a general account of applied mathematics would simply be negligence.

Conclusion

I have discussed just two of the many philosophical issues that arise in connection to inconsistent mathematics. Both the issues discussed in this chapter revolve around applications of inconsistent mathematics. The first concerned drawing conclusions about ontological commitments from the indispensability of mathematics. When we find ourselves forced to admit the indispensability of inconsistent mathematical theories, a counterintuitive conclusion looms: sometimes we ought to believe that inconsistent objects exist. The second issue concerns the provision of an adequate account of applied mathematics—one that provides an adequate account of applications of inconsistent mathematics.

Apart from the much-discussed crisis in set theory, there has been very little work in philosophy on inconsistent mathematical theories, presumably because such mathematics is thought not to occupy a central position in mathematics itself; inconsistent mathematics is thought to be at best a curiosity or a pathological limiting case, and at worst something to be avoided at all costs. I hope this chapter has gone some way to establishing that inconsistent mathematics is interesting in its own right and that including it in our stock of examples will help shed light on major issues in mainstream philosophy of mathematics.[20]

Notes

1. Russell, for example, suggests that "[w]e tend to believe the premises because we can see that their consequences are true, instead of believing the consequences because we know the premises to be true" [1907, p. 273].

2. See Giaquinto (2002) for a nice account of this episode and its fallout.

3. The consistent treatments of such figures (Penrose, 1991; Penrose and Penrose, 1958) do not do justice to the cognitive dissonance one experiences when viewing the figures in question *as inconsistent*.

4. In modern calculus, we'd say that the limit of $a + \delta$, as δ goes to zero, is a. But in the early days of calculus, a rigorous theory of limits was not available. Newton and Leibniz were stuck with equations like the one above. It's also worth noting that inconsistency is usually thought to be a property of formal theories and the early calculus was a long way from anything that would count as a formal theory. To claim that the early calculus was inconsistent, then, also involves some substantial claims about the interpretation of that theory *as inconsistent*. This is a big issue and much more needs to be said in order to establish

beyond doubt that the early calculus was inconsistent. (For example, the early practitioners may have been groping towards one of the modern consistent interpretations of the calculus.) But it does seem that *prima facie*, at least, both the natural interpretation and Newton's changing quantity interpretations of the early calculus were inconsistent. See Mortensen (1995) for more on this.

5. Later in the 1960s work on non-standard analysis (and infinitesimals) by Robinson (1966) provided a separate consistent interpretation of the calculus, and, arguably, one closer to the spirit of the original. A little later Conway (1976) provided yet another way to rehabilitate infinitesimals.

6. See Colyvan, 2001a, 2008a; Putnam, 1971 and Quine, 1981 for details of the indispensability argument.

7. Throughout this chapter, I will take an inconsistent object to be an object that has inconsistent properties assigned to it by the theory positing it.

8. This is a logic where there is some Q such that $P \wedge \neg P \nvdash Q$. That is, in paraconsistent logics not everything follows from a contradiction.

9. And we can also deal with the related worry that there would seem to be only one inconsistent theory. As we shall see shortly, in a paraconsistent setting, we can make sense of different inconsistent theories.

10. For example, since an infinitesimal $\delta \neq 0$, it follows that either $\delta \neq 0$ or the fundamental theorem of calculus holds. But since $\delta = 0$, by disjunctive syllogism, we have the fundamental theorem of calculus.

11. Of course, it's hard to think about conservativeness when inconsistent theories are in the mix. If we take a theory Δ to be a conservative extension of Γ, then conservativeness amounts to (roughly) that any statement formulated in the vocabulary of Γ and derivable from $\Delta + \Gamma$ is derivable from Γ alone. But if Δ is inconsistent, then it can never be conservative, so long as the logic in question is explosive (i.e. supports *ex contradictione quodlibet*). But sense can be made of conservativeness in such settings, if the logic is paraconsistent.

12. You might think that in order for a theory to count as one of our best theories (and thus relevant to the indispensability argument), it needs to be consistent. This would rule out such cases as I'm considering here right from the start. It is hard to motivate such a privileged position for consistency, though (Bueno and Colyvan, 2004; Priest, 1998). Consistency is one among many virtues theories can enjoy, but it does not seem to trump all other virtues in the way this response would require. Indeed, if I am right that scientists take inconsistent theories seriously, anyone wishing to argue that such theories are never candidates for our best theories (so no ontological conclusions can be drawn from them) would seem to be at odds with scientific practice and thereby flying in the face of philosophical naturalism. See Colyvan (2008b) for further objections and responses to the indispensability argument I've outlined here.

13. There are interesting connections here with debates about ontological vagueness (or vagueness in the world) and Russell's dismissal of it as "the fallacy of verbalism" (Colyvan, 2001c). There are also related debates about whether vagueness might give us reason to believe that the world is inconsistent (Beall and Colyvan, 2001).

14. There is also a related puzzle, often called the unreasonable effectiveness of mathematics (Colyvan, 2001b; Steiner, 1995, 1998; Wigner, 1960), of understanding how an apparently *a priori* discipline such as mathematics can provide the tools so often required by empirical science.

15. Of course, nothing I've said here tells us why the latter is so useful, but the strategy here is to deal with any special issues arising from the inconsistency.

16. I'm thinking here of the role of applications in helping legitimise the Dirac delta function, the early complex numbers, and, of course, the calculus (Kline, 1972).

17. See Bueno and Colyvan (forthcoming) for more details and a defence of the account.

18. The situation here is not unlike using continuous mathematics to model discrete phenomena (e.g. differential equations in population ecology). The discrete phenomena are treated as continuous in the limit, modelled using continuous mathematics, then inferences drawn in the mathematics, and the results interpreted discretely.

19. Something like this may have been going on with at least some of the applications of the early calculus: the underlying (unmathematised) theories of change, for example, were inconsistent in precisely the ways revealed when these theories were represented using the inconsistent calculus.

20. I'd like to thank Otávio Bueno, Stephen Gaukroger, Øystein Linnebo, and an anonymous referee for comments on earlier drafts or for discussions that helped clarify my thinking on the issues addressed in this paper. I am also grateful to audiences at the Baroque Science Workshop at the University of Sydney in February 2008, at the conference on New Waves in Philosophy of Mathematics at the University of Miami in April 2008, and at the 4th World Congress of Paraconsistency at the University of Melbourne in July 2008. Finally, I'd like to thank Chris Mortensen for impressing on me the significance of inconsistent mathematics. Work on this paper was funded by an Australian Research Council Discovery Grant (grant number DP0666020).

References

Batterman, R. W. 2010. "On the Explanatory Role of Mathematics in Empirical Science", *British Journal for the Philosophy of Science, 61*(1): 1–25.

Beall, J.C. and Colyvan, M. 2001. "Looking for Contradictions", *The Australasian Journal of Philosophy, 79*(4): 564–569.

Bueno, O. and Colyvan, M. 2004. "Logical Non-Apriorism and the Law of Non-Contradiction", in G. Priest, J.C. Beall, and B. Armour-Garb (eds.), *The Law of Non-Contradiction: New Philosophical Essays*. Oxford: Oxford University Press, pp. 156–175.

Bueno, O. and Colyvan, M. forthcoming. "An Inferential Conception of the Application of Mathematics", *Noûs*.

Colyvan, M. 2001a. *The Indispensability of Mathematics*. New York: Oxford University Press.

Colyvan, M. 2001b. "The Miracle of Applied Mathematics", *Synthese*, 127: 265–278.

Colyvan, M. 2001c. "Russell on Metaphysical Vagueness", *Principia*, 5(1–2): 87–98.

Colyvan, M. 2008a. "Who's Afraid of Inconsistent Mathematics?", *Protosociology*, 25: 24–35.

Colyvan, M. 2008b. "The Ontological Commitments of Inconsistent Theories", *Philosophical Studies*, 141: 115–123.

Colyvan, M. 2008c. "Vagueness and Truth", in H. Dyke (ed.), *From Truth to Reality: New Essays in Logic and Metaphysics*. London: Routledge, pp. 29–40.

Conway, J.H. 1976. *On Numbers and Games*. New York: Academic Press.

Field, H. 1980. *Science Without Numbers: A Defence of Nominalism*. Oxford: Blackwell.

Gaukroger, S. 2008. "The Problem of Calculus: Leibniz and Newton On Blind Reasoning", paper presented at the Baroque Science Workshop at the University of Sydney in February 2008.

Giaquinto, M. 2002. *The Search for Certainty: A Philosophical Account of Foundations of Mathematics*. Oxford: Clarendon Press.

Gödel, K. 1947. "What Is Cantor's Continuum Problem?", reprinted (revised and expanded) in P. Benacerraf and H. Putnam (eds.), *Philosophy of Mathematics Selected Readings*, second edition. Cambridge: Cambridge University Press, 1983, pp. 470–485.

Kline, M. 1972. *Mathematical Thought from Ancient to Modern Times*. New York: Oxford University Press.

Leng, M. 2002. "What's Wrong With Indispensability? (Or, The Case for Recreational Mathematics)", *Synthese*, 131: 395–417.

Leng, M. 2010. *Mathematics and Reality*. Oxford: Oxford University Press.

Meyer, R.K. 1976. "Relevant Arithmetic", *Bulletin of the Section of Logic of the Polish Academy of Sciences*, 5: 133–137.

Meyer, R.K. and Mortensen, C. 1984. "Inconsistent Models for Relevant Arithmetic", *Journal of Symbolic Logic*, 49: 917–929.

Mortensen, C. 1995. *Inconsistent Mathematics*. Dordrecht: Kluwer.

Mortensen, C. 1997. "Peeking at the Impossible", *Notre Dame Journal of Formal Logic*, 38(4): 527–534.

Mortensen, C. 2004. "Inconsistent Mathematics", in E.N. Zalta (ed.), *The Stanford Encyclopedia of Philosophy* (Fall 2004 edition), URL= <http://plato.stanford.edu/archives/fall/2004/entries/mathematics-inconsistent/>.

Mortensen, C. 2008. "Inconsistent Mathematics: Some Philosophical Implications", in A.D. Irvine (ed.), *Handbook of the Philosophy of Science Volume 9: Philosophy of Mathematics*. North Holland: Elsevier.

Penrose, L.S. and Penrose, R. 1958. "Impossible Objects, a Special Kind of Illusion", *British Journal of Psychology*, 49: 31–33.

Penrose, R. 1991. "On the Cohomology of Impossible Pictures", *Structural Topology*, 17: 11–16.

Pincock, C. 2004. "A New Perspective on the Problem of Applying Mathematics", *Philosophia Mathematica* (3), 12: 135–161.

Pincock, C. 2007. "A Role for Mathematics in the Physical Sciences", *Noûs*, 41: 253–275.

Priest, G. 1997. "Inconsistent Models of Arithmetic Part I: Finite Models", *Journal of Philosophical Logic*, 26(2): 223–235.

Priest, G. 1998. "What Is So Bad About Contradictions?", *The Journal of Philosophy*, 95(8): 410–426.

Priest, G. 2000. "Inconsistent Models of Arithmetic Part II: The General Case", *Journal of Symbolic Logic*, 65: 1519–1529.

Putnam, H. 1971. *Philosophy of Logic*. New York: Harper.

Quine, W.V. 1981. "Success and Limits of Mathematization", *Theories and Things*. Cambridge, MA.: Harvard University Press, pp. 148–155.

Robinson, A. 1966. *Non-standard Analysis*. Amsterdam: North Holland.

Russell, B. 1907. "The Regressive Method of Discovering the Premises of Mathematics", reprinted in D. Lackey (ed.), *Essays in Analysis*. London: George Allen and Unwin, 1973, pp. 272–283.

Steiner, M. 1995. "The Applicabilities of Mathematics", *Philosophia Mathematica* (3), 3: 129–156.

Steiner, M. 1998. *The Applicability of Mathematics as a Philosophical Problem*. Cambridge, MA: Harvard University Press.

Wigner, E.P. 1960. "The Unreasonable Effectiveness of Mathematics in the Natural Sciences", *Communications on Pure and Applied Mathematics*, 13: 1–14.

Why Do We Believe Theorems?

ANDRZEJ PELC

Introduction

Most common mathematical practice deals with proofs of theorems. As authors, mathematicians[1] invent proofs and try to write them down rigorously; as readers, they try to verify and understand proofs of other mathematicians; as referees and journal editors, they assess the interest and value of proofs, and as teachers, they explain proofs to novices.

Why are mathematicians so concerned with proofs? Or, reformulated in a more direct way: why do we prove theorems? This question serves as the title of the paper by Rav [1999]. The most obvious answer is: we prove theorems to convince ourselves and others that they are true. While this is often, indeed, the direct reason for proving a theorem, the convincing power is far from being the unique role of proofs in mathematical practice. Often the new ideas and techniques conveyed by a proof are much more important than the theorem for which the proof was originally invented. Rav formulates it succinctly:

> Proofs are for the mathematician what experimental procedures are for the experimental scientist: in studying them one learns of new ideas, new concepts, new strategies—devices which can be assimilated for one's own research and be further developed. [1999, p. 20]

and illustrates this epistemic function of proofs by well-chosen examples.

Nevertheless, the role of proofs as a means of convincing the mathematical community of the validity of theorems is very important. While proofs can also serve other purposes, only proofs can directly serve this purpose. Since in this paper we discuss the reasons for believing theorems, rather than discussing the epistemic value of proofs, our perspective on proofs is much more restrictive than that of [Rav, 1999]. It dealt with the question 'Why do we prove theorems?' and so was concerned with all aspects of proofs, both as ways to convince that the proved theorem is correct, and as repositories of mathematical knowledge. Our question (the title of this paper is deliberately modeled on the title of Rav [1999]) is 'Why do we believe theorems?' Thus we are only interested in the 'convincing' role of proofs.

Before announcing our position concerning reasons for believing theorems, we need to explain two crucial terms that will be used throughout the paper: *proof* and *derivation*. In choosing these particular terms, we follow [Rav, 1999]. (The expressions *informal proof* instead of *proof* and *formal proof* instead of *derivation* could be also used, as they convey the sense we want to give to those terms, but we find Rav's terminology more convenient.) By *proofs* we will mean the arguments used in mathematical practice in order to justify the correctness of theorems. Since this simply reports the meaning adopted by the mathematical community, there is no formal definition of this term. On the other hand, the term *derivation* is used in its formal logical sense. Recall that a formalized language is described first. It has an appropriate alphabet (containing, among other things, logical symbols). Terms and formulae of this language are defined as specific strings of symbols of this alphabet. Then a formalized theory T is defined in this language by specifying a—possibly infinite—recursive set[2] of formulae called *axioms* that form the basis of the theory. A finite set of transformations called *inference rules* is specified. These transformations permit us to obtain a new formula from a finite set of formulae. One example of an inference rule is *modus ponens*. Finally, a (linear) *derivation* is a finite sequence of formulae such that every term of this sequence is either an axiom or is obtained from a set of earlier terms of this sequence by applying one of the inference rules. A formula of the above formalized language is called a *theorem* of the theory T if it is the last formula in some derivation. The crucial characteristic of this formal definition is that, because of the fact that the set of axioms is recursive (although possibly infinite), there is a mechanical way to verify whether a given sequence of formulae of the formalized language is a derivation or not. (This should not be confused with the possibility of a mechanical method of verifying whether a given formula is a theorem of T or not. The well-known undecidability theorem states that the latter cannot be verified by a mechanical method for sufficiently strong theories.) This feature of derivations is the reason for their important role in the construction of formalized theories. Once an (alleged) derivation is presented, there is a mechanical way to verify its correctness, and in the case of a positive verdict, to assert that the last formula of the derivation is indeed a theorem of the theory T.

Since the formulation of Hilbert's program, most of the discussions concerning reasons for confidence in mathematical theorems are conducted along the axis between formalists and their opponents. In a nutshell, the first group asserts that *derivations* underlying proofs actually carried out in mathematical practice are the reason to believe those theorems, while the second group dismisses or significantly weakens the role of those derivations in building confidence in theorems, and points to other features of proofs, such as the analysis of the meaning of mathematical notions, as well as to social factors, as responsible for belief in mathematical statements. Hence, while the radical interpretation of Hilbert's program, as expressed in the famous Problem Number 2 stated in Hilbert's address[3] in Paris in 1900, has been shown to be infeasible by the later negative results of Gödel, a more moderate interpretation, trying to harvest confidence gained from some link between proofs and derivations, is still present and lively in discussions of the topic.

This controversy between formalists and their opponents is well exemplified by a discussion between Jody Azzouni and Yehuda Rav in a sequence of papers ([Rav, 1999], [Azzouni, 2004], and [Rav, 2007]). In this discussion Azzouni presents a moderate formalist point of view which he calls the *derivation-indicator* view, while Rav opposes his arguments. Azzouni's view is well summarized in the abstract of his [2004], where he says:

> A version of Formalism is vindicated: Ordinary mathematical proofs indicate (one or another) mechanically checkable derivation of theorems from the assumptions those ordinary mathematical proofs presuppose. The indicator view explains why mathematicians agree so readily on results established by proofs in ordinary language that are (palpably) not mechanically checkable.

The present paper is no exception to the rule that reasons for mathematical beliefs are discussed in terms of the controversy between formalists and non-formalists. We choose Azzouni [2004] as a protagonist of the formalist point of view because his position is moderate:[4] we want to show that even such a moderate position is impossible to maintain. However, our opposition to the views expressed by Azzouni [2004] is presented from a very different standpoint from that of Rav [2007] and is stronger. To summarize our position, we will argue that:

> No link between derivations and proofs can contribute to increasing confidence in theorems in the present state of knowledge.

> When Rav says:

> Consequently, when it comes to the nature of the *logical* justification of mathematical arguments in proofs, with Azzouni putting his faith in *formal derivations*, even if just indicated, and further maintaining

that on *this* basis proofs are recognized by mathematicians as valid, as they see them hence being *mechanically checkable*, against such formalist-mechanist claims I have objections to voice, both on technical and on historical grounds. As opposed to various formalist views, I hold that mathematical proofs are *cemented* via arguments based on the *meaning* of the mathematical terms that occur in them, which by their very conceptual nature cannot be captured by formal calculi. [2007, p. 294] (emphasis in original)

we think that his critique is too mild. Indeed, we will argue that not only do mathematical proofs contain ingredients that cannot be captured by formal calculi but, in fact, no link between those proofs and the hypothetical derivations underlying them can be established in the case of many theorems. Thus, while Rav argues that there is added epistemic value in proofs with respect to derivations, we make a much stronger claim: in the case of many theorems, derivations cannot contribute any epistemic value at all.

The Impossibility of Gaining Confidence from Derivations

We will argue that, in the present state of knowledge, it is impossible to gain any confidence in most mathematical theorems from hypothetical derivations underlying their proofs. Hence we will argue against the statement of Azzouni who says:

it's derivations, derivations in one or another algorithmic system, which underlie what's characteristic of mathematical practice: in particular, the social conformity of mathematicians with respect to whether one or another proof is or isn't (should be, or shouldn't be) convincing. [2004, p. 83]

Our impossibility argument concerning such a gain of confidence is based on considering lengths of hypothetical derivations underlying proofs.

It should be stressed that our impossibility claim is quite strong. We want to argue that not only do mathematicians not in practice use derivations to get or increase confidence in their results, but that in the present state of knowledge it is *theoretically impossible* to achieve such a gain of confidence in the case of most interesting mathematical theorems.

Since we want to argue against the statement of Azzouni quoted above, we need to try to understand what he means by saying that derivations 'underlie what's characteristic of mathematical practice.' First we must ask: what exactly is an algorithmic system and derivations in which algorithmic system? Azzouni [2004] provides a kind of answer to the first question, saying in the footnote to the above quote: 'An algorithmic system is one where the recognition procedure for proofs is mechanically implementable. By no means are algorithmic systems restricted to language-based axiom systems.' He is more vague as far as the choice of a specific system is concerned:

> ... the picture doesn't require mathematicians, in any case, when studying a subject-matter, to remain within the confines of a single (algorithmic) system—indeed, if anything, mathematicians are required to transcend such systems by embedding them in larger ones. The derivation indicated (by the application of new tools to a given subject matter) can be a derivation of the weaker system the mathematician started with, or it can be a derivation of a stronger system (some of) the terms of which are taken to pick out the same items supposedly referred to in the weaker system. [2004, p. 93]

Nevertheless, for a given theorem in mind, we have to focus on a particular 'algorithmic system' (while agreeing that this system could be modified several times in the course of developing the proof of the theorem). Derivations in an axiomatized theory, such as Zermelo-Fraenkel set theory with the Axiom of Choice (ZFC) are in principle mechanically checkable (when written correctly); hence we will use the system ZFC as an example of an algorithmic system in which derivations are 'indicated' by usual proofs, to use Azzouni's terminology. ZFC is a good example because many mathematicians believe that most of the mathematical lore could be *in principle* formalized in this system. Actually, defining other mathematical notions (such as functions, relations, algebraic systems) in terms of sets is a standard mathematical practice which can be considered as a step in the direction of such a formalization. (Further steps are usually not made.)

Next we should see what Azzouni means by saying that proofs 'indicate' derivations. This point is somewhat obscure, as a precise definition of this relation between proofs and derivations is never given in his paper. To be sure, Azzouni is well aware of the fact that 'mechanically recognizable derivations ... (generally) aren't ones (that can be) exhibited *in practice*' [2004, p. 95] (emphasis original). However, the very fact of his stressing that derivations are not exhibited *in practice* points to the important distinction between practical considerations and a *theoretical* accessibility of such derivations. This is also implied by his statement 'it's *derivation* which provides the skeleton for (the flesh of) *proof*' (p. 95, emphasis original). In order to provide such a skeleton, a derivation should be *at least theoretically* accessible for scrutiny, and hence it should be *at least theoretically* possible to *record and verify* it. Since Azzouni stresses mechanical checkability as an important feature of derivations, it is therefore fair to assume that, whatever could be the precise meaning of the phrase 'proofs indicate derivations', it should be *theoretically possible* to check such indicated derivations mechanically.

Now comes a crucial remark that forms the basis of our impossibility argument. Since Azzouni (and other formalists) claim that confidence concerning theorems is somehow linked to derivations underlying proofs, the burden of justifying that such (indicated) derivations can (at least theoretically) be recorded and mechanically checked is on *their* side, in every case when such a gain of confidence is claimed. A caveat is in order here. Formalists do not need actually to *exhibit* an indicated derivation, but they should justify that such a derivation can theoretically be recorded and mechanically checked. This theoretical possibility is a *necessary* condition for such a gain of confidence and hence should be justified by anybody claiming the gain. There is an important asymmetry with respect to the 'burden of justification'. In order to counter this formalist claim for a given theorem, there is no need to show that such a mechanically checkable derivation is theoretically impossible; it is enough to show that the formalist side has not justified the theoretical possibility of mechanical checking of such a derivation. We will argue that this is the case, and hence that the claimed confidence gain cannot be harvested.

Before proceeding with our impossibility argument, a disclaimer is in order. Our argument refers to 'complex' 'deep' theorems, as opposed to simple corollaries from definitions, such as the unicity of the neutral element in groups, or simple geometric theorems, *e.g.*, that the sum of angles of a triangle totals π. For the latter two examples it is not hard to imagine a derivation in ZFC; it is probably even possible (although very likely a terribly boring task) to write it explicitly. Notice, however, that confidence in these simple results is extremely high anyway; so the additional potential gain would be small, if not non-existent. Of course, the above notions of 'complex' 'deep' theorems are very fuzzy, which is the reason we put them in quotation marks. However, most mathematicians would agree that Fermat's Last Theorem, proved by Wiles [1995], is both complex and deep (no need of quotation marks in this case). Hence we will use this theorem, referred to as *FLT*, as an example in our impossibility argument.

We should also stress that the choice of ZFC as the underlying axiomatic system in which derivations are indicated by usual proofs is only given as an example. It could be replaced by any normally used system such as Peano Arithmetic, Kelley-Morse theory of classes, *etc.*, whose axioms and inference rules people find intuitively obvious. Of course one might create an artificial system, *e.g.*, one containing FLT as an axiom (in which the derivation of FLT would have length one), but such manipulations cannot possibly increase confidence in this theorem.

The starting point of our argument is the following controversy between Rav [1999] and Azzouni [2004]. Rav writes:

In reading a paper or monograph it often happens—as everyone knows too well—that one arrives at an impasse, not seeing why a certain claim B is to follow from claim A, as its author affirms. Let us symbolise the author's claim by '$A \to B$'. (The arrow functions iconically: there is an informal logical path from A to B. It does *not* denote formal implication.) Thus, in trying to understand the author's claim, one picks up paper and pencil and tries to fill in the gaps. After some reflection on the background theory, the meaning of the terms and using one's general knowledge of the topic, including eventually some symbol manipulation, one sees a path from A to A_1, from A_1 to $A_2, \ldots$, and finally from A_n to B. This analysis can be written schematically as follows:

$$A \to A_1, A_1 \to A_2, \ldots, A_n \to B.$$

Explaining the structure of the argument to a student or non-specialist, the other may still fail to see why, for instance, A_1 ought to follow from A. So again we interpolate $A \to A'$, $A' \to A_1$. But the process of interpolations for a given claim has no theoretical upper bound. In other words, how far has one to analyse a claim of the form 'from property A, B follows' before assenting to it *depends on the agent*. There is no theoretical reason to warrant the belief that one ought to arrive at an atomic claim $C \to D$ which does not allow or necessitate any further justifying steps between C and D. This is one of the reasons for considering proofs as *infinitary objects*. Both Brouwer and Zermelo, each for different reasons, stressed the infinitary character of proofs. [1999, p. 14]

and Azzouni responds:

What's going on? Why should anyone think that a *finitary* piece of mathematical reasoning, a step in a proof, say, corresponds to something infinitary (if, that is, we attempt to translate it into a derivation)? [2004, p. 97]

While we agree with the main thought of Rav [1999], it seems that the use of the word 'infinitary' may have been exaggerated. It is this word and not the argument itself that seems to have provoked the controversy. What we consider as the crux of Rav's reasoning is that there is no clearly defined *upper bound* on the number of justifying steps. If the argument is presented in this way, there is no need to invoke any infinitary character of the proof, that raised Azzouni's objections. Our impossibility argument will be based on this important distinction. (In the sequel we never make any claims about an infinitary character of proofs or derivations. Like Azzouni, we believe that these are finite objects.)

For any theorem T whose derivation (or many derivations) in ZFC is indicated (according to Azzouni's terminology), by its 'ordinary' published proof, let $L(T)$ denote the length of the shortest possible derivation (not only among those derivations indicated by the particular considered proof, but among all possible derivations of T in (ZFC). Such derivations exist since, by assumption, they are indicated by the ordinary proof. Hence the integer $L(T)$ is well defined. We conjecture that in the case of most 'complex' 'deep' theorems T, and in particular in the case of FLT, it is impossible to provide (and justify) *any* upper bound on $L(T)$. We agree that this is a bold statement but we would like to challenge a skeptical reader to provide (and justify!) any upper bound on $L(\text{FLT})$. Should it be a million? a quadrillion? $10^{10^{10}}$? What would be the justification of the choice of any such number? Actually, of any number at all? To be sure, we have not *proved* that it is impossible to give such an estimate, we have only conjectured it. Nevertheless, and this time it is not a conjecture, but a statement of a fact: nobody (including formalists claiming confidence gains from derivations underlying proofs) has ever given any such estimate. This simple observation is the first crucial claim in our argument. Our conjecture is stronger and says that such an estimate is impossible to establish, but it will not be used in the argument: due to the asymmetry with respect to the 'burden of proof', the fact that such an estimate has not been given will be enough for our reasoning.

The next step of our argument requires the definition of a very large integer number M, an *extremely* large number indeed. We are not concerned with determining the size of this number. The only thing that matters for our purposes is that no derivation of length larger than M could ever be actually written down or verified mechanically, even in theory. We will proceed with a series of estimates leading to the definition of M. First consider the period of time known as the *Planck time* (*cf.* [Halliday *et al.*, 1996]). This interval of time, call it t_p, of length of order 10^{-43} seconds, is a lower bound on the duration of any observable physical event, and hence it is also a lower bound on the time needed for any conceivable processing device to record or verify one step of a derivation. Next, switching to the macroscale, consider a future state of the universe precluding any information-processing activity. A good candidate for such a state would be the heat death of the universe, *i.e.*, when the universe has reached maximum entropy. Let τ denote an upper bound (in seconds) on the time until this state of the universe is reached. Let $Q = \tau/t_p$. Hence Q is an upper bound on the number of steps of a derivation that can ever be recorded or verified by a single processing device. It is of course possible that many processing devices cooperate in verifying some derivation in parallel, each processing unit working on a different part of the derivation. Hence we need a third estimate, an upper bound on the number of elementary particles in the observable universe. Call this number P. Clearly, any processing device must be composed of at least one particle, hence the number of such conceivable devices is at most P. Finally let $M = P \cdot Q$. It follows that the maximum number of derivation steps that could ever be recorded or verified is bounded by M. In other words, a derivation of length larger than M could never be verified, even if the entire observable universe were converted into a machine totally devoted to the verification of this single derivation.

It should be noted that all our estimates leading to the definition of M are grossly exaggerated: the time of recording or verifying one step of a derivation is much larger than t_p, any computing activity in the universe would stop sooner than after τ seconds, and the number of processing devices in a hypothetical verifying machine is much smaller than P. As a consequence, our claim concerning the number M could be even made about a much smaller number. However, we chose to use the above exaggerated estimates because, for the purpose of our argument, we only need to indicate some number bounding the length of a theoretically verifiable derivation, and the number M has the advantage of being rather simple to define and of not leaving any doubt concerning the validity of our claim about it.

Consider a derivation of length larger than M. The hypothetical existence of such a derivation of a theorem T could not possibly contribute to our confidence in T because we could never have any kind of access (even theoretically) to all the terms of such an extremely large sequence of formulae, and hence we could never verify that it is indeed a correct derivation. Trying to gain any confidence concerning the validity of theorem T from the existence of such a hypothetical derivation is (metaphorically speaking) similar to getting insight into a black hole.

Let us now call a theorem T *reachable*, if $L(T) \leq M$. In other words, a theorem is reachable, if and only if there exists its derivation in ZFC of length at most M. It follows from what was said above that reachability of a theorem is a necessary condition for gaining any confidence in it from some hypothetical derivation of this theorem.

Now we are ready for the third, final step of our argument. Consider the theorem FLT. Since no upper bound on $L(\text{FLT})$ has been justifiably provided, the answer to the question of whether FLT is reachable is unknown. Given the fact that the number M is so enormous, one would be tempted to give the answer 'yes' to this question, *i.e.*, to establish M as an upper bound on $L(\text{FLT})$. However, as observed before, this has never been done (and seems impossible to do, although we do not need this stronger conjecture in our argument). To avoid misunderstandings: we are not claiming that FLT is not reachable, we are only arguing that the answer to this question is unknown.

We can now conclude that, since in the case of FLT (and the reasoning would be similar in the case of many other 'complex' 'deep' theorems) we are unable to decide—in the present state of knowledge—whether FLT is reachable, there can be no gain of confidence from indicating any derivation of FLT. Indeed, as observed at the beginning of our argument, the side claiming such a gain has the burden of justifying that for some derivation of FLT it is *theoretically possible* to check it mechanically. This however would imply reachability of FLT, which, as previously argued, is not known to be true.

We would like to present the following analogy. A biologist studying animals' behavior or a trainer working with animals considers such notions as fear, anger, pain, or sexual attraction, and tries to explain various actions of animals (attacking, running away, mating, *etc.*) in these terms. However, these actions could also be explained at some lower level, the level of biochemical reactions in the animal's brain, or even at a subatomic level, involving statistical information about motions of particles. It is clear that these lower levels (especially the subatomic one) are not very useful in explaining and predicting the animal's behavior: they are too detailed. Likewise, a proof of a theorem should be explained at a 'macroscopic' level, involving mathematical objects and principal techniques, rather than at a 'subatomic' derivation level. However, we would like to observe that the analogy ends at this point. Indeed, the subatomic level in the case of living organisms can be (indirectly) observed. Hence one may argue that (with suitable hypothetical tools) we may be able to gain additional biological knowledge studying such a microscopic level. In the case of mathematical theorems, the accessibility of this 'subatomic' level, *i.e.*, the level of derivations, is quite different. The core of the problem is their *mechanical verifiability*. As previously argued in the case of particular theorems these objects may be too large to be written, perceived (in any possible sense), let alone scrutinized for mechanical verification. So while in biology it is unlikely but theoretically possible that the subatomic level can provide additional insight into animals' behavior, such a possibility (even theoretical) need not exist in mathematics in the case of many theorems, for a simple but crucial reason: the size of objects at the 'subatomic' level (*i.e.*, at the level of derivations) may be too large to be examined with the aim of verification by a human mind or by any technological device.

Finally it should be stressed that, as previously mentioned, our impossibility argument concerns the present state of knowledge. In other words, we argued that it is *now* impossible (even theoretically) to gain additional confidence concerning the validity of theorems such as FLT by 'indicating' their derivations. Indeed, we based our argument on the observation that the question of whether FLT is reachable has not been settled until now. This impossibility at present is enough to refute Azzouni's [2004] claim, which is made about such gains of confidence at present. It is of course hard to predict how the situation could change with additional insights in the future. If, for example, short derivations of all published mathematical theorems were discovered one day (a very unlikely but not completely impossible scenario) then the situation would change dramatically, and Azzouni's claim might become substantiated. This, however, does not concern our discussion whose purpose was to refute the possibility of such confidence gains *now*, a claim made by the formalist side.

Acknowledgments

Thanks to Jody Azzouni for enlightening discussions concerning the subject of this paper and to anonymous referees whose important remarks permitted us to correct the arguments and improve presentation. The remaining flaws remain, of course, solely the responsibility of the author. This research was partially supported by NSERC discovery grant and by the Research Chair in Distributed Computing at the Université du Québec en Outaouais.

Notes

1. The term 'mathematician' is used here in a large sense and encompasses all scientists adopting the methodology of proofs taken from 'mainstream' mathematics. Hence apart from, say, algebraists or topologists, it includes, *e.g.*, mathematical logicians and theoretical computer scientists but excludes users of mathematics such as engineers or physicists, who are interested in mathematical results but not in the way they are obtained.

2. This is a technical term whose informal meaning is that there exists a mechanical method of verifying whether a given sentence is a member of the set or not.

3. 'I wish to designate the following as the most important among the numerous questions which can be asked with respect to the axioms: *To prove that they are not contradictory, that is, that a finite number of logical steps based upon them can never lead to a contradictory result.*' Hilbert [1900] (emphasis in original).

4. For example, Azzouni's abstract quoted above goes on to say: 'Mechanically checkable derivations in this way structure ordinary mathematical practice without its being the case that ordinary mathematical proofs can be "reduced to" such derivations.'

References

AZZOUNI, J. [2004]: 'The derivation-indicator view of mathematical practice', *Philosophia Mathematica* (3) **12,** 81–105.

HALLIDAY, D., M. RESNICK, and J. WALKER [1996]: *Fundamentals of Physics.* New York: Wiley.

HILBERT, D. [1900]: 'Mathematische Probleme', *Nachrichten Königl. Gesell. Wiss. Göttingen, Math.-Phys. Klasse,* 253–297; English trans. Mary W. Newson, 'Mathematical problems', *Bull. Amer. Math. Soc.* **8** (1902), 437–479.

RAV, Y. [1999]: 'Why do we prove theorems?' *Philosophia Mathematica* (3) **7,** 5–41.

————. [2007]: 'A critique of a formalist-mechanist version of the justification of arguments in mathematicians' proof practices', *Philosophia Mathematica* (3) **15,** 291–320.

WILES, A. [1995]: 'Modular elliptic curves and Fermat's Last Theorem', *Annals of Mathematics* **141,** 443–551.

Mathematics in the Media

Mathematicians Solve 45-Year-Old Kervaire Invariant Puzzle

ERICA KLARREICH

A trio of mathematicians has solved a 45-year-old mystery about the shape of high-dimensional spaces by showing that all "framed" shapes of dimension higher than 126—shapes that are impossible to visualize, but that can be described by equations—are related in a certain fundamental way to spheres.

Their finding has startled mathematicians, many of whom expected the opposite to be true. Over the past four decades, many mathematicians have tried—and failed—to settle the question.

"I don't think anyone had any idea this was going to be solved in the next 20 years," said Mark Hovey, a mathematician at Wesleyan University. "The new method shows a path up a previously inaccessible mountain."

The new finding, by Michael Hopkins of Harvard University, Michael Hill of the University of Virginia in Charlottesville, and Douglas Ravenel of the University of Rochester in New York, concerns manifolds, shapes that are made by gluing together pieces of ordinary—though possibly high-dimensional—Euclidean space.

The research bridges two different ways to study these shapes. One method, called topology, is concerned with the way a shape is connected up—that is, which of the shape's qualities remain unchanged if the shape is stretched and distorted without being torn. The other method, called differential topology, studies shapes that are smooth enough to allow the use of ideas from calculus, such as tangent lines and derivatives.

In 1956, John Milnor, now at the State University of New York at Stony Brook, astonished mathematicians by showing that in dimension 7, there exist manifolds that have the same topology as an ordinary seven-dimensional sphere, but have a different differential structure. In other words, these manifolds—now known as exotic spheres—are each connected up in the same way as a sphere, but are so bumpy that calculus works differently on each one.

The natural next questions were, in what other dimensions do exotic spheres exist? Where they do exist? And how many different ones are there?

In the 1960s, Milnor and the late Michel Kervaire, and then William Browder of Princeton University, took a stab at answering this question. They showed that in almost all dimensions, every manifold—or more precisely, any "framed" manifold, a technical restriction—can be converted into an exotic sphere by performing a process called surgery. In surgery, a part of the manifold gets drilled out, and a new piece is glued in along the boundary of the excised piece. Thus, the researchers showed, if mathematicians can get a handle on which manifolds are related to each other via surgery, they can figure out how many exotic spheres there are.

The work of Milnor, Kervaire and Browder left an important gap. They didn't know what happens in dimensions of the form $2^n - 2$, for whole-number values of n: that is, dimensions 2, 6, 14, 30, 62, 126, 254, 510, etc. In these dimensions, it was conceivable that some manifolds might exist that could not be converted into spheres via surgery—manifolds with such complicated topological twists and turns that even surgery could not trim away their unusual features.

Building on algebra developed by the late Turkish mathematician Cahit Arf, Kervaire defined an invariant of a framed manifold—that is, a number determined by the manifold's topology—that measures whether the manifold could be surgically converted into a sphere. This number, called the Arf-Kervaire invariant, evaluates to 0 if the manifold can be converted to a sphere, and 1 otherwise. The researchers showed that in any given dimension, there are only two possibilities: either all manifolds have Arf-Kervaire invariant equal to 0, or half have Arf-Kervaire invariant 0 and the other half have Arf-Kervaire invariant 1.

In the 1970s and 1980s, mathematicians discovered that framed manifolds with Arf-Kervaire invariant equal to 1—oddball manifolds not surgically related to a sphere—do in fact exist in the first five dimensions on the list: 2, 6, 14, 30 and 62. A clear pattern seemed to be established, and many mathematicians felt confident that this pattern would continue in higher dimensions.

"It's as if there is an emerald mine, and you've pulled out five emeralds," Ravenel said. "You hope there are infinitely many more, if you can just get the shovels to dig them out."

Researchers developed what Ravenel calls an entire "cosmology" of conjectures based on the assumption that manifolds with Arf-Kervaire invariant equal to 1 exist in all dimensions of the form $2^n - 2$. Many called the notion that these manifolds might not exist the "Doomsday Hypothesis," as it would wipe out a large body of research. Earlier this year, Victor Snaith of the University of Sheffield in England published a book about this research, warning in the preface, ". . . this might turn out to be a book about things which do not exist."

Just weeks after Snaith's book appeared, Hopkins announced on April 21 that Snaith's worst fears were justified: that Hopkins, Hill and Ravenel had proved that no manifolds of Arf-Kervaire invariant equal to 1 exist in dimensions 254 and higher. Dimension 126, the only one not covered by their analysis, remains a mystery.

The new finding is convincing, even though it overturns many mathematicians' expectations, Hovey said.

"After so many false proofs over the years, you start being suspicious, but this is completely different," he added. "Their new method is very powerful. It's clear that it is going to work."

The trio had not set out to prove the Doomsday Hypothesis, Ravenel said. Rather, they had been working for several years on a different problem, when they realized almost a year ago that it was connected to the Arf-Kervaire invariant question.

"Everyone else was knocking on the front door of the Arf-Kervaire problem, but we were investigating a house down the street," he said. "Then we found a tunnel that led to the Arf-Kervaire problem."

At first, the three researchers thought that they would have to do an extremely intricate calculation to make the leap from their research to the Arf-Kervaire problem. As a warm-up, they did what Hopkins called a "toy example," to see if that type of calculation worked in a simplified setting. At a week-long conference in Mexico, Hopkins showed the calculation to his former thesis advisor, Mark Mahowald—an emeritus professor at Northwestern University—who had studied the Arf-Kervaire problem for decades.

"By the end of that week, after a lifetime of believing that these things existed [in all dimensions of the form $2^n - 2$], he believed that we were going to get it and that these things would exist in at most six dimensions," Hopkins said. "It's hard to explain what it feels like when the person whose mind you admire most has spent most of his career thinking about a problem, and then you solve it."

In February, the three researchers had what Hopkins called "a radically new idea" that simplified the calculations and allowed them to reach their goal. "It was a thrilling, obsessive sprint where we thought of nothing else for six months," Hopkins recalled. "Now I understand why authors are always dedicating their books to their family."

Once the researchers were sure their surprise finding was correct, they kept it under wraps for a couple more months, saving their big announcement for a major conference in Edinburgh on geometry and physics last

April, honoring the famous mathematician Sir Michael Atiyah. To heighten the surprise, Hopkins even gave his talk a fake title so attendees would have no clue what they were about to hear.

While some mathematicians may have been disappointed to have an entire cosmology of conjectures wiped out in a single stroke, the glass may be seen as half empty or half full. It's true that, with the possible exception of dimension 126, there are no more emeralds in the mine—no more manifolds so bumpy and twisty that they cannot be surgically converted to an exotic sphere. Then again, Hill observes, since in these high dimensions every manifold can be converted to an exotic sphere, "in some sense there are twice as many exotic spheres as we thought."

Even though the new finding has killed off an entire cosmology of ideas, it is more a beginning than an end. "Finding something that wasn't predicted using the standard tools is really exciting," Hill said. "It shows that the great pictures we have so far are incomplete."

And while many mathematicians may have been hoping that there would be infinitely many emeralds in the mine, in some ways it's even more exciting to find out that there are at most six, Hopkins said. These few emeralds will join the pantheon of famous collections of geometric objects that happen only a few times, such as the Platonic solids, of which there are only five—cube, tetrahedron, octahedron, dodecahedron and icosahedron.

"After one of my talks on our work, Igor Frenkel [of Yale University] said to me, 'God has given you a beautiful gift. When there are only six of something, each one is very beautiful'," Hopkins said.

At this point, Hopkins understands why there are just five or six special dimensions "on the level of a car mechanic, in the way Mr. Scott on Star Trek knew how to keep the Enterprise running—I know how the machine works, in my bones," he said. "We don't have enough distance yet to give a philosophical reason why there are at most six. But I think over the next few years, our proof will simplify greatly, and we'll have some very natural constructions of the list of six."

In cracking the Doomsday Hypothesis, the researchers created a whole new set of techniques that will inform many unsolved problems, predicts Nicholas Kuhn, of the University of Virginia in Charlottesville.

While the new finding has cleared up the relationship between exotic spheres and surgery on manifolds, it remains an open—and exceedingly difficult—problem to classify fully the exotic spheres. "I don't expect this to be solved in my grandchildren's lifetime," Ravenel said. "It's a very hard problem that has kept the subject going for decades."

Darwin: The Reluctant Mathematician

JULIE REHMEYER

For all his other talents, Charles Darwin wasn't much of a mathematician. In his autobiography, he writes that he studied math as a young man but also remembers that "it was repugnant to me." He dismissed complex mathematical arguments and wrote to a friend, "I have no faith in anything short of actual measurement and the Rule of Three," where the "Rule of Three" was an extremely simple mathematical calculation.

But history played a joke on the great biologist: It made him a contributor to the development of statistics.

It was the wildflower common toadflax that got the whole thing started. Darwin grew the plant for experiments, and he carefully cross-fertilized some flowers and self-fertilized others. When he grew the seeds, he found that the hybrids were bigger and stronger than the purebreds.

He was astonished. Although he had always suspected that inbreeding was bad for plants, he had never suspected it could have a significant effect within a single generation.

So he repeated the experiment with seven other kinds of plants, including corn. He had a clever, and at that time novel, idea. Since slight differences in soil or light or amount of water could affect the growth rates, he planted the seeds in pairs—one cross-pollinated seed and one self-pollinated seed in each pot. Then he let them grow and measured their heights.

Sure enough, on average, the hybrids were taller. Among his 30 corn plants, for example, the purebreds were only 84 percent as tall as the hybrids. But Darwin was savvy enough not to simply trust the average heights of so few plants. "I may premise," Darwin wrote, "that if we took by chance a dozen or score of men belonging to two nations and measured them, it would I presume be very rash to form any judgments from such small numbers on their average heights." Could it be, he wondered, that the height differences in the plants were just random variation?

Darwin noted, though, that men's heights vary a lot within a single country, whereas the heights of his plants didn't. His result might be more meaningful, but he wanted to be able to quantify how meaningful.

Doing that, however, required Darwin's hated mathematics.

So he turned to his cousin, Francis Galton, who just happened to be a leader in the emerging field of statistics. Galton had recently invented the standard deviation, a way of quantifying the amount of random variability in a set of numbers.

But Galton wasn't all that much use. He could calculate the standard deviation, but he couldn't use that number to tell Darwin how likely it was that the height difference wasn't just random. Furthermore, he was pretty sure it was too few plants to tell. "I doubt," he wrote, "after making many trials, whether it is possible to derive useful conclusions from these few observations. We ought to have measurement of at least fifty plants in each case, in order to be in a position to deduce fair results."

And there the matter rested, in frustrating uncertainty, for 40 years.

Resolving the impasse, it turned out, required some beer. The Guinness brewing company hired a young University of Oxford graduate, William Sealy Gosset, to develop statistical techniques to cheaply monitor the quality of its beer. The method Gosset developed was so powerful that it transformed statistics and continues to be a workhorse to this day.

Ironically, though, Gosset wasn't allowed to publish the method under his own name, because Guinness wanted to keep it a secret that statistics could help make better beer. But publish it he did, under the pseudonym "Student." The technique has hence become known as the "Student's t-test."

The Student's t-test did just what Galton didn't know how to do: Given the standard deviation Galton had calculated, it told how likely it was that the difference in the heights between the hybrids and the purebreds were just random. The answer? The chance was about one in 20. By statistical standards, that's significant, but barely so.

It took another 10 years and the intervention of another statistical genius for the next breakthrough on the problem. As a college student, Sir Ronald Aylmer Fisher learned about Gregor Mendel's work in genetics and Darwin's work in evolution, but the theory connecting the two hadn't yet been developed. Fisher set out to create the statistical foundation to make the connection possible. Darwin's experiment with hybrids was just the kind of problem Fisher needed to be able to solve.

He noticed something that Galton had missed: Galton had ignored Darwin's clever method of pairing the plants. He had calculated the standard deviation of the plants as a single, large group.

Fisher repeated the analysis but calculated the standard deviation of the difference in heights between the pairs of plants in each pot. Suddenly, instead of a one in 20 chance that the result didn't mean anything, he calculated about a one in 10,000 chance. In other words, it was nearly certain that the hybrids really did grow taller than the purebreds.

Fisher noted that the Student's t-test had one possible flaw: It assumed that the plant heights would vary in a predictable way (according to a normal distribution, to be precise). Just in case that assumption was wrong, he devised another way of analyzing the data and confirmed the result. "He was very clever in the way he did it," says Susan Holmes of Stanford University. Only in the 1980s did statisticians realize the full potential of Fisher's method and develop it into the subject of "exact testing."

Fisher's analysis was only possible because Darwin had designed his experiment so well. In fact, Fisher was often frustrated with the quality of other people's experiments. "To call in the statistician after the experiment is done," he said, "may be no more than asking him to perform a postmortem examination: he may be able to say what the experiment died of."

David Brillinger, a statistician at the University of California, Berkeley, says that Darwin's method of pairing is now common practice. "Darwin was a leader in a subfield of statistics called experimental design," he says. "He knew how to design a good experiment, but what to do with the numbers was something else."

Darwin himself came around eventually in his attitude toward mathematics. While he wrote in his autobiography of his youthful distaste for math, he also wrote that he wished he had learned the basic principles of math, "for men thus endowed seem to have an extra sense."

Loves Me, Loves Me Not (Do the Math)

STEVEN STROGATZ

"In the spring," wrote Tennyson, "a young man's fancy lightly turns to thoughts of love." And so in keeping with the spirit of the season, this week's column looks at love affairs—mathematically. The analysis is offered tongue in cheek, but it does touch on a serious point: that the laws of nature are written as differential equations. It also helps explain why, in the words of another poet, "the course of true love never did run smooth."

To illustrate the approach, suppose Romeo is in love with Juliet, but in our version of the story, Juliet is a fickle lover. The more Romeo loves her, the more she wants to run away and hide. But when he takes the hint and backs off, she begins to find him strangely attractive. He, on the other hand, tends to echo her: he warms up when she loves him and cools down when she hates him.

What happens to our star-crossed lovers? How does their love ebb and flow over time? That's where the math comes in. By writing equations that summarize how Romeo and Juliet respond to each other's affections and then solving those equations with calculus, we can predict the course of their affair. The resulting forecast for this couple is, tragically, a never-ending cycle of love and hate. At least they manage to achieve simultaneous love a quarter of the time.

The model can be made more realistic in various ways. For instance, Romeo might react to his own feelings as well as to Juliet's. He might be the type of guy who is so worried about throwing himself at her that he slows himself down as his love for her grows. Or he might be the other type, one who loves feeling in love so much that he loves her all the more for it.

Add to those possibilities the two ways Romeo could react to Juliet's affections—either increasing or decreasing his own—and you see that there are four personality types, each corresponding to a different romantic style.

My students and those in Peter Christopher's class at Worcester Polytechnic Institute have suggested such descriptive names as Hermit and Malevolent Misanthrope for the particular kind of Romeo who damps out his own love and also recoils from Juliet's. Whereas the sort of Romeo who gets pumped by his own ardor but turned off by Juliet's has been called a Narcissistic Nerd, Better Latent Than Never, and a Flirting Fink. (Feel free to post your own suggested names for these two types and the other two possibilities.)

Although these examples are whimsical, the equations that arise in them are of the far-reaching kind known as differential equations. They represent the most powerful tool humanity has ever created for making sense of the material world. Sir Isaac Newton used them to solve the ancient mystery of planetary motion. In so doing, he unified the heavens and the earth, showing that the same laws of motion applied to both.

In the 300 years since Newton, mankind has come to realize that the laws of physics are always expressed in the language of differential equations. This is true for the equations governing the flow of heat, air and water; for the laws of electricity and magnetism; even for the unfamiliar and often counterintuitive atomic realm where quantum mechanics reigns.

In all cases, the business of theoretical physics boils down to finding the right differential equations and solving them. When Newton discovered this key to the secrets of the universe, he felt it was so precious that he published it only as an anagram in Latin. Loosely translated, it reads: "It is useful to solve differential equations."

The silly idea that love affairs might progress in a similar way occurred to me when I was in love for the first time, trying to understand my girlfriend's baffling behavior. It was a summer romance at the end of my sophomore year in college. I was a lot like the first Romeo above, and she was even more like the first Juliet. The cycling of our relationship was driving me crazy until I realized that we were both acting mechanically, following simple rules of push and pull. But by the end of the summer my equations started to break down, and I was even more mystified than ever. As it turned out, the explanation was simple. There was an important variable that I'd left out of the equations—her old boyfriend wanted her back.

In mathematics we call this a three-body problem. It's notoriously intractable, especially in the astronomical context where it first arose. After Newton solved the differential equations for the two-body problem (thus explaining why the planets move in elliptical orbits around the sun), he turned his attention to the three-body problem for the sun, earth and moon. He couldn't solve it, and neither could anyone else. It later turned out that the three-body problem contains the seeds of chaos, rendering its behavior unpredictable in the long run.

Newton knew nothing about chaotic dynamics, but he did tell his friend Edmund Halley that the three-body problem had "made his head ache, and kept him awake so often, that he would think of it no more." I'm with you there, Sir Isaac.

Note

In the first story above, Romeo's behavior was modeled by the differential equation $dR/dt = aJ$. This equation describes how Romeo's love (represented by R) changes in the next instant (represented by dt). The amount of change (dR) is just a multiple (a) of Juliet's current love (J) for him. This equation idealizes what we already know—that Romeo's love goes up when Juliet loves him—by assuming something much stronger. It says that Romeo's love increases in direct linear proportion to how much Juliet loves him. This assumption of linearity is not emotionally realistic, but it makes the subsequent analysis much easier. Juliet's behavior, on the other hand, was modeled by the equation $dJ/dt = -bR$. The negative sign in front of the constant b reflects her tendency to cool off when Romeo is hot for her. Given these equations and an assumption about how the lovers felt about each other initially (R and J at time $t = 0$), one can use calculus to inch R and J forward, instant by instant. In this way, we can figure out how much Romeo and Juliet love (or hate) each other at any future time. For this elementary model, the equations should be familiar to students of math and physics: Romeo and Juliet behave like simple harmonic oscillators.

References

For models of love affairs based on differential equations, see Section 5.3 in Strogatz, S. H. (1994) *Nonlinear Dynamics and Chaos*. Perseus, Cambridge, MA.

For Newton's anagram, see page vii in Arnol'd, V. I. (1988) *Geometrical Methods in the Theory of Ordinary Differential Equations*. Springer, New York.

Chaos in the three-body problem is discussed in Peterson, I. (1993) *Newton's Clock: Chaos in the Solar System*. W.H. Freeman, San Francisco.

For the quote about how the three-body problem made Newton's head ache, see page 158 in Volume II of Brewster, D. (1855) *Memoirs of the Life, Writings, and Discoveries of Sir Isaac Newton*. Thomas Constable and Company, Edinburgh.

The Mysterious Equilibrium of Zombies and Other Things Mathematicians See at the Movies

Samuel Arbesman

If you were a normal moviegoer sitting through the opening credits of "Casino Royale," you might notice the snappy visuals, or you might begin dwelling on how this new Bond is a lot less refined than the previous ones, what with all the brawling and fisticuffs.

On the other hand, if you were a mathematician, you would also be noticing that the animated pattern of playing-card clubs, as they hypnotically subdivide on screen, starts to look an awful lot like a fractal.

In any decade there are really only a handful of movies about math ("Proof" comes to mind, as well as "A Beautiful Mind"), but a surprising number of movies that end up embodying math, even if it's accidental. "Six Degrees of Separation" is based on the math of social networks. Thrillers have a special propensity for edgy twists on game theory. And what is a disease-outbreak movie if not an illustration of mathematical epidemiology, with puffy suits? To see movies through their math, sometimes, is to watch a whole different drama.

Troubled Bridge

In the opening minutes of the new film "Harry Potter and the Half-Blood Prince," the Millennium Bridge—an elegant modern footbridge in the center of London—is sensationally destroyed. While casual watchers might recognize this as a nice point of reference to the real world, those who think about math are actually bothered by it, since the bridge buckles while moving up and down.

The Millennium Bridge did indeed have a dangerous wobble to it, but not that one. In 2000, the then-new bridge had to be closed due to a spontaneous side-to-side swing that surprised the engineers who had designed it, and continues to intrigue mathematicians. One explanation, put forth by Steven Strogatz (my graduate school advisor) along with some colleagues, involved the spontaneous synchronization of human behavior, where a slight initial wobble is amplified as people unconsciously begin walking in synchrony to keep their balance, which in turn causes even more people to walk in unison. The most recent theory, proposed by John Macdonald, actually eliminates the need for people to adjust their pace, at least when the bridge initially begins to move.

The bridge has since been modified to prevent the wobbling. And thus far, no math has been used to model the precise forces required for dark wizards to destroy the bridge.

Holy Bat-hematics

In the final scenes of "The Dark Knight" (spoiler alert!), the Joker gives the following choice to the passengers of two ferries: they can either blow up the other boat and save themselves, or themselves be blown up. If no one decides within a certain amount of time, both ferries are destroyed.

The typical moviegoer pretty much thinks one thing: Batman better show up now. But the mathematician immediately recognizes the Joker's trap as a variation on the classic problem of the prisoner's dilemma, where two individuals, each isolated in a prison cell, are given a choice: betray their friend and go free, or cooperate by saying nothing, and be given a short prison sentence. If each betrays the other, however, they will get a longer prison sentence.

This seminal problem in game theory has an important property: while cooperation is a more socially beneficial strategy, it is actually a more "stable" strategy for each person to betray the other, since this makes each better off independent of the whims of his friend. This behavior is known as a Nash equilibrium and is named after John Nash, well-known from the more obviously mathematical film, "A Beautiful Mind."

Characteristically, the Joker is playing a warped variant of the game, in which cooperation is a very bad choice, since everyone will die. Of course, as science blogger Jake Young has noted, the ferry passengers must take into account something that most participants in the game could safely ignore: the nontrivial probability that they will be saved by a superhero.

The Math of Zombies

You think you know what zombie flicks are about: the undead's insatiable desire for human flesh. True, but they're also about epidemiology. One zombie bite infects a healthy person, who infects others, and so on until the few healthy humans left are holed up in Wal-Mart with a shotgun, aiming for the head.

The problem of zombies intrigued Philip Munz of Carleton University and his colleagues at the University of Ottawa, who recently wrote a scientific paper quantifying various properties of zombie epidemics. Standard modeling techniques for disease outbreaks weren't quite sufficient, the authors found. "The key difference between the models presented here and other models of infectious disease," they wrote, "is that the dead can come back to life."

After a thorough, if tongue-in-cheek, analysis, the authors found that the optimal method for halting such epidemics involves killing zombies early and often—the rare scientific paper that satisfies both the splatter-film aficionado and the Centers for Disease Control.

More like 6.6 Degrees

In the film "Six Degrees of Separation," Stockard Channing's character at one point makes the following observation: "I read somewhere that everybody on this planet is separated by only six other people. Six degrees of separation between us and everyone else on this planet. The President of the United States, a gondolier in Venice, just fill in the names."

Back in the 1960s, the psychologist Stanley Milgram attempted to measure the number of connections separating people in Nebraska and Kansas from those around Boston, and found a typical distance of about six steps. Since then, this concept of a tightly connected world has captured the public imagination, and in math and science has spawned a large field known as social network theory. Jure Leskovec and Eric Horvitz, two computer scientists, recently calculated that, at least in online networks (they used MSN Messenger interactions), the number of steps separating any two individuals on average is about 6.6.

Of course, this area of research is also applicable to Six Degrees of Kevin Bacon, the game in which players compete to link any given actor to Kevin Bacon, through mutual co-stars, in the minimal number of steps. When the calculations are done, it turns out that Bacon is actually not the most connected actor. Dennis Hopper is, with Harvey Keitel running second. Kevin Bacon is 507th.

Balance Theory, with Pistols

Speaking of Harvey Keitel, at the end of the movie "Reservoir Dogs" (another spoiler!), Keitel and two other main characters find themselves in a triangle, all pointing guns at one another. This situation, known as a Mexican standoff (and borrowed from "The Good, the Bad, and the Ugly") is the province of an area of social network analysis known as balance theory. Which social relationships are stable? Well, not this one.

If you and someone else hate the same third person, but like each other, balance theory says you're golden—all three can persist without changing their opinions. On the other hand, if all three of you despise the others, it's an unstable triad, as well as a wildly common plot point for crime movies. While there are numerous resolutions—one person changes his preference toward another, a relationship tie is cut—another route back to stability, albeit a messy one, is the gunning down of at least one person. Presumably not a mathematician, who would already be out of there.

Strength in Numbers:
On Mathematics and Musical Rhythm

Vijay Iyer

I've been making music for most of my life, but I also spent a number of years studying mathematics, physics, and cognitive science. Because of my background, I am often asked about a connection between science and music.

It has been shown that playing music "sharpens" the brain; despite clichés about rock drummers, music apparently makes you smarter. This suggests a reversal of the usual assumption I encounter. Science and math won't necessarily make me (or anyone) a better musician or composer, but playing music early in life often makes you better at science and math.

Chicken-or-egg questions aside, I have benefited from prolonged immersion in both disciplines, and I like to let them talk to each other. One of my artistic goals is to find new, unique or surprising moments of beauty. I sometimes employ elaborate means to get there, because I want to reach for things beyond my grasp. So, like many composers before me, I resort to formal techniques to generate pre-compositional material. The mathematical ideas I work with in music are easy enough for a twelve-year-old to understand, but they help me find sounds and rhythms that I might never have made otherwise.

I should stress at the outset that such methods are not the point of my music. Music is not a mere display of technique, any more than conversation is a display of vocabulary. Music is made by and for people; it's about sensation, emotion, and connection. Technique is just a means to achieve those aims.

I want to show you one example involving Fibonacci numbers. Fibonacci was a 13th-century Italian mathematician who, among other achievements, brought the Indian-Arabic number system to Europe. He also wrote about this set of numbers that now bears his name. I became intrigued by these numbers some years ago, and have used them to structure much of my work ever since.

The Fibonacci sequence begins like this: 1, 1, 2, 3, 5, 8, 13, 21, 34, 55, 89, 144... Each number in the sequence is the sum of the previous two numbers, and it continues ad infinitum. Now if you look at the ratios of two successive Fibonacci numbers, and keep going up the sequence, you get: 1, 2, 1.5, 1.667, 1.6, 1.625, 1.615, 1.619, 1.618... As you go up the sequence, you see that this ratio gets closer and closer to a famous irrational number called the "golden ratio": 1.6180339887... To name a few uses, Plato employed the golden ratio to describe two of his Platonic solids; Kepler made it the basis of his Kepler triangle; Roger Penrose applied it to his Penrose tiles.

A "golden" rectangle, one whose side lengths form the golden ratio, has an interesting trait: the ratio of its side lengths is equal to the ratio of the sum of both side lengths to the longer side length. Perhaps because of this basic property, the ratio has been observed frequently in dimensional proportions across many different contexts - in architecture from the Pyramid of Giza and the Parthenon to constructions by Le Corbusier and Mies Van der Rohe; images by artists from Da Vinci and Albrecht Dürer to Juan Gris, Mondrian, and Dalí; and rhythmic durations and pitch ratios in works by composers from Bartok and Debussy to John Coltrane and Steve Coleman. (In fact it was Coleman who introduced me to this whole idea, when I first met him in 1994.)

What interests me now about Fibonacci numbers is their scaling property. Because they get successively closer to the golden ratio, the ratio 5:3 is somehow "similar" to the ratio 8:5, which is "similar" to the ratio 13:8, or 144:89, or 6765:4181. But what could I mean by as vague a term as "similar"? This is a question I explore musically, with my trio's version of "Mystic Brew," a 70s soul-jazz classic by Ronnie Foster, famously sampled by the pioneering hip hop group A Tribe Called Quest in their song "Electric Relaxation."

The harmonic rhythm in Foster's original is asymmetric in what you could call a Fibonacci way: a short chord and then a long chord, three beats plus five beats, totalling eight beats. It's standard four-four time, with one added feature: if you were to step to the beat, you'd hear a chord when you take your first step, and then another chord while your knee is aloft between the second and third steps. This is a rhythm that you hear in all kinds of places – another example being the memorable opening chords in Michael Jackson's "Billie Jean."

In our version we work with that asymmetry, moving it through Fibonacci-like transformations. To explore what the similarity of Fibonacci ratios feels like, we perform an asymmetric "stretch" that maintains the same "golden" balance over the entire measure. This stretch is not quite the usual kind; we don't transform simply by multiplying, as you might when shifting from duple to triple meter, or when doubling a cooking recipe. Rather, we try to preserve an "impression" of the original – the short-and-long-ness of it – to see if we can achieve that feeling of similarity.

Suppose you had a round pie and eight guests; you know how to divide that pie into eight equal pieces, and you know exactly what that pie would look like with three pieces missing. Now, suppose five more friends unexpectedly show up. You have the same pie and thirteen guests. How do you divide a circle into thirteen by eye? A decent short cut would be to imagine it divided into eight with three pieces missing, and cut that shape. Then, divide the smaller section you've just cut into five equal pieces, and the larger section into eight. Your result would be close enough that guests probably wouldn't complain.

This is something like the technique we use here, only instead of a pie, we divide a length of time. The 3 beats and 5 beats of the original are transformed to a faster 5 beats and 8 beats (totaling 13), which then becomes a still-faster 8 beats and 13 beats (totaling 21). Each transformed measure is roughly the same length and, importantly, the second chord lands at roughly the same time, about $3/8$ of the way through (or $5/13$, or $8/21$).

The goal is that you *perceive* the "short-long" division of the cycle the same way in each case. Thankfully the ear is forgiving; because we expect and even *crave* continuity in our perception, our listening brains help smooth things out. Like the gracious guests eating your slices of pie, the ear doesn't complain about small differences. In this case, the overall motion proceeds seemingly undeterred – including a sense of regular pulse - while the music's inner mechanism seems to quicken.

And a bonus finding is that, through accenting these changes just right, you can make these "irregular" or "artificial" rhythms sound simple and natural - like a buoyant, composite version of the original's asymmetric $4/4$. I never would have guessed that an irregular 21-beat cycle could feel as powerful or joyous as it feels to me in this treatment.

Does this finding reveal anything about the golden ratio? I can't be certain, but perhaps because that proportion is so familiar to us, it may even help us hear the continuity through the transformations. For the listener, that special "short-long" balance might override the non-duple structure, tricking your ear into hearing the familiar amidst the unfamiliar.

Musical mini-discoveries like this are the reason that I use such formalisms in my work. I am not trying to make any point about the beauty of mathematics or to flex any intellectual muscle; I am trying to make music that hits me viscerally and in surprising, unobvious ways.

As abstruse as some of this may seem, there are specific cultural origins for these techniques. As the son of immigrants from India, I'm very inspired by Karnatak (south Indian classical) music. It is a tradition of religious song, very intricately organized: melodically nuanced and rhythmically dazzling, full of systematic permutations. I'm also inspired by the African roots of African-American music; I've spent time playing, absorbing, and studying West and Central African musical traditions, in order to understand their profound and widespread influence on nearly every vernacular music we have in the west, from New Orleans jazz to German techno. (In fact, the original Fibonacci-like asymmetry referred to above in "Mystic Brew" and "Billie Jean" is a common trait found throughout music of the Black Atlantic; its presence in these songs is no accident, because asymmetries are valued in those traditions.)

These non-western musical traditions are just as deeply ordered with rhythm as western music is with harmony. But there's a qualitative difference between rhythm and harmony: when you organize rhythms, you structure a listener's experience in time. Rhythm is the first thing we perceive about music. Rhythm hits us viscerally; it reaches into our bodies, stimulates our brain stems, and in many cases, provokes sympathetic motion.

Why? One reason is that the rhythms of music are not so different from the inherent timescales of human bodies. Think of the rapid clip of our speech, the bounce of our walk, the slow ebbing of our breath. And then think of Charlie Parker's conversational saxophone solos, Ray Brown's loping basslines, or Billie Holiday's cries and sighs. Musical rhythm resembles human bodies in motion because music *is* the sound of bodies in motion.

So when you impose rigorous order on musical rhythm – as has been done around the world for centuries - you are organizing human motion. You create a dialogue between the physical and the ideal: embodied human action in a structured environment. The process gives us something to strive for, to work through, to achieve with virtuosity and grace. This is the case with music, sport, dance, ritual, games, art – indeed, with most areas of human endeavor, improvised and otherwise.

The dialectic between soul and science, body and number, freedom and discipline, self and non-self – dare I say it? That's culture in a nutshell. It is this very dialogue - this sustained interaction between ourselves and the world around us - that I wish to make audible through music.

Math-hattan

NICK PAUMGARTEN

For ten years, Glen Whitney, a mathematician, worked as an algorithm manager at the giant quantitative hedge fund Renaissance Technologies, on Long Island. He was the man—or one of many—in the so-called "black box." During that time, Renaissance did extremely well, as did Whitney, and so when he left the firm, last year, he had the wherewithal to devote himself to his favorite shower-time epiphany—that what the world needs (if not most, then at least a lot) is a museum devoted to math. The equation goes something like this: for the variables Expertise (E), Computational Power (CP), Capital (C), Risk (R), Altruism (A), Obsession (O), Indifference (I),

$$\frac{\Delta O + [E(CP+C)+A] - R}{I} = \text{Math Museum}$$

The idea behind the museum, which doesn't yet have a home, is that math is ubiquitous, supercool, underappreciated, poorly taught, and even more poorly learned. To drum up interest in his museum, Whitney has been leading free tours of Manhattan neighborhoods, lingering over the math-y bits. On a recent morning, about two dozen civilians (each head count seemed, disconcertingly, to yield a different sum) met up with him on the new cantilevered grandstand outside Alice Tully Hall, at Lincoln Center. Whitney, slim, forty, and dressed for math (bow tie, Dockers, leather walking shoes), made a couple of mild jokes ("People who go on math tours are very prompt") and then began explaining how the hyperbolic paraboloids of the grandstand employed straight lines to form a curved surface.

Next he jaywalked across Sixty-sixth Street ("This is the most efficient route") to Lincoln Center Plaza, where, over the screech of a construction crew's saws, he explained a centrally symmetrical bench (don't ask), a row of semicircular arches, and what he called "the best spot in the city for perfect one-point perspective"—a view east through the cloistered walk-way on the north side of Avery Fisher Hall. A Philip Johnson clock just off Columbus Avenue led to a disquisition on Pythagoras, octaves, calendars, eclipses, and time. Leaning out into the traffic, he said, "Now I am going to switch from the sublime motion of the heavens to the mundane motion of the cars"—and gave a little talk, his voice occasionally cracking, on the tim-ing patterns of the city's thousands of stoplights. As to the eternal question of whether it's faster to take Third Avenue or Park Avenue uptown, the former having staggered lights, the latter not, he equivocated: "It depends on your nature. If you want speed right now, you take Park. If you like to understand a system and maximize it, you take Third. I'm a system guy."

A man, perhaps a cabbie in math-tour disguise, said, "I like Park Avenue. It has a tunnel."

As the tour moved down Broadway, Whitney calculated the advantage of taking the hypotenuse while travelling through a grid (you gain two blocks for every nine). He stopped at a fire hydrant, produced a wrench, and explained the rationale behind pentagonal lug nuts (they are wrench-proof) and reverse-threaded screws (they are righty-tighty-lefty-loosey-proof). As the group moved to Whole Foods for a discussion of queueing theory, Sylvain Cappell, a math professor at N.Y.U., and a member of the math museum's advisory council, indulged a mathophobe with back-of-the-classroom mutterings about the ham-sandwich theorem (any two slices of bread and chunk of ham can be bisected on a single plane), the optimiza-tion of prepared foods, and the curiously frequent incidence, at least in the nineteen-eighties, of high-school calculus teachers who were also football coaches. The tour ended in Columbus Circle—within sight of the uni-sphere, at the foot of Trump Tower—with a discussion of Eratosthenes, Christopher Columbus, and the measurement of the earth. No question, math was cool.

In Whitney's view, the standard progression in math education—alge-bra, geometry, trig, pre-calculus, calculus—is random and baseless, a lin-ear conceit that creates a false sense of increasing difficulty. Mathematical ignorance is insidious, and it manifests itself in many ways. "The purest example is the lottery," he said. "The lottery is a tax on the mathematically illiterate."

Another example, impure in the extreme, is the national debt. On his way to the train station, Whitney stopped off at the National Debt Clock, on Forty-fourth Street, which tracks the deficit in real time. At that second, the clock read $11,518,960,404,062, and a second later the number was tens of thousands of dollars higher. The family share of it held steady, at $96,700. "People just don't seem to care," he said. "Why? Because it's im-possible to understand it." He has tried the analogical approach favored by astronomers. "If Jupiter is the size of a basketball, then Earth is a pea. O.K., so let's suppose the national debt is a basketball. Your annual salary would be unseen by the most powerful microscope in the world."

When he was at Renaissance, he said, the firm processed more than a terabyte of data every day. That's a lot. The trick, he said, was to "find the nuggets . . . the very faint phenomena amidst the cacophony of static." That's where the math came in.

Contributors

Kathleen Ambruso Acker received her doctorate in mathematics education from American University and works as an independent researcher in mathematics education. Her research interests include teaching with technology, teaching students with learning disabilities, mathematics education as it applies to home-schooled students, and the history of mathematics. Presently she works for American University teaching graduate students enrolled in the Teach for America program.

David L. Alderson is assistant professor of operations research at the Naval Postgraduate School, Monterey, California. Information on his work can be found on his Web site, http://faculty.nps.edu/dlalders/.

Samuel Arbesman is a postdoctoral research fellow in the Department of Health Care Policy at Harvard Medical School and is affiliated with the Institute for Quantitative Social Science at Harvard University. He is a regular contributor to the Ideas section of the *Boston Globe*.

Philip L. Bowers is professor and chair of the Mathematics Department at Florida State University, where he has taught and nurtured his passion for mathematics for over twenty-five years. Aside from his research interest in geometry and topology, he has a keen appreciation of the historical and philosophical aspects of the subject, as well as an enduring interest in both the technical and philosophical aspects of modern theoretical physics.

Carlo Cellucci is professor of logic at the University of Rome–La Sapienza. He is the author of four books, *Teoria della dimostrazione* (*Proof Theory*, 1978), *Le ragioni della logica* (*The Reasons of Logic*, 1998), *Filosofia e matematica* (*Philosophy and Mathematics*, 2002), and *Perché ancora la filosofia* (*Why Still Philosophy*, 2008). He is currently completing another book, *Remaking Logic: What Is Logic, Really?*

Barry A. Cipra is a freelance mathematics writer based in Northfield, Minnesota. He has been a contributing correspondent for *Science* magazine and a regular writer for *SIAM News*, the monthly newsletter of the Society for Industrial and Applied Mathematics. He is the author of *Misteaks . . . and How to Find Them Before the Teacher Does: A Calculus Supplement*, published by A.K. Peters, Ltd.

Mark Colyvan is professor of philosophy and director of the Sydney Centre for the Foundations of Science at the University of Sydney, Australia. He is the author of *The Indispensability of Mathematics* (Oxford University Press, 2001), co-author (with Lev Ginzburg) of *Ecological Orbits: How Planets Move and Populations Grow* (Oxford University Press, 2004), and author of a number of articles in and around the philosophy of mathematics, logic, formal decision theory, and their applications.

Chandler Davis, though he failed in 1992 to stop the University of Toronto from making him emeritus, remains a mathematician and editor—recently, co-editor—of *The Mathematical Intelligencer*. For some of his nonmathematical prose, see *It Walks in Beauty*, edited by J. Lukin (Aqueduct, 2010).

Philip J. Davis holds a doctoral degree from Harvard University and is currently professor emeritus in the Division of Applied Mathematics at Brown University. He is known for his work in numerical analysis and approximation theory, as well as his investigations and writings on the history and philosophy of mathematics.

Keith Devlin is a mathematician at Stanford University. He is a Fellow of the American Association for the Advancement of Science. He has written twenty-eight books and published more than eighty research articles. The recipient of the Pythagoras Prize, the Peano Prize, the Carl Sagan Award, and the Joint Policy Board for Mathematics Communications Award, he is also "the Math Guy" on National Public Radio.

Alicia Dickenstein is professor of mathematics in the School of Exact and Natural Sciences of the University of Buenos Aires and a principal researcher of the Scientific Research Council of Argentina (CONICET). She has been an invited visiting professor and guest scientist at many institutions throughout the world. Dickenstein is currently working in the area of algebraic geometry and its applications, a subject on which she has co-organized several international conferences. Her publications include not only numerous research articles in prestigious journals and chapters in books on her specialty, but also textbooks on mathematics for children ages nine through twelve.

John C. Doyle is the John G. Braun Professor of Control & Dynamical Systems, Electrical Engineering, and BioEngineering at Caltech, Pasadena, California. Information on his work can be found on his Web site, http://www.cds.caltech.edu/~doyle/.

Freeman Dyson is a retired professor at the Institute for Advanced Study in Princeton, New Jersey. He began his career as a pure mathematician in England but switched to physics after moving to the United States. A volume of his selected papers in mathematics and physics was published by the American Mathematical Society in 1996.

Harold M. Edwards is professor emeritus of mathematics at New York University. He has received the Whiteman and Steele prizes of the American Mathematical Society and is the author of eight books: *Advanced Calculus, Riemann's Zeta Function, Fermat's Last Theorem, Galois Theory, Divisor Theory, Linear Algebra, Essays in Constructive Mathematics*, and *Higher Algebra*.

Timothy Gowers is a Royal Society 2010 Anniversary Research Professor at the University of Cambridge. He works in analysis and combinatorics. For his work in these areas he was awarded a European Mathematical Society prize in 1996 and a Fields Medal in 1998.

Judith V. Grabiner is the Flora Sanborn Pitzer Professor of Mathematics at Pitzer College, one of the Claremont Colleges in California. She is the author of *The Origins of Cauchy's Rigorous Calculus* (MIT Press) and *The Calculus as Algebra: J.-L. Lagrange, 1736–1813* (Garland Press), as well as a Teaching Company DVD course titled "Mathematics, Philosophy, and the 'Real World.'" She has received several Lester Ford and Allendoerfer Awards from the Mathematical Association of America and the Deborah and Franklin Tepper Haimo Award for Distinguished College or University Teaching from the MAA in 2003.

Mary W. Gray is professor and chair of the Department of Mathematics and Statistics at American University, Washington, D.C. She has received the Presidential Award for Excellence in Science, Engineering and Mathematics Mentoring and is a Fellow of the American Statistical Association. A lawyer as well as a statistician, Dr. Gray has over 100 publications in statistics, economic equity, law, and mathematics education.

Branko Grünbaum received a doctoral degree from Hebrew University in Jerusalem in 1957. He is professor emeritus at the University of Washington, where he has taught since 1966. His book, *Convex Polytopes* (1967, 2003), has been very popular, as was the book *Tilings and Patterns* (co-authored with G. C. Shephard), published in 1986. He hopes that the volume *Configurations of Points and Lines* (2009) will revive interest in this exciting topic, which was neglected during most of the twentieth century. Grünbaum's interests are mostly in various branches of combinatorial geometry. He has been a Guggenheim Fellow and a Fellow of the AAAS. The American Mathematical Society awarded him a Leroy P. Steele prize, and the Mathematical Association of America awarded him the Lester Ford and Carl Allendoerfer prizes.

Brian Hayes is senior writer for *American Scientist* magazine and a former editor of both *American Scientist* and *Scientific American*. He is also the author of *Infrastructure: A Field Guide to the Industrial Landscape* (W. W. Norton, 2005) and *Group Theory in the Bedroom and Other Mathematical Diversions* (Hill and Wang, 2008). He has been journalist-in-residence at the Mathematical Sciences Research Institute in Berkeley and a visiting scientist at the Abdus Salam International Centre for Theoretical Physics in Trieste. He is the winner of a National Magazine Award.

Orit Hazzan is associate professor in the Department of Education in Technology and Science of the Technion–Israel Institute of Technology. In her current research on computer science and software engineering education she partially relies on her research background in mathematics education. In addition to about 100 papers Hazzan has published in refereed journals and conference proceedings, she is also a co-author of two books: *Human Aspects of Software Engineering* (Charles River Media, 2004) and *Agile Software Engineering* (Springer, 2008).

Theodore P. Hill is professor emeritus of mathematics at Georgia Tech and has held visiting appointments in Costa Rica, Germany (Gauss Professor), Holland (NSF-NATO Fellow), Israel, Italy, and Mexico. He studied at West Point (BS), Stanford (MS), Göttingen (Fulbright Scholar), and the University of California–Berkeley (MA, PhD). His primary research interests are in mathematical probability, especially optimal-stopping theory, fair-division problems, and Benford's law.

Howard T. Iseri is professor of mathematics at Mansfield University of Pennsylvania. In mathematics, he likes to find ways to make geometric concepts accessible and intuitive. Outside of math, he is obsessed with the idea that his bikes, feet, and kayaks are too slow.

Vijay Iyer is a pianist, composer, bandleader, and producer in New York City. His fourteen albums include *Historicity*, winner of the *Village Voice* and *Downbeat Magazine* critics' polls and Germany's Echo Award. In addition to his work in jazz, Iyer has composed orchestral and chamber works; scored for film, theater, radio, and television; collaborated with poets and choreographers; and joined forces with artists in hip-hop, rock, experimental, electronic, and Indian classical music. He teaches at New York University and the New School. His writings appear in *Music Perception*, *Journal of Consciousness Studies*, *Current Musicology*, *Critical Studies in Improvisation*, *Jazz-Times*, *Wire*, *The Guardian*, and the anthologies *Uptown Conversation*, *Sound Unbound*, and *Arcana IV*.

Behzad Jalali is director of educational services in the Department of Mathematics and Statistics of the American University, Washington, D.C. His research focus is on teaching mathematics to students with disabilities, as well as on the role of language in mathematics learning.

Tim Johnson is the UK Research Council's Academic Fellow in Financial Mathematics at Heriot-Watt University and the Maxwell Institute for Mathematical Sciences in Edinburgh, Scotland. He studied physics as an undergraduate, then worked in the oil industry, where he realized guessing was not a good policy when uncertainty and money are involved, and so became a mathematician. His research is centered on the field of stochastic optimal control.

Ann Kajander is associate professor of mathematics education at Lakehead University, Thunder Bay, Ontario, Canada. She has taught mathematics at both the secondary and post-secondary levels and has received a teaching award from Lakehead University. She has published two books for teachers, the more recent being *Big Ideas for Growing Mathematicians*.

Erica Klarreich is a mathematics and science writer based in Berkeley, California. She holds a doctorate in mathematics from Stony Brook University and a certificate in science writing from the University of California, Santa Cruz. Her articles have appeared in *Nature*, *New Scientist*, *American Scientist*, *Science News*, and other publications.

Uri Leron is emeritus holder of the Churchill Family Chair of Education in Science and Technology at the Technion–Israel Institute of Technology. He was a ring theorist before switching his research interests to mathematical thinking and learning. He was awarded a residency at the Rockefeller Center in Bellagio, Italy, to study the implications of cognitive psychology and evolutionary psychology for mathematical thinking in general, and for understanding the sources of ubiquitous and recurring errors in particular.

Miroslav Lovric is a professor in the Department of Mathematics and Statistics at McMaster University, Hamilton, Ontario. His areas of research interest include differential geometry, applied mathematics and mathematics education. Besides publishing his research, Lovric wrote a textbook on vector calculus, and is currently working on a textbook for life sciences mathematics. In 2001 Lovric was awarded the national 3M teaching award.

Melvyn B. Nathanson is professor of mathematics at the City University of New York (Lehman College and the Graduate Center) and the author of more than 150 research papers and books in mathematics. Before becoming a mathematician he was an undergraduate major in philosophy at the University of Pennsylvania and a graduate student in biophysics at Harvard University.

Michael Nielsen is a Toronto-based writer currently completing a book titled *Reinventing Discovery*, about the use of collective intelligence in science. He was formerly a theoretical physicist and is co-author of the standard text in quantum computing.

Nick Paumgarten has been a staff writer for the *New Yorker* since 2005. From 2000 to 2005 he was the deputy editor of The Talk of the Town, to which he regularly contributes. He has also written features on subjects ranging from sports talk radio to backcountry skiing. Before coming to the *New Yorker*, Paumgarten was a reporter and senior editor at the New York *Observer*. Paumgarten lives in Manhattan.

Andrzej Pelc received a doctorate in mathematics in 1981 from the University of Warsaw, Poland. He is currently professor of computer science and director of the Research Chair in Distributed Computing at the Université du Québec en Outaouais, Gatineau, Quebec, Canada. He has published more than 250 papers in computer science and mathematics. Pelc is the recipient of the 2003 Prize for Excellence in Research from the Université du Québec en Outaouais.

David Pimm has been a professor of mathematics education since 2003 at the University of Alberta, Canada. Much of his academic career prior to 2003 was spent at the Open University in the United Kingdom, in the Faculty of Mathematics. He is the author of two books, *Speaking Mathematically* and *Symbols and Meaning in School Mathematics*, and editor or co-editor of seven others, including most recently *Mathematics and the Aesthetic: New Approaches to an Ancient Affinity* (Springer, 2006), with Nathalie Sinclair and William Higginson, in the CMS series.

Julie Rehmeyer is a math and science writer in Berkeley, California. She did graduate work in mathematics at MIT and taught math and the classics at St. John's College. She has written for *Science News, Wired, Discover, New Scientist*, and many other publications.

Nathalie Sinclair is associate professor at Simon Fraser University in the Faculty of Education. She co-edited the book *Mathematics and the Aesthetic: New Approaches to an Ancient Affinity* (Springer) and authored two books related to school mathematics: *Mathematics and Beauty: Aesthetic Approaches to Teaching Children* (Teachers College Press) and *The History of the Geometry Curriculum in the United States* (IAP). In addition to her interests in the aesthetic dimension of mathematics and in history, she studies the use of digital technology in mathematics teaching and learning, with a special interest in dynamic geometry software.

Steven Strogatz is the Jacob Gould Schurman Professor of Applied Mathematics at Cornell University. In 2007 he received the JPBM Communications Award, a lifetime achievement award for the communication of mathematics to the general public. His books include *Sync* (Hyperion, 2003) and *The Calculus of Friendship* (Princeton University Press, 2009).

Robert Thomas is a Fellow of St John's College and professor in the Department of Mathematics, both at the University of Manitoba, Winnipeg, Canada. He teaches mathematics, chiefly to engineering students; does mathematical research; and edits *Philosophia Mathematica*, the journal on the philosophy of mathematics published by Oxford Journals. His article title is due to Chandler Davis.

William P. Thurston is professor of mathematics and computer science at Cornell University. He is a pioneer in the field of low-dimensional topology. For the depth and originality of his contributions to mathematics he was awarded the Fields Medal in 1982.

David Wagner is associate professor in the Faculty of Education at the University of New Brunswick. His research focuses on human interactions in mathematics and mathematics learning situations, and how these interactions are structured in speaking and writing in these environments.

Anne Watson is professor of mathematics education at the University of Oxford, where she has worked in teacher education since 1996. Before that she was a secondary mathematics teacher in challenging schools, maintaining a lifelong commitment to helping more students achieve in mathematics, summarized in *Raising Achievement in Secondary Mathematics* (Open University Press).

Walter Willinger is a member of the Information and Software Systems Research Lab at AT&T Labs-Research, Florham Park, New Jersey. Information on his work can be found on his Web site, http://www.research.att .com/people/Willinger_Walter/index.html.

Henryk Woźniakowski is professor of computer science at Columbia University, New York, and professor of applied mathematics at the University of Warsaw, Poland. His main research interest is computational mathematics, in particular the computational complexity of continuous problems.

Acknowledgments

For the last five or six years I planned a volume similar to this one, but life circumstances in each case stalled the project before completion. The breakthrough came after a fortuitous conversation with Steven Strogatz in the fall of 2009. From there I consulted closely with Vickie Kearn, the Princeton editor of the book. She is the perfect collaborator on such a project, and every time I use the third person pronoun "we" in the introduction I refer to Vickie just as much as to me. Thanks also to Vickie's assistant, Stefani Wexler, for her efforts to obtain many reprinting permissions.

In finalizing the content I benefited from several suggestions offered, once again, by Steven Strogatz and Vickie Kearn, and one by Aaron Weinberg. Thanks to all other people who have seen the table of contents and opined on it. I also thank the contributors who made available their original figure and photo files. Useful comments made by anonymous reviewers contributed to the final content of the introduction; I am thankful to them. Of course, the responsibility for the shortcomings in the final result is entirely mine.

The staff at the Cornell University Libraries (particularly the Mathematics Library) were always kind and expedient in ordering materials not immediately available.

Many thanks to the people who assigned me adequate teaching and research duties over the last few years, especially David W. Henderson and Maria Terrell—but also Katherine Gottschalk, Paul Sawyer, Stanley Seltzer, Richard Durrett, Elise West, Sarah Hale, and Carol Shilepsky. Directly or indirectly, they made it possible for me to work on this book.

I thank Kristen Grace and Brendan Wyly for indirect support, in difficult personal circumstances for me.

Finally, I thank my daughter, Ioana Emina. She accompanied me to libraries, sometimes patiently but more often impatiently, making sure that we did not spend too much time in a noise-free environment. Her cheerful indifference to my scrupulousness in tracing books and periodicals helped me put my task in proper perspective. I dedicate this book to her.

Credits

"The Role of the Untrue in Mathematics" by Chandler Davis. First published in *Mathematical Intelligencer 31* (3) pp. 4–8 (2009). Reprinted by kind permission of Springer Science and Business Media.

"Desperately Seeking Mathematical Proof" by Melvin Nathanson. First published in *Mathematical Intelligencer* 31, no. 2 (2009): 8–10. Reprinted by kind permission of Springer Science and Business Media.

"An Enduring Error" by Branko Grünbaum. First published in *Elemente der Mathematik* 64:2(2009), pp. 89–101. Reprinted by permission.

"What is Experimental Mathematics?" by Keith Devlin. Copyright the Mathematical Association of America 2010. All rights reserved. Reprinted by permission.

"What Is Information-Based Complexity?" by Henryk Woźniakowski. First published in *Essays on the Complexity of Continuous Problems*, ed. Erich Novak, Ian H. Sloan, Joseph Traub, and Henryk Woźniakowski, 89–95. Zurich: European Mathematical Society, 2009. Reprinted by permission.

"What Is Financial Mathematics?" by Tim Johnson. First published in *Plus Online Magazine*. Reprinted by kind permission of the author.

"If Mathematics Is a Language, How Do You Swear in It?" by David Wagner. First published in *The Montana Mathematics Enthusiast* 6, no. 3 (2009): 449–58. Reprinted by permission.

"Birds and Frogs" by Freeman Dyson. First published in *Notices of the American Mathematical Society* 56.2(2009): 212–23. Reprinted by kind permission of the American Mathematical Society.

"Mathematics Is Not a Game But . . ." by Robert Thomas. First published in *Mathematical Intelligencer* 31, no. 1 (2009): 4–8. Reprinted by kind permission of Springer Science and Business Media.

"Massively Collaborative Mathematics" by Timothy Gowers and Michael Nielsen. Reprinted by permission from Macmillan Publishers Ltd: *Nature 461*, pp. 879–81. Copyright 2009.

"Bridging the Two Cultures: Paul Valéry" by Philip J. Davis. First published in *Svenska Matematikersamfundet Medlemsutskicket,* May 15, 2009. Reprinted by permission.

"A Hidden Praise of Mathematics" by Alicia Dickenstein. First published in *Bull. Amer. Math. Soc. (N.S.)* **46** (2009), 125–129. Reprinted by kind permission of the American Mathematical Society.

"Mathematics and the Internet: A Source of Enourmous Confusion and Great Potential" by Walter Willinger, David Alderson, and John C. Doyle. First published in *Notices of the AMS* 56: 5(2009), pp. 586–99. Reprinted by kind permission of the American Mathematical Society.

"The Higher Arithmetic: How to Count to a Zillion without Falling Off the Number Line" by Brian Hayes. First published in *American Scientist* 97.5(2009): 364–68. Reprinted by permission.

"Knowing When to Stop: How to Gamble If You Must—The Mathematics of Optimal Stopping" by Theodore P. Hill. First published in *American Scientist* 97.2(2009): 126–33. Reprinted by permission.

"Homology: An Idea Whose Time Has Come" by Barry A. Cipra. First published in *SIAM News, 42.10*(2009). Reprinted by permission of the author.

"Adolescent Learning and Secondary Mathematics" by Anne Watson. First published in *Proceedings of the 2008 Annual Meeting of the Canadian Mathematics Education Study Group/Groupe canadien d'étude en didactique des mathématiques*, 21–29. Burnaby, BC: CMESG/GCEDM, 2009. Reprinted by permission of the author.

"Accommodations of Learning Disabilities in Mathematics Courses" by Kathleen Ambruso Acker, Mary W. Gray, and Behzad Jalali. Originally published in *Notices of the AMS* 56.9(2009): 1072–80. Reprinted by kind permission of the American Mathematical Society.

"Audience, Style, and Criticism" by David Pimm and Nathalie Sinclair. First published in *For the Learning of Mathematics* 29, no. 2 (2009): 23–28. Reprinted by permission.

"Aesthetics as a Liberating Force in Mathematics Education?" by Nathalie Sinclair. First published in *ZDM Mathematics Education* 41 (2009): 45–60. Reprinted by kind permission of Springer Science and Business Media.

"Mathematics Textbooks and Their Potential Role in Supporting Misconceptions" by Ann Kajander and Miroslav Lovric. First published by Taylor and Francis Ltd. (http://www.informaworld.com) in *International Journal of Mathematical Education in Science and Technology* 40, no. 2 (2009): 173–81. Reprinted by permission of the publisher.

"Exploring Curvature with Paper Models" by Howard T. Iseri. Reprinted with permission from *Understanding Geometry for a Changing World*, copyright 2009 by the National Council of Teachers of Mathematics. All rights reserved.

"Intuitive vs Analytical Thinking: Four Perspectives" by Uri Leron and Orit Hazzan. First published in *Educational Studies in Mathematics* 71, no. 3